Adhesion Measurement Methods

Theory and Practice

Robert Lacombe

Taylor & Francis
Taylor & Francis Group
Boca Raton   London   New York

A CRC title, part of the Taylor & Francis imprint, a member of the
Taylor & Francis Group, the academic division of T&F Informa plc.

Published in 2006 by
CRC Press
Taylor & Francis Group
6000 Broken Sound Parkway NW, Suite 300
Boca Raton, FL 33487-2742

© 2006 by Taylor & Francis Group, LLC
CRC Press is an imprint of Taylor & Francis Group

No claim to original U.S. Government works
Printed in the United States of America on acid-free paper
10 9 8 7 6 5 4 3 2 1

International Standard Book Number-10: 0-8247-5361-5 (Hardcover)
International Standard Book Number-13: 978-0-8247-5361-0 (Hardcover)

Library of Congress Cataloging-in-Publication Data

Catalog record is available from the Library of Congress

Taylor & Francis Group
is the Academic Division of Informa plc.

Visit the Taylor & Francis Web site at
http://www.taylorandfrancis.com

and the CRC Press Web site at
http://www.crcpress.com

Preface

This volume has arisen out of a short course on adhesion measurement methods given in conjunction with a series of symposia on surface related aspects of materials science technology. The conference Web site* caught the attention of John Corrigan, who at that time was an acquisition editor for Marcel Dekker. As I had long been contemplating writing a volume on adhesion measurement to use as supporting material for a short course on adhesion measurement, John did not have to work very hard to convince me that it would be a good idea to write a volume on this topic. In addition, John and I felt that such a volume would fill an important gap in the engineering science literature because there was no single text devoted to adhesion measurement technology notwithstanding the fact that an enormous body of literature existed on the subject in scientific journals and edited volumes. Having thus decided to engage in the project, I concluded that the main purpose of the volume would be to provide a useful reference work and handbook for the practicing engineer/scientist who has a need to confront problems of adhesion either in support of manufacturing operations or in the development of new products. Thus, this book is meant to be used and kept handy for any and all of the following purposes:

1. As a monograph/reference work to be used either for self-study or to become aware of what has been done in the realm of developing methods and useful tools for measuring the adhesion of coatings and thin films.
2. As supplementary reading material for courses on materials science, mechanics of materials, or engineering design of laminate structures at the advanced undergraduate or graduate level.
3. As a handbook for looking up useful information and formulae on adhesion-related matters, such as driving force formulae for various modes of delamination, methods for estimating stress buildup, and material property data in support of "back-of-the-envelope" calculations.
4. As an introductory reference work for accessing the vast scientific and engineering literature on adhesion measurement. A substantial bibliography of some 40 key reference works plus over 500 articles and books is organized topicwise to provide a convenient introduction to the veritable ocean of information available in the literature.

The contents of the book are organized into seven chapters and five appendices in the following order: Chapter 1 gives a brief introduction to the subject, including

---

* Those interested in surface-related phenomena such as adhesion, cleaning, corrosion, and the like can go the conference Web site at www.mstconf.com; there you will find the programs and abstract listings of some 24 previous symposia on these topics, as well as up-to-date information on current and future symposia.

an attempt to define the term *adhesion* for the purpose of providing a definition that is both accurate and useful in practice.

Chapter 2 provides an overview of the most common adhesion measurement methods plus a few exotic methods to round out the mix. From the point of view of this work, adhesion measurement techniques fall into one of the three following categories: qualitative, semiquantitative, and fully quantitative techniques. Each of the methods discussed has its uses and drawbacks, and the intent is to make this as clear as possible. Something akin to a *Consumer Reports* format is adopted to help the reader interested in selecting a method with which to address current adhesion problems. Anyone just getting involved with adhesion-related issues should find this chapter helpful.

Chapter 3 lays the foundation required to step up to the problem of implementing fully quantitative adhesion measurement methods. To do this, however, one has to confront headlong the thermal-mechanical behavior of the materials with which one is dealing. This comes about from the simple fact that all adhesion measurement methods in some way or another apply an external load to the structure tested and then draw conclusions based on the observed deformations or mechanical reaction forces observed. The most general formalism available for dealing with this type of behavior is the continuum theory of solids, which is treated in some detail in this chapter. Inescapably, the level of mathematical treatment rises considerably over that given in Chapter 2. Every attempt is made to avoid excessive rigor and abstract formalism, which does more to flaunt the level of erudition as opposed to shedding light on the technical matters at hand. Thus, those who subscribe to the *International Journal of Solid Structures* will most likely find the mathematical level quite pedestrian, whereas members of the laity could find the discussion fairly heavy sledding. This should in no way, however, prevent anyone from using the results presented in succeeding chapters to practical advantage.

Chapter 4 deals with the discipline of fracture mechanics, which draws directly on all the supporting material presented in Chapter 3. Fracture mechanics is the ultimate organizing tool for performing fully quantitative adhesion measurements. It provides the concepts of stress intensity factor and strain energy release rate, which are two of the most useful quantitative measures of adhesion strength. Thus, from the point of view of fracture mechanics, a delamination is nothing more than a particular kind of crack occurring at an interface in a bimaterial structure.

Chapter 5 attempts to draw all of the above material together and make it seem more coherent and relevant by providing several specific and detailed examples of adhesion measurement in action. Thus, extensive examples of the peel test, the scratch test, and the pull test are presented. It is hoped that the reader will gain significant insight and intuition into how adhesion testing is carried out in practice and perhaps find some answers to specific problems of interest.

Chapter 6 deals with the problem of measuring the residual or intrinsic stress in a coating or other laminate structure. Whoever reads through the previous three chapters will quickly realize that residual stress is one of the key factors governing the delamination behavior of coatings and laminates and is a critical parameter in most fracture mechanics formulae for stress intensity factors and strain energy release rates. Thus, a fairly comprehensive overview of most of the useful stress measurement methods is provided. Use of one or more of these methods can be considered as providing an indispensable foundation for developing effective adhesion measurement procedures.

Chapter 7 concludes with more examples taken from the author's direct experience in wrestling with adhesion problems in the microelectronics industry. Silicon chips and ceramic multichip modules used to package these chips into useful devices give rise to a welter of adhesion-related problems because the number of interfaces involved is so varied and extensive that most structures can be looked at as one extensive interface. The presentation here is informal and intended to provide insight and intuition as to how adhesion problems and adhesion measurement happen in the "real world."

Several appendices are provided to make the volume more useful as a day-to-day handbook and handy reference for looking up simple formulae and material property data for performing back-of-the-envelope type calculations. Rudimentary calculations for estimating the stress expected in a coating or the driving force for delamination can be very helpful for making decisions regarding which processes or materials one should employ for fabricating a specific device. Appendix A provides an overview of vector and tensor calculus for those with a need to brush up on the topic both regarding performing elementary calculations and in understanding more fully the mathematical developments in Chapters 3 and 4. Appendix B gives a quick overview of the most useful aspects of strength of materials theory, which essentially amounts to computing all the ways a beam can bend. This is important material because bending beams figure heavily in many adhesion measurement schemes and stress measurement experiments. Appendix C gives an extended table of material property data required for nearly all the formulae in Appendices B and D. This type of data tends to be scattered far and wide in a variety of texts and reference works, which by Murphy's second law are never at hand when most needed. Thus, having these data located next to the formulae that require them should prove convenient. Appendix D provides a list of the most useful formulae from fracture mechanics that can be applied to the most common failure modes observed in coatings and laminates. Finally, a two-part bibliography is given that should prove handy in obtaining useful references from the vast technical literature.

In closing, I would like to state that this is very much a work in progress. It makes no attempt at being fully comprehensive or definitive in any sense. Rather, the goal is to provide a volume extensive enough to be useful but not so vast as to be more of a burden than a help. In this regard, it is clear that much of relevance had to be left out. It is hoped that future editions will correct this problem to some extent. I invite any and all constructive criticisms and suggestions to correct errors or improve the presentation. I am most readily reached at the e-mail address given below and will give due attention to any and all who respond.

Finally, this work would not be complete without giving recognition to my long-time friend, colleague, and co-worker Dr. Kashmiri Lal Mittal (Kash to his friends). Kash has certainly been a key inspiration in completing this work. A cursory look at the references will make it clear that his contributions to the adhesion literature have been monumental. In addition to editing the *Journal of Adhesion Science and Technology*, he has published 81 (and counting) edited volumes dealing either directly or indirectly with problems of surface science and adhesion. His many comments and corrections to this work have been indispensable. My thanks go also to John Corrigan, who gave the initial impetus that got this project going. Many thanks, John, wherever you are.

Robert H. Lacombe
MST Conferences, LLC
3 Hammer Drive
Hopewell Junction, NY 12533-6124
E-mail: rhlacombe@compuserve.com

The Author

Robert Lacombe, Ph.D., received his doctoral degree in macromolecular science from Case Western Reserve University and was a postdoctoral fellow at the University of Massachusetts, working on problems of polymer solution thermodynamics. Following his stint in academia, he was an IBM physicist for 20 years and worked on problems relating to thin film wiring technology for both semiconductor chips and multichip modules. After IBM, he worked for an independent consulting firm on hybrid nonintrusive inspection and evaluation techniques, as well as problems of materials compatibility of both semiconductor and microelectronic packaging devices. He is an expert in the area of stress buildup in laminate structures and using the techniques of fracture mechanics in solving problems of delamination and cracking in composite devices. He has been a leader in the areas of materials characterization and published some of the first mechanical response data on monolayer nanostructures in the early 1980s. In addition, he pioneered innovative uses of large-scale computation using finite element methods and applied this expertise directly to problems affecting product development and manufacturing. Dr. Lacombe is Chairman of MST Conferences and has organized over 20 international symposia covering the topics of adhesion and other surface-related phenomena since 1998. He is credited with over 40 publications and patents. Dr. Lacombe is a leading innovator in the field of subsurface inspection methods dealing with flaws in structural parts of air, auto, marine, and aerospace vehicles. As part of his activities with MST Conferences, he teaches a semiannual course on adhesion measurement methods in conjunction with the symposia.

Table of Contents

1 Introduction

Can you measure it? Can you express it in figures? Can you make a model of it? If not, your theory is apt to be based more upon imagination than upon knowledge.

Lord Kelvin

1.1 OVERVIEW

The science of adhesion has the curious distinction of being at one and the same time both a "sticky" subject and a "slippery" one. The sticky aspect is obvious and needs no further comment; it is the slippery aspect that will draw much of our attention in this volume. The essential problem arises from the fact that adhesion is a basic property of surfaces, and according to the famous physicist Wolfgang Pauli: "God created matter; surfaces were invented by devil." Surfaces are indeed devilish entities, especially for those who seek a quantitative understanding of their behavior. This arises from the fact that, for nearly all macroscopic objects, the surface area forms a very small portion of the bulk, is further subject to highly asymmetric forces, and is strongly prone to contamination and a large variety of defects.

All of this makes it very tricky to characterize the nature of a real surface. For instance, one can have a glass surface that, for all practical purposes, appears to be absolutely clean, but the presence of even a monolayer of contaminant can change its behavior from water wettable to hydrophobic. This in turn has dramatic consequences if one is trying to get a particular coating to adhere to the glass. From a practical point of view, obtaining good adhesion in a manufacturing process requires very serious attention to the state of cleanliness of the surface to which something must be bonded.

Furthermore, simply cleaning the surface may not be enough because contaminants may intrude from the ambient atmosphere. A notorious example of this arises in brazing processes where, for instance, one is trying to adhere a metal cap to a ceramic substrate using a solder formulation. Every precaution can be taken to ensure that the mating surfaces are scrupulously clean, but a contaminant such as silicone oil can condense onto the mating surfaces in question, preventing proper wetting of the solder and thus causing a defective joint to be formed. In many instances, silicone oil may be used for a variety of purposes in neighboring sections of a manufacturing plant. A common heating/ventilation system can then carry microscopic amounts of the volatile liquid throughout the entire plant, and because only a monolayer of coverage is enough to disrupt the wetting properties of a surface, the consequences can be catastrophic for the brazing process. Such problems can and do happen rather more frequently than one might imagine in the world of industrial processing, which emphasizes the slippery aspect of adhesion phenomena.

1

The scope of problems alluded to here makes the science of adhesion a very broad one that covers the range of disciplines from surface characterization to strength of materials. Indeed, many volumes have been written on the subject. However, the focus here is on one key aspect of the study of adhesion phenomena: the science of adhesion measurement. The basic goal is twofold. First, this volume is to serve as a manual for the relatively uninitiated worker in either academia or industry who has a practical need to perform adhesion measurements either as part of an experimental program or possibly as part of some quality control procedure. A broad range of well-known experimental techniques are reviewed and critiqued regarding their respective strengths and weaknesses. References are given both to the technical literature and to applicable commercial instrumentation. Emphasis is given to those methods that supply at least a semiquantitative estimate of adhesion strength and can be implemented in any laboratory with minimum expenditure of resources. Thus, the practical user can look to this volume as a ready guide and resource for getting a handle on personal adhesion measurement needs.

The second goal of this work is to provide a more fundamental understanding of the adhesion measurement process and its consequences regarding the design and manufacture of structures that have a critical dependence on the adhesion of disparate material layers. The multilevel wiring layers in microelectronic structures immediately come to mind as an appropriate paradigm. Such systems often involve the buildup of material stresses that, to be effectively handled, require an understanding of continuum theory and fracture mechanics. Thus, a substantial section of this volume is devoted to the elements of continuum theory and fracture mechanics most relevant to the problem of adhesion measurement. It should be mentioned also that the topics of material behavior and fracture mechanics are of great concern even for the practical individual mainly concerned with subduing a specific adhesion problem that is plaguing a particular manufacturing process. The point is that, if material stresses get high enough, no amount of adhesion improvement will solve the problem of device failure. If the interface proves to be too strong, then the delamination failure simply proceeds as a fracture crack in the bulk material. Solving such problems unavoidably requires understanding and control of the material stresses. Again, microelectronic structures provide a prime example in which such situations can arise. In response to this type of dilemma, an entire section of this work treats the topic of stability maps, which provide a means of navigating through such problems. This is a topic of significant interest for both the practical user as well as anyone interested in the more fundamental and theoretical aspects of adhesion measurement and control.

1.2   WHAT IS ADHESION AND CAN IT BE MEASURED?

There has been considerable debate in the technical literature concerning the question of what adhesion is and if it can be measured, and it is certainly not the intent here to expound further on what has been discussed many times. However, to provide a certain measure of perspective and to establish a defensible position on this matter, it is worthwhile at least to discuss this problem briefly.

One of the earliest overviews of this issue was given by Mittal,[1] and we take his discussion as a starting point. The basic argument comes down to what one takes as the definition of *adhesion*. Now, from a commonsense point of view, one would like to think that adhesion is a simple matter of how well two different materials tend to stick together, and that adhesion measurement is some indication of the force required to separate them. Although this approach may suffice for the person in the street, it runs into serious difficulty when one tries to arrive at a more scientific definition of adhesion that can be useful for engineering purposes. A truly useful definition of the term adhesion needs to have the properties as given in definition A.

1.2.1 DEFINITION A: CRITERIA FOR A TRULY USEFUL DEFINITION OF THE TERM ADHESION

If we say that X is the adhesion of material A to material B, then it should have the following characteristics:

1. X has the same meaning for all practitioners who would stick A to B.
2. X is unambiguously measurable by one or more commonly understood methods.
3. Knowing X allows the practitioner to predict the loading conditions that will cause material A to delaminate from material B.

Many would agree that the above is certainly a worthy definition of the term adhesion, but it unfortunately runs into a number of difficulties in practice because of the slippery aspects of the science of adhesion discussed in this chapter. In particular, the following types of problems are likely to arise in practice:

- **Problem 1: Universality of the Definition Regarding Selection of Materials to be Adhered and Nature of Active Practitioners** — Definition A runs into a variety of problems regarding the universality condition mentioned as characteristic 1. Let us assume that material A is a coating material to be applied to a thick rigid substrate comprised of material B. For simplicity, we assume that these materials are universally available to practitioners P_1 and P_2 in reasonably pure form. However, practitioner P_1 wants to coat A onto B as a thin film less than 1 μm thick, whereas practitioner P_2 wants to coat A onto B as a thick coating more than 25 μm thick. All other conditions are equal. Let us also assume that our practitioners are equally scrupulous in cleaning the substrate material B, and it is reported that A adheres well to B. Nonetheless, it can easily happen that practitioner P_1 will find good adhesion of A to B whereas P_2 will experience delamination. Among a wide variety of problems that can occur, one of the most common is the presence of thin film stresses. If the two practitioners use different techniques to coat A onto B, then the two coatings can develop different material morphologies, which can lead to substantially different stress levels in the two coatings. The more highly stressed coating will be much more likely to delaminate than the other.

Furthermore, even assuming that the two practitioners used identical coating techniques, practitioner P_2 will be much more likely to experience delamination because of the thicker coating as, for a given level of film stress, the driving force for edge delamination will scale linearly with film thickness. A second problem that can occur is that the two coatings may be subjected to different loading conditions when in use. For example, practitioner P_1's coating may see predominantly shear loads, whereas that of practitioner P_2 experiences predominantly tensile loads. Because most coatings tend to resist shear delamination (commonly referred to as mode II) much better than tensile delamination (commonly referred to as mode I), practitioner P_2 will be much more likely to experience delamination problems than his counterpart P_1. Thus, definition A falls short regarding the universality criterion in that it may give good results for those putting down thin coatings and poor results for those trying to make thick coatings.

- **Problem 2: Definition A Enables the Use of Simple and Unambiguous Measurement Procedures** — From the point of view of the practical practitioner, the use of simple and unambiguous measurement procedures is quite likely the most important property desired in any truly usable definition of the term adhesion. However, the two qualifications of simple and unambiguous tend to be mutually contradictory in that a truly simple test is not likely to be unambiguous because, for the sake of simplicity, a number of important details will be either omitted or glossed over. The truly unambiguous test will specify in great detail the conditions of sample preparation, including cleaning procedures, control of material properties, precise specification of loading conditions, and control of the ambient environment. Observing all of these caveats will tend to undermine the goal of achieving simplicity. Clearly, any truly usable definition of adhesion will have to seek an appropriate balance between these two criteria.
- **Problem 3: Definition A Has True Predictive Power** — It is at once clear that the property that definition A has true predictive power is at odds with characteristic 2 because to obtain a truly predictive measure of adhesion, clearly the utmost attention must be given to all the details that will ensure an unambiguous result. This will sacrifice the goal of achieving simplicity. In addition, to be truly predictive the measurements have to be fully quantitative and consistent with detailed calculations. This implies at minimum the use of fracture mechanics methods and continuum theory of materials.

So where does this leave us? It is clear that although definition A is what we would like to have for a definition of the term "Adhesion", it is clearly unworkable in the real world. In this volume, therefore, we back off somewhat from definition A and revert to definition B. However, before we frame definition B, we reflect on just what we are after vis-à-vis what is practicably obtainable. As this volume is primarily intended for those who have a practical need to perform adhesion measurements in support of the need to fabricate useful structures, our definition focuses on the need to be readily measurable. This puts the property of simplicity in the forefront as complex measurement procedures have a relatively low probability of implementation in a manufacturing process in which time and resource expenditure are carefully regulated commodities. In addition, our working definition has to be hierarchal because there are different levels of quantitativeness that can be specified depending specifically on the end use of the adhesion measurement. Thus, we frame definition B:

1.2.2 DEFINITION B: ADHESION

We say the adhesion of material A to material B is such and such based on the following criteria:

1. The adhesion of A to B is a relative figure of merit indicating the tendency of A to stick or bind to B derived from an observation or measurement that can be entirely qualitative, semiquantitative, or fully quantitative.
2. The precise meaning of the term is entirely dependent on the details of the measurement technique employed and the experimental and environmental conditions under which the measurement was made. This leads to a hierarchy of definitions. Thus, qualitatively we might say A has good adhesion to B based on the observation that A was never observed to separate from B under a variety of common loading conditions. A semiquantitative statement of the adhesion of a coating of material A onto a substrate of material B might indicate that 2% of the coating was removed during a "Scotch tape" test. Finally, a fully quantitative statement might conclude that the adhesion strength of A to B is 10 J/m^2 based on a double-cantilever beam experiment carried out at 50°C under 40% relative humidity.

1.3 COMMENTS ON NOMENCLATURE AND USAGE

Definition B can be considered as defining the *practical adhesion* of one material to another. As defined, it is clearly measurable and has definite significance within the context of the specific measurement technique employed. This definition stands in contradistinction to the term *fundamental adhesion*, which would imply the work required for separation of material A from B assuming they were joined across a perfect mathematical plane and that the separation avoided any tearing out of either material. The fundamental adhesion of one material to another is an important theoretical concept and is clearly definable in principle. The problem for our purposes is that the fundamental adhesion is exceedingly difficult, if not impossible, to measure under most conditions and even if measured is not likely to be of any immediate use. At this point, we simply recognize that practical adhesion is some complex function of fundamental adhesion that we fortunately do not need to know under most circumstances. Furthermore, we recognize that the fundamental adhesion between two materials can be modified by physicochemical means and is a powerful tool for modifying practical adhesion. For the most part, however, this topic is beyond the scope of this volume.

At this point, a number of comments on usage are in order. We have defined the noun adhesion from definition B, and we largely follow common dictionary usage and use the term *adhere* as the verb form. Also, when a coating separates from its substrate we use the term *delamination*. In the mechanics literature, one sometimes comes across the term *decohesion*. A coating is said to decohere from its substrate. The genesis of this term arises from the fact that the locus of failure lies in the substrate material a few microns below the true interface. Thus, it makes sense to talk about cohesive failure. One can also say the coating delaminated from the surface because that is the net effect, and failure occurred so close to the interface that, given the analytical tools at hand, the cohesive nature of the failure could not be detected. In all cases, one has to rely on the context of the measurements and observations to decide which term is most appropriate.

References

1. Mittal, K.L. Adhesion Measurement: Recent Progress, Unsolved Problems, and Prospects, in *Adhesion Measurement of Thin Films, Thick Films, and Bulk Coatings, ASTM STP-640* (K.L. Mittal, Ed.), American Society for Testing and Materials, Philadelphia, PA, 1978, pp. 5–17.

2 Overview of Most Common Adhesion Measurement Methods

No experiment is so dumb that it should not be tried.

Attributed to Walther Gerlach in reply to Max Born's comments that his work on magnetic fields with strong spatial gradients would not be worthwhile. This work turned out to be pivotal to the success of the now-famous Stern-Gerlach experiment, which demonstrated spatial quantization for the first time and has been one of the experimental linchpins for quantum theory ever since.[1]

2.1 PREAMBLE

This chapter overviews some of the most commonly used, and consequently the most useful, adhesion measurement methods. The approach follows that of a *Consumer Reports*-type examination in which both the strengths and the weaknesses of each technique are examined, and specific recommendations are made as to which methods are most suitable for which applications. Definition B of the term "adhesion" as presented in the Chapter 1 is assumed throughout. In addition, the discussion is guided by the following criteria for the "ideal" adhesion test:

- *Quantitative*:
  Gives numerical data that can be unambiguously interpreted.
  Data analysis straightforward and clear.
- *Ease of sample preparation*:
  Samples quickly and easily prepared with readily available equipment.
  If sample preparation is too complex, the test will not likely be implemented.
- *Results relevant to real world*:
  Final data must have relevance to final use conditions.

It should be made clear from the outset that none of the techniques described below meets all of the above criteria for the ideal adhesion test. However, each of these tests provides useful and reliable data when applied in a conscientious manner. Finally, it should be stated that this survey in no way covers all of the adhesion measurement methods reported in the literature. Mittal[2] informally counted over 300 reported methods as of early 1994. Most of the reported methods, however, are essentially variations on one of the techniques described in what follows.

Adhesion measurement methods can be divided into two broad categories: destructive or nondestructive. The overwhelming majority fall into the destructive class, by which a loading force is applied to the coating in some specified manner and the resulting damage subsequently observed. Nondestructive methods typically apply a pulse of energy to the coating/substrate system and then try to identify a specific portion of the energy that can be assigned to losses occurring because of mechanisms operating only at the interface. Inferences are then made regarding the bonding strength between the coating and the substrate.

2.2 PEEL TEST

2.2.1 INTRODUCTION

Within the realm of destructive adhesion tests, there are two major classes: those dealing with relatively soft flexible coatings and those dealing with hard brittle coatings. By far the most common test for flexible coatings is the peel test. Anyone who has removed wallpaper from an old house already has considerable practical experience with the basic rudiments of this test. When dealing with polymer-based paints, for example, the peel test readily suggests itself as the preferred experiment for testing the adhesion of the coating. Such coatings on curing and drying tend to build up a significant level of internal stress, which increases dramatically near the edge of the coating. If the level of adhesion between the coating and the substrate is not sufficient, then the coating will delaminate and peel back. The now-released coating can be grasped with a tweezer, and an ersatz peel test is performed.

Thus, the peel test automatically suggests itself as an adhesion test for flexible paint coatings. The main task for the experimenter is to standardize and quantify the peel test experiment so that the results can be used either to establish a quantitative ranking among the coatings tested or to set a numerical specification for adhesion strength that can be subsequently used as a quality control standard. The problem of quantifying the test is readily solved by the use of an appropriate tensile test apparatus in conjunction with suitable hardware for applying the peel force load and maintaining the required peel angle. Figure 2.1 exhibits a number of common configurations.

Figure 2.1a illustrates the common 90° peel test; it is the favored test for flexible coatings on rigid substrates. This is by far the most prevalent and most thoroughly studied of all the peel tests. Figure 2.1b illustrates the 180° version of Figure 2.1a. This configuration offers advantages when space is cramped; Chapter 5 presents this scenario for the peel test performed in a calorimeter. It is also clear that the peel test can be performed at any angle between 0 and 180°. For most practical purposes, there is little need to consider angles other than 90 or 180° unless there are geometric constraints imposed by the sample or test apparatus. From an analytical point of view, however, varying the peel angle can provide information on the effect of mode mixity on peel strength.

When performing a peel test, the interfacial region is subjected to both tensile (mode I) and shear (mode II) loads. The ratio of these two loading types is loosely referred to as the loading *mode mixity*. The importance of knowing the mode mixity

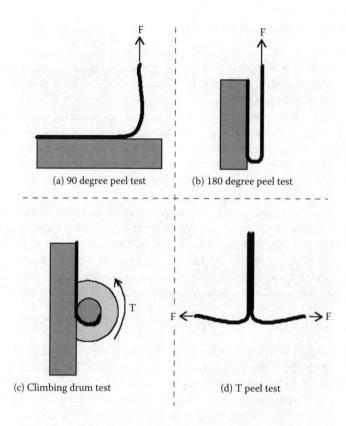

FIGURE 2.1 Four standard peel test configurations: (a) 90° peel test, the most commonly used configuration; (b) 180° test preferred when available space precludes the 90° test; (c) climbing drum or peel roller test, which has the advantage of controlling the radius of curvature of the peel strip; (d) T-peel test, preferred when testing adhesion of two flexible strips. A symmetric peel is obtained when identical strips are separated. Can also be used in a nonsymmetric configuration in which the two strips are not identical.

stems from the fact that the apparent adhesion strength of many coatings is sensitive to the mode mixity. For example, glues tend to be much stronger in shear than in tension, which implies that they will exhibit much higher adhesion strength in a predominantly mode II test as opposed to a mode I test. A very common example of this phenomenon is exhibited by Velcro fasteners. When loaded in shear, the Velcro fastener exhibits a very tenacious grip. However, when pulled at a right angle (i.e., tensile or mode I loading) the fastener readily separates.

Figure 2.1c illustrates the climbing drum test used in testing the adhesion of rubbers in the tire industry. One advantage of this version is that the radius of curvature of the peeling film is fixed by the drum radius, which simplifies later numerical analysis of the data. Finally, the T-peel test shown in Figure 2.1d can be used to test the adhesion between two flexible films.

2.2.2 ADVANTAGES OF THE PEEL TEST

The peel test in its many configurations meets many of the criteria of the ideal adhesion test. Sample preparation is typically reasonably simple and straightforward. This single fact more than anything else accounts for the overall popularity of this test. Also, the peel force gives a semiquantitative measure of the coating adhesion to the substrate, which can be readily used for ranking or quality control purposes.

A further advantage of this test is the fact that the rate of delamination and the locus of failure can be controlled fairly precisely. This stems from the fact that a very high stress concentration exists at the point where the coating just lifts off the substrate. This tends to narrowly focus the failure region very close to the geometric interface between coating and substrate, which is the region of most interest in any adhesion test. Because the rate of delamination can be precisely controlled by the test equipment, studies of the rate dependence of adhesion strength can be easily carried out. This can be very important when studying coatings that exhibit a strong molecular relaxation behavior (i.e., glass transition and related phenomena) near the test temperature. Finally, the peel test readily lends itself to use under conditions of controlled temperature and environment (for example, temperature and humidity conditions).

2.2.3 DISADVANTAGES OF THE PEEL TEST

As noted, the peel test works quite well when used as a method for ranking the adhesion of a coating when the substrate has been subjected to a number of different surface treatments. However, when trying to ascertain whether the coating will survive a given set of end-use conditions, several problems arise. The main issue is the fact that the peel test subjects the coating to very high strain levels near the peel bend, which most coatings never see under common end-use conditions. The strain in the coating at the peel bend can easily approach 25% or higher, whereas real coatings delaminate under nearly strain-free conditions. Thus, the load state imposed by the peel test does not reasonably approximate the load conditions that cause failure in the field; therefore, conclusions arrived at on the basis of peel testing can be highly misleading when trying to anticipate the actual service behavior of a particular coating.

A particularly illuminating example of how far off you can be was given by Farris and Goldfarb.[3] These authors tested the adhesion of polyimide films to aluminum and demonstrated apparent peel strength adhesion values in the range of 500 to 900 J/m^2. However, the same coatings self-delaminated at an adhesion strength of 23 J/m^2 when the coating thickness was increased to 120 μm. Thus, it is clear that peel test results can present highly misleading estimates of the actual adhesion strength of a coating when subjected to realistic end-use conditions.

Further limitations of the peel test stem from the fact that it is applicable only to tough flexible coatings. Attempts have been made to circumvent this limitation by applying peelable backing coatings on top of the coating to be tested and then peeling the composite laminate. One major problem with this approach is the fact that the locus of failure at the peel front can become unstable. The delamination front can wander between the backing layer and the test coating or between the coating under test and the substrate, making interpretation of the results unclear.

A number of other drawbacks and limitations apply to the peel test, including difficulty in initiating a peel strip for coatings with strong adhesion and controlling sample-to-sample variability. These problems can typically be dealt with by developing appropriate experimental techniques.

2.2.4 SUMMARY AND RECOMMENDATIONS

The peel test is quite likely the method of choice when dealing with tough, flexible coatings on rigid substrates as it meets many of the criteria of the ideal adhesion test. First and foremost is the consideration of sample preparation. In this regard, it is typically quite straightforward to fabricate convenient-size coupons of the substrate material and apply the coating of interest to them. Suitable care should be taken to clean the substrate and apply the adhesion promoters of interest. Further technical details, such as providing a release layer so that the peel strip can be easily initiated, should not be overlooked. Last, care should be exercised in interpreting the final data. Peel test data can be reliable for ranking the effectiveness of adhesion promoters or for quality control measurements. As noted, one should not rely on peel test measurements as a guide to performance of the coating under end-use conditions because the load state imposed by the peel test does not generally reproduce actual load conditions in the field. A detailed example of a successful implementation of the peel test for studying the adhesion of high-temperature coatings to silicon substrates is given in Chapter 5.

2.3　TAPE PEEL TEST

2.3.1 INTRODUCTION

The tape peel test is a rough-and-ready variant of the standard peel test. Its main advantage is ease of sample preparation. The predominant disadvantage is that the results of the test will tend to be qualitative only. Attempts at systematizing the test, however, have given semiquantitative results. In a typical application, a strip of specially fabricated tape is applied to the coating to be tested in a predefined manner. The main concern is to be as consistent as possible to achieve reproducible results. The tape is subsequently peeled off in a prescribed fashion, and the test surface is then inspected for whatever resulting damage that may have occurred. At the purely qualitative level, the experiment gives a "go/no go" type of result. indicating whether the adhesion of the coating is acceptable. A number of techniques have been invented to give a semiquantitative result by quantifying the level of partial damage that may have happened to the coating. An example of this for the case of ink coatings is discussed in Section 2.3.2.

The main problem with obtaining truly quantitative results with the tape peel test is that one now has to deal with four different materials: the substrate, the coating, the tape adhesive, and the tape backing material. Satas and Egan[4] reported the effect of the backing layer and the adhesive layer on the peel strength of pressure-sensitive tapes. Their data showed that, depending on the tape backing material, the peel force can vary by as much as a factor of 2 for a given layer thickness.

In a separate study Aubrey et al.[5] investigated the effect of adhesive molecular weight, adhesive layer thickness, backing film thickness, peel rate, and peel angle on the peel strength of polyester backing polyacrylate adhesive pressure-sensitive tapes. They demonstrated that all of these factors have a significant effect on the measured peel force. In particular, the peel force showed dramatic dependence on peel rate with three fundamentally different modes of peeling. At low rates, the peel force is controlled by flow of the tape adhesive and is strongly rate dependent. At high rates, little viscous deformation occurs, and the peel force is largely rate independent. At intermediate peel rates, the peel force exhibits cyclic instability driven by alternate storage and dissipation of elastic energy. The net result is a type of "stick-slip" peeling. Thus, without even considering the coating and substrate properties, we already have a considerable degree of complexity introduced just by the properties of the tape alone. If we now introduce further degrees of freedom arising from the mechanical response of the coating and substrate, we see that the problem of deriving a truly quantitative analysis of the tape peel test rapidly becomes prohibitive.

2.3.2 ADVANTAGES OF THE TAPE PEEL TEST

Despite the difficulties mentioned in obtaining quantitative results with the tape peel test, it can still be a useful and effective measurement method in certain applications. This is best illustrated by the study of ink coatings by Calder et al.[6] These authors succinctly summarized the case for using the tape peel test in a cogent manner[7]:

> There is a body of experience in the industry that confirms that the tape test is a reasonable predictor of how the ink will remain in place, intact on the substrate under many actual use conditions.

> The test is fast and can be performed at press side. It is obviously important to know rather quickly whether an ink has adequate adhesion when the film is being printed at 600 ft/min.

In their experiments on ink coatings, the authors applied the subject ink coating to the relevant paper or foil substrate. Tapes were applied to the ink coating after a specified drying time and removed rapidly by a 90° peel test. The degree of adhesion of the ink coating is then rapidly evaluated using a light spectrophotometer and is reported as percentage coating removal as compared to a standard untested sample. Appropriate calibration methods are used to ensure repeatability. Figure 2.2 illustrates some representative data from this type of experiment. The solid line shows the apparent adhesion versus time for an ink with a relatively "soft" binder matrix, and the dashed line illustrates the same behavior for an ink with a "hard" binder matrix. The data clearly reveal that the soft binder gives an ink with stronger adhesion at short times (lower percentage coating removal), and that both ink types level off to substantially the same adhesion level at longer times. This is the type of information that can be of practical use in the printing industry, in which different printing techniques have different requirements for ink adhesion.

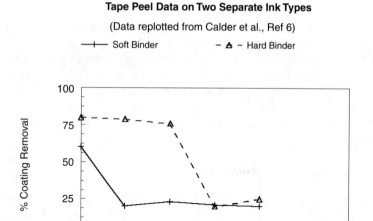

Tape Peel Data on Two Separate Ink Types

(Data replotted from Calder et al., Ref 6)

FIGURE 2.2 Tape peel test data on printing ink. The relative adhesion strength is measured by the amount of light that can pass through the ink/paper layer after the tape is peeled off. The ordinate gives the percentage of light passing through the sample as a percentage referred to an appropriate standard. The abscissa represents the drying time. A high value indicates more light passing through and thus poorer adhesion of the ink. A low value indicates little ink removal and thus good adhesion. The solid line represents an ink using a "soft" binder system and the dashed line a "hard" binder system. The data clearly show that the soft binder gives better adhesion at short times than the hard one, and that both systems behave much the same at long drying times.

A different type of application of the tape peel test in the photographic film industry was given by Grace et al.[8] These investigators used the tape peel test in conjunction with a time-resolved salt bath technique for investigating the adhesion of silver coatings to poly(ethylene terephthalate) (PET) films. In the salt bath test, silver-coated PET films were immersed in a salt bath, and the time required for the silver to lift off was noted. These results were then correlated with standard tape peel testing in a manner similar to that mentioned above. The essential result of this investigation was the demonstration that the salt bath test was able to better discriminate different levels of adhesion of the silver coatings than the tape peel test alone. The tape test basically gave a good/not good type of result, whereas coatings tested in the salt bath would survive for different lengths of time, thus giving a more continuous scale of adhesion performance. In particular, coatings that the tape peel test indicated were good were shown to delaminate at intermediate times in the salt bath test. However, films shown to be poor by the tape peel test were also poor by the salt bath test. Thus, the tape peel test supported the salt bath experiments but did not give the same degree of resolution of adhesion strength.

2.3.3 Disadvantages of the Tape Peel Test

As pointed out, the tape peel test at best can give a semiquantitative estimate of the adhesion of a coating. The results of the test tend to be confounded by the mechanical response and other failure modes of the tape backing and the tape adhesive as well as similar behavior of the coating and the substrate material. With so many potential complicating factors, the interpretation of tape peel test data is very difficult if more than a simple qualitative estimate of adhesion strength is required. The use of calibration methods and reference samples is mandatory to ensure a reasonable level of repeatability.

2.3.4 Summary and Recommendations

Although the tape peel test is limited to a qualitative or at best semiquantitative evaluation of adhesion, it has a number of advantages that make it attractive in specific applications. In particular, in situations for which a simple rapid test is required, such as for testing printing inks, or for which a straightforward go/no go evaluation is sufficient, this test may be perfectly adequate. In some cases, it may be the only reasonable test available. However, great caution is recommended in evaluating tape peel data. and the results should not be overinterpreted in terms of trying to understand the fundamental adhesion of a coating because a large number of confounding factors come into play with this test.

2.4 PULL TEST

2.4.1 Introduction

As with the peel test, the pull test is a general method for assessing both qualitatively and semiquantitatively the adhesion of coatings to a variety of substrates. Also, as with the peel test, it enjoys a number of advantages and suffers from several disadvantages. The advantages include the following:

- Applicable to a wide variety of coatings and substrates including
  - brittle coatings
  - flexible coatings
- Gives both qualitative and semiquantitative results
- Relatively easy sample preparation

The disadvantages of this method include

- Difficult data analysis, especially for quantitative measurements
- Rapid uncontrollable failure mode
- Wide scatter in data
- Need for bonding adhesive or solder

An excellent evaluation of the pull test as applied to paint coatings was given by Sickfeld.[9] This author investigated two basic pull test configurations, as illustrated in Figure 2.3. The symmetric configuration is preferable for testing coatings on

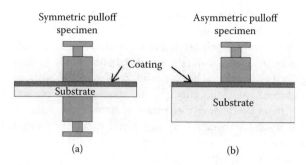

FIGURE 2.3 Schematic of two pull test configurations. The symmetric configuration is preferred when the substrate is thin or flexible. The asymmetric mode is preferred when the substrate is thick or very rigid. *Note:* Glue/adhesive layer not shown.

relatively thin flexible substrates, whereas the asymmetric sample is preferred for thick rigid substrates. Similar to the tape peel test, the pull test involves two additional materials besides the coating and substrate under investigation. The test stud itself is fabricated out of a high-modulus metal or ceramic material and for the case of paint coatings can be considered almost perfectly rigid. This is not the case for testing stiff brittle coatings such as diamond or ceramics; in these cases, the stud material must be carefully figured into the analysis. In addition, an adhesive is required to attach the test stud to the coating under test. For paint coatings, this is typically an epoxy glue, and its properties will always enter the analysis.

After the test stud is appropriately cemented to the coating under test, it is pulled off under controlled conditions using a tensile test apparatus. A number of complications now enter the picture as follows:

- Unless the load is applied very carefully, there can be an off-axis component that can impose a bending moment to the sample in addition to the tensile load.
- Even assuming pure tensile loading, any real sample will not be uniformly bonded, and the applied stress field will seek out any defects or bonding weaknesses.
- Failure will be initiated at the weakest point in the structure and propagate at acoustic velocities to complete separation.
- Failure can occur either adhesively at any of the three sample interfaces or cohesively in any of the four bulk materials. Mixed-mode interfacial and cohesive fracture is the most common failure mode.

Given the list of complexities, it should come as no surprise that typical pull test data show a wide range of variation. Multiple tests must be run at any given condition, and data-censoring techniques should be applied to ferret out unwanted failure modes.

2.4.2 Advantages of the Pull Test

The main advantage of the pull test is its wide-ranging applicability to all manner of coatings, from relatively soft flexible polymer coatings to hard brittle coatings such as diamond. In addition, as pointed out by Sickfeld,[9] there are two types of information that can be obtained from this test. The first is qualitative, deriving from an analysis of the resulting pull off fracture surface. Some idea of the integrity of the coating can be obtained by noting whether failure tends to be mainly cohesive in the coating itself or interfacial between the coating and the substrate. In particular, Sickfeld was able to study the effect of moisture and solvent immersion on the failure mode of paint coatings. For the case of immersion of the coating in water, subsequent pull testing showed nearly interfacial failure between the coating and substrate. On the other hand, coatings immersed in gasoline or oil demonstrated a mixed interfacial/cohesive type of failure. This type of data can be very valuable when evaluating a particular coating for use under particular service conditions.

A second advantage is the quantitative information derived from the pull test. It has been hypothesized that, in most cases, the failure mode in a given pull test experiment is determined by a preexisting distribution of flaws in the sample. Thus, as discussed in the preceding section, the applied stress field seeks out the largest, most vulnerable flaw in the sample. Failure initiates at this point, and the initial flaw rapidly propagates at acoustic velocities to ultimate separation of the pull stud and the sample surface. In effect, it is assumed that all samples will have some kind of inherent flaw distribution no matter how carefully they were prepared, and the stress field deriving from the pull test will inevitably find the most vulnerable flaw; failure will initiate and propagate from that point. It has further been found that Weibull statistics are very effective in analyzing this type of data.

Pawel and McHargue[10] used the pull test to analyze the adhesion of iron films to sapphire substrates. These investigators ion implanted both nickel and chromium impurities at the interface between an iron film and sapphire substrate. Subsequent pull testing and Weibull analysis unequivocally demonstrated that the chromium interlayer substantially improved the adhesion of the iron film over the untreated and nickel-treated cases.

Finally, there are situations that are perfectly disposed toward the pull test, such as evaluating the durability of pins on a microelectronic packaging substrate. For large mainframe machines, such substrates can carry over 100 silicon chips and require over 1000 pins to distribute power and signal data to a supporting carrier board. The reliability of these pins is critical to the proper function and performance of the total chip/substrate assembly, and each pin must meet very stringent reliability and performance criteria. The pull test is the natural performance evaluation procedure for this application. Coupled with the appropriate Weibull analysis, the pull test provides a crucial engineering and quality control tool for the design and fabrication of such structures.

2.4.3 DISADVANTAGES OF THE PULL TEST

One of the main disadvantages of the pull test is the wide variability in typical test data. This problem has been documented most cogently by Alam et al.[11] These investigators attempted to evaluate the adhesion of chemical vapor deposition (CVD) diamond coatings to tungsten substrates. Because of a variety of conditions affecting their sample preparation, including nonuniformity of film thickness, diamond quality, film cohesion, and surface preparation, they observed considerable variability in their pull test data. In the authors' own words: "The measured adhesion values showed larger variations from point to point across the sample surface and from identically prepared samples than variations as a function of the film processing parameters." Thus, the data derived from pull testing in this case were mainly qualitative. Every form of sample failure was observed, including clean interfacial delamination, partial delamination with partial film cohesive failure, cohesive failure in the epoxy adhesive coupled with delamination of the epoxy from the coating, and pure cohesive failure of the diamond coating. With such a wide range of failure modes, it was no wonder that the data showed a high degree of variability. With the use of statistical analysis, however, the authors were able to show that substrate preparation, gas flow, and gas pressure were the most important processing parameters.

2.4.4 SUMMARY AND RECOMMENDATIONS

In conclusion, it is clear that the pull test can be an effective tool for evaluating both the qualitative and semiquantitative durabilities of a wide variety of coatings. The main advantage of this technique is its versatility and applicability to a wide range of coating/substrate systems. It can be applied to both soft flexible coatings as well as hard brittle ones. Pull test equipment is commercially available and can be set up in any laboratory with a tensile testing apparatus. The use of advanced statistical analysis such as Weibull analysis can be helpful in providing semiquantitative information concerning the durability of various coatings.

The wide variation in typical pull test data remains one of the main weaknesses of this technique. Multiple tests must be done on a given sample coupled with statistical analysis to obtain reliable quantitative data. However, there are a number of specific instances, such as pin testing on microelectronic substrates, for which the advantages of pull testing make it the most natural choice for reliability testing.

2.5 INDENTATION DEBONDING TEST

2.5.1 INTRODUCTION

Figure 2.4 shows a schematic representation of the indentation test. In this test, an indenter with a hemispherical point (i.e., with a tip radius that is on the order of the thickness of the coating tested) is thrust into the coating under carefully controlled conditions. The dominant effect of this maneuver is to greatly compress the coating material directly under the indenting tip. Surprisingly, however, a concomitant delamination of the coating can also occur starting at the edge of the indenter and extending for a distance that can be several times the indenter tip radius.

An early example of the use of this technique for testing the adhesion of epoxy to copper in circuit boards was given by Engel and Pedroza.[12] These investigators worked with epoxy coatings in the range from 25 to 300 μm on copper metal substrates. Using an indenter tip of approximately 0.2-mm radius, they observed peripheral delamination around the central indentation out to a radius of up to 2 mm.

One simple way of understanding the mechanics of what is happening is given in Figure 2.5. As the indenter penetrates the epoxy coating, material is extruded to

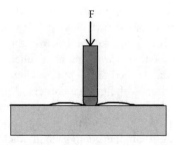

FIGURE 2.4 Idealized representation of the indentation debonding test. An indenter with a spherical tip is pressed into the coating under carefully controlled loading conditions. A high compressive stress is developed under the indenter; however, depending on the elastic/plastic response of the substrate, a tensile field is developed at the periphery that can cause the coating to delaminate.

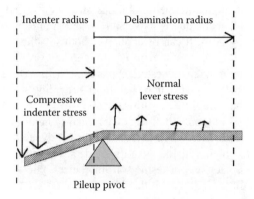

FIGURE 2.5 Simplified description of indentation debonding test. The coating is treated as a simple beam setup. The compressive force of the indenter loads the left ligament of the beam while also pushing underlying material to the periphery, where it piles up and forms a pivot. Material to the right of the pivot point experiences normal tension because of the lever effect of the coating acting as a simple beam.

the periphery of the indenter, causing a pileup at the edge. In this case, the underlying copper material is also pushed to the indenter edge because copper is a highly plastic material, which also contributes to the pileup of material. The excess mound of both epoxy and copper at the edge of the depression can be thought of as forming a sort of pivot for the epoxy coating to act in the fashion of a simple lever, as shown in Figure 2.5. The epoxy film will have significant rigidity on the length scale of 1 mm, so the normal stress generated by this levering effect can be quite significant and lead to delamination if the coating adhesion is not sufficient. A further contributing mechanism is the shear stress generated by the extrusion of the epoxy material from under the indenter. Thus, a combination of flow shear stress coupled with normal lever stress can operate to delaminate the coating starting at the indenter edge. Engel and Pedroza[13] also used a simple plate model of the coating to estimate the radial strain in the coating and referred to this as the peel strain. Such a strain can be an energy source for driving the indentation delamination.

A far more rigorous analysis of the stresses driving the delamination process in the indentation test was given by Jayachandran et al.[14] These authors treated the case of a poly(methylmethacrylate) polymer coating on a rigid substrate. Having access to extensive data characterizing the constitutive behavior of the poly(methylmethacrylate), they were able to carry out highly detailed numerical studies of the indentation process using the finite element method, including full details of large deformation and viscoplastic strain phenomena. Because these authors assumed a perfectly rigid substrate, their results cannot be compared directly to those of Engel and Pedroza. What was found is that there was indeed a massive shear flow and pileup of material created by the indenter all the way to the edge and beyond. However, because of the assumption of a rigid substrate, only a small tensile normal stress was predicted beyond the indenter edge. Thus, the normal stress in the epoxy/copper system is mostly caused by the pileup of the copper at the indenter edge, which forms a pivot on which the epoxy coating acts as depicted in Figure 2.5.

The indentation debonding method has also been applied to hard refractory coatings, as demonstrated by the work of Weppelmann et al.[15] These authors investigated the titanium nitride (TiN)/silicon system both theoretically and experimentally. Experimentally, they used a diamond indenter in conjunction with a digital interference microscope to follow the sample deformation precisely. Theoretically, they were able to develop a simple formula for the strain energy release rate for delamination caused by the radial strain induced by the indentation process. Using their experimental results, they were able to estimate an adhesion strength of approximately 1.2 J/m^2 for the TiN/silicon system.

2.5.2 ADVANTAGES OF THE INDENTATION DEBONDING TEST

The indentation test has a number of clear advantages, which can be summarized as follows:

- Applicable to a wide variety of coating/substrate systems
- Ease of sample preparation
- Both qualitative and quantitative results obtained
- Commercial equipment readily available

The indentation test is readily implemented both in the laboratory and on the production line for a wide variety of coatings. As mentioned, it has been applied to both soft flexible coatings on metals and hard brittle coatings on silicon. Engel and Pedroza[12] commented on the use of this test for quality control in testing the adhesion of epoxy on copper in circuit boards. Other than preparing the coating, no special preparation of the test sample is necessary. The test is clearly applicable to a wide variety of substrates and has been applied to testing scratch-resistant coatings on curved plastic lenses. As mentioned, the indentation test can be analyzed to give quantitative results in addition to a simple qualitative estimate of the coating durability. Finally, commercial off-the-shelf equipment is readily available in the form of indentation test equipment and powerful microscopes with digital interferometers for evaluating both substrate damage and deformation.

2.5.3 DISADVANTAGES OF THE INDENTATION DEBONDING TEST

The main disadvantages of the indentation test can be summarized as follows:

- Complex mode of loading involving large compressive stress and high shear strains
- Difficult quantitative analysis and poorly understood precise mechanism of delamination

The very high compressive load induced by the indentation test coupled with the high shear flow associated with soft coatings may make the relevance of the indentation test questionable for some coating systems. In particular, coatings subjected to large temperature swings may delaminate at edges or other discontinuities under loading conditions that are far different from those induced by the indentation test. A further drawback for hard coatings is the fact that, in addition to a large compressive stress, a very significant hoop stress is also generated by this test, which can lead to radial cracking in the sample substrate as well as the coating. Thus, multiple failure modes can greatly complicate the interpretation of the data when one is primarily interested in the coating adhesion.

2.5.4 SUMMARY AND RECOMMENDATIONS

The indentation debonding test clearly passes many of the criteria required for an ideal adhesion test. Primary among these are the ease of sample preparation and applicability to a wide variety of coating/substrate systems. Ready availability of commercial equipment makes this test a favorite in many industries that have to deal with quality control issues involving coatings. The main problem to be aware of is whether the loading conditions created by this test are reasonably close to those that the coating under test must endure in practice. In general, this is a very relevant test for coatings that must endure abrasive conditions and contact with potentially penetrating surfaces. Great care should be taken, however, if the coating in question will be subjected to large thermal strains that can be induced by large temperature gradients or thermal expansion mismatch between the coating and substrate.

2.6 SCRATCH TEST

2.6.1 INTRODUCTION

Figure 2.6 gives a highly schematic representation of the scratch test, which can be thought of as an extension of the indentation test with the added feature that the indenter is translated along the sample surface as well as into the coating. An informative overview of the early history of this technique was given by Ahn et al.[16] Apparently, Heavens[17] and Heavens and Collins[18] were the first to employ this

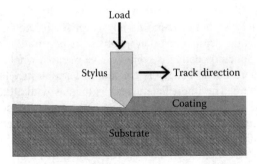

FIGURE 2.6 Essentials of the scratch test. Loading conditions are similar to the indentation test in that the stylus pressure is ramped up according to some specified program; however, the stylus is simultaneously driven forward at a fixed rate.

technique to study the durability of metallic films evaporated onto glass. Subsequently, Benjamin and Weaver[19] performed an elementary mechanics analysis of this method and derived the following simple formula for the shear force to be overcome by the scratch stylus:

$$F = \frac{AH}{\sqrt{(R^2 - A^2)}}$$

$$(2.1)$$

$$A = \sqrt{\frac{W}{\pi H}}$$

where

A = Radius of stylus contact circle
R = Radius of stylus tip
W = Applied load normal to coating surface
H = Indentation hardness of substrate
F = Shearing force resisting lateral motion of stylus

The hope was that the load W required to remove the coating could be taken as a measure of the coating adhesion by relating it to the generated shear force given in Equation (2.1). However, a number of complications were noted by later workers,[20] who discovered several difficulties, including the following:

- Delamination of the coating can be observed even before the stylus removes all traces down to the substrate. In addition, the film can be thinned to the point at which it becomes translucent and cannot be removed.
- Complex material properties such as the elastoplastic behavior of the coating and substrate determine the nature of the scratch track.
- Multiple modes of failure are observed, including mechanical failure in the bulk of the coating or substrate in addition to interfacial delamination.

Given the complications observed in the mechanically simpler indentation test, none of these remarks should come as any surprise. It should be clear that an experiment such as the scratch test that involves the penetration and dragging of a stylus through an adhered coating is going to give rise to a whole range of complex thermomechanical response behaviors, including viscoplastic flow, bulk fracture, and interfacial failure. The immediate upshot is that, as with all the other adhesion tests discussed, the scratch test will be at best a semiquantitative technique. However, this does not preclude the usefulness or effectiveness of this method for providing insight into the adhesion and durability of coatings in a variety of applications. In particular, Ahn et al.[21] demonstrated that the scratch test can readily reveal poor adhesion in a coating because in this case lateral delamination of the coating can be observed to occur along the length of the scratch track.

In an attempt to put the scratch test on firmer footing, Oroshnik and Croll[22] came up with the concept of threshold adhesion failure (TAF). These investigators noticed that, in the thin aluminum films they were investigating, small patches of delamination could be observed in the scratch track well before the scratch stylus penetrated to the underlying substrate. It occurred to them that the load at which this patchy delamination occurred could be used as a measure of the coating adhesion. They proposed the following definition:

> Threshold Adhesion Failure occurs if, within the boundaries of a scratch and over its 1-cm path, removal of the film from its substrate can be detected by transmitted light with a microscope (×40 magnification) at even one spot, no matter how small.

This definition coincides well with Definition B (*see* Chapter 1) for adhesion and is certainly serviceable for the purposes at hand. Oroshnik and Croll described the method by which TAF is obtained for a given coating. The load on the stylus is increased incrementally as it moves over the sample surface up to the point at which spots of delamination are just detected. The load is then incrementally decreased until the delamination events just disappear, at which point the load is again increased to the point at which the delaminations again appear. This procedure of successively incrementing and decrementing the stylus load is repeated until the apparent threshold load for producing delaminations is reliably boxed in between upper and lower load conditions. Figure 2.7 shows a schematic of the type of data obtained from this procedure. Note from this figure how the data tend to settle at a fixed level of apparent adhesion strength.

Oroshnik and Croll went on to discover that, even though using a given stylus, the TAF data were highly reproducible and consistent, no two stylus tips were identical, and each gave different TAF results. Using microscopic interferometry, these authors discovered that the stylus tips they were using were neither spherical nor with an unambiguous radius. In fact, data were presented showing measurements taken with different styli on a single film for which the TAF load differed by nearly a factor of 2. Furthermore, it was shown that the Benjamin and Weaver result for the shear force given by Equation (2.1) was not verified by the data, which was not surprising given the nonspherical nature of the stylus tips used. It is in fact well established that the most critical factor controlling the scratch test is the nature of

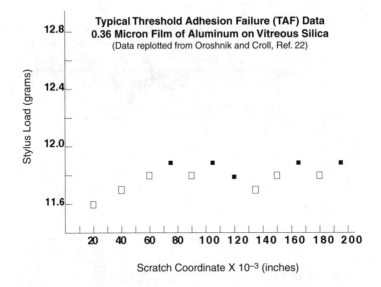

FIGURE 2.7 Typical threshhold adhesion data. Filled squares indicate load level at which delamination first appears while the load is increasing. Open squares indicate load level at which delamination first ceases while the stylus load is decreased. At the critical load for delamination, the load increasing and the load decreasing data should closely overlap.

the stylus tip. Different results will be obtained depending on the stylus material (steel, tungsten carbide, or diamond) and the precise topography of the tip region that first contacts the coating surface.

Later investigators expanded on the scratch test not only to include more sophisticated instrumentation for controlling the load program of the stylus and advanced microscopy for observing the scratch track, but also to couple the scratch method with acoustic spectroscopy, by which the sound vibrations generated by the stylus are detected and recorded at the same time that the scratch track is formed. Typical of the more modern approach is the work of Vaughn et al.[23] These investigators performed adhesion measurements on both copper and diamondlike carbon (DLC) films coated onto PET substrates. These experiments gave an ideal opportunity to evaluate the performance of the scratch test on two radically different types of coatings on the same substrate material. The highly plastic copper coatings gave completely unremarkable stylus load-versus-scratch length plots in which the load simply increases monotonically, with scratch length showing no particular discontinuities where the film began to delaminate. Further, these coatings gave no discernible acoustic signal at coating failure events such as delamination. By contrast, the rigid brittle DLC coatings showed a sharp drop in the stylus load versus scratch length curve at points where the DLC coating fractured in the typical "herringbone"-style cracks that can appear in the scratch track because of the high tensile stress just behind the advancing stylus tip. Figure 2.8 shows a schematic representation of the type of data obtained from the DLC coatings.

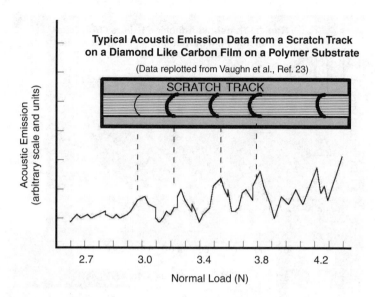

FIGURE 2.8 Correlation of acoustic emission data with microscopy of scratch track. Note that the appearance of a "herringbone" crack as the load increases coincides exactly with a local peak in the acoustic emission plot.

An innovative version of the scratch test was implemented by Sarin,[24] who basically turned the experiment on its side and performed a microscratch test on a cross section of the coating substrate sample at 90° to the coating surface (Figure 2.9). The sample is sliced across a section of interest, and the edge is precisely polished by standard metallographic methods. The scratch test is then carried out in the standard manner only on a microscale because of the very narrow cross section under investigation. One advantage of this approach is that, under certain conditions the failure mode can be isolated at the coating/substrate interface, and the precise locus of failure is immediately apparent by direct microscopic investigation. Sarin, in fact, found that for a titanium carbide (TiC) film coated onto a silicon nitride/titanium carbide (Si_3N_4 + TiC) ceramic composite, he was indeed able to generate a crack along the coating/substrate interface, thus obtaining a true delamination failure mode. However, when the same type of experiment was performed on a TiC coating on tungsten carbide-cobalt (TC-Co) substrate, cracks were observed to start in the substrate and then propagate into the coating. Sarin's findings illustrate what tends to be a general rule for all adhesion measurement methods in that any given method will be found to work well on a given specific sample system but will generally give unexpected results when the sample is changed even in a relatively minor way.

Further attempts at quantifying the scratch test have been reviewed by Bull,[25] who examined the scratch test as applied to a number of qualitatively different coating/substrate systems. This author basically found that the following criteria must be met to achieve a truly quantitative assessment of adhesion strength using the scratch test:

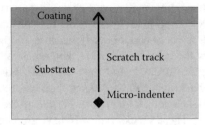

90 degree micro-scratch test
(Adapted from V. K. Sarin ref. 24)

FIGURE 2.9 Innovative variation of scratch test in which the sample is sectioned, and using a microindenter, the scratch track proceeds from within the sample substrate toward the coating/substrate interface at a 90° angle.

1. There must be a well-defined delamination mode present under the prevailing test conditions.
2. Knowledge of the sample stress state leading to delamination failure must be available either through direct measurement or calculation.

With these conditions in mind, Bull examined a variety of coating/substrate systems that basically fell into the categories of hard and soft. Soft systems generally give rise to a high level of viscoplastic deformation, and clean failure modes may not be clearly observable. Hard systems, on the other hand, often exhibit brittle fracture in the substrate, the coating, or both. Combinations of soft and hard materials can give rise to all of the previously mentioned failure modes plus interfacial delamination. Table 2.1 gives a highly qualitative overview of the general trends. This table illustrates the general observation that the scratch test works best when at least one component of the coating/substrate system is a relatively hard material. On reflection this is intuitively reasonable. You clearly would not use the scratch test to quantify the adhesion of molasses to chewing gum.

TABLE 2.1
Qualitative Summary of Failure Modes in the Scratch Test as a Function of Coating/Substrate Hardness

Substrate hardness ⇒ Coating hardness ⇓	Soft	Medium	Hard
Soft	Plastic deformation extrusion	Coating thinning scrape off	Coating thinning scrape off
Medium	Plastic deformation extrusion	Delamination	Delamination fracture
Hard	Plastic deformation extrusion	Delamination fracture	Delamination fracture

However, Bull did illustrate two situations for which the scratch test gives promising results for hard coatings on hard substrates. The first case involves what can basically be called buckling spallation. For the system TiN on stainless steel, a thin coating of the TiN can be made to spall off the substrate. The scratch stylus creates a compressive stress ahead of itself because of the deformation of the substrate, which causes the coating to buckle. High tensile stresses in the coating cause the buckled coating to crack and then flake off. For thicker coatings, the bending required for buckling does not occur because the coating is too stiff. However, compressive shear cracks can form ahead of the indenter through the thickness of the coating. These cracks typically have sloping sides that can act as an inclined plane or wedge. The forward motion of the indenter can then drive the edge of the coating up the ramp created by the crack, causing the interface between the coating and substrate to separate, which leads to spalling of the coating. Bull referred to this mode of delamination as *wedge spallation.*

Finally, even though the scratch test has proven most effective for hard brittle coatings, work by Bull et al.[26] showed that this test can also give valuable information concerning the adhesion and durability of polymer coatings such as epoxies. The application in this case was the evaluation of epoxy coatings for covering the inner hull of coal-carrying vessels. The epoxy is intended to protect the metal hull from abrasion by the coal and corrosion caused by the brackish marine environment. Lumps of coal settling against the inner hull have an abrasive effect; thus, the scratch test was deemed an appropriate method for evaluating the performance of the epoxy coatings. A number of different coating formulations were investigated, and all showed varying tendencies to either crack or delaminate under the action of the scratch stylus. One interesting failure mode arose because of the tensile load imposed on the coating behind the moving stylus. This would cause a through crack to form behind the stylus, which would then drag the coating along, opening up the crack and further buckling the coating in front in much the same manner as a rug buckles when pushed laterally at one of its edges. In addition, it was found that coatings with extender pigmentation tended to crack more readily than the unfilled coatings. However, the filler also retarded adhesion degradation as determined by aging studies. Thus, scratch testing can be used to determine the optimal tradeoff between coating toughness and adhesion by comparing scratch results on coatings with varying levels of extender pigmentation.

2.6.2 Advantages of the Scratch Test

As with the indentation test, the main advantage of the scratch test lies in the relative ease of sample preparation. One simply prepares coupons of a convenient size out of the relevant coating and substrate materials in the same manner as on the manufacturing line or in the build shop. In addition, the newer commercially available equipment can be fitted with a number of auxiliary tools, such as microscopes, acoustic spectrometers, and surface-profiling attachments. Because the stylus can also act as an indenter, the coating hardness and elastic properties can be determined. Thus, a single instrument can give valuable information on surface topography, mechanical properties, and modes of deformation and delamination. Semiquantitative information can be obtained by recording the stylus load at failure; thus, scratch testing can be used to rank the durability of a series of coating formulations. In some cases, a fully quantitative estimate of the fracture energy of the coating/substrate interface can be obtained if care is taken to measure carefully all relevant mechanical properties and carry out the appropriate fracture mechanics calculations.

2.6.3 DISADVANTAGES OF THE SCRATCH TEST

There are two essential disadvantages to using the scratch test as an adhesion measurement tool: This test is essentially limited to hard brittle coatings even though some exceptions, such as brittle epoxy coatings, may be successfully investigated. The softer metals and most polymer coatings tend to viscoplastically flow and deform around the scratch stylus, causing mounding at the edges of the scratch track and pileup in front of the stylus. In addition, these coatings do not give a distinct acoustic signal at the failure point, thus negating the use of acoustic spectroscopy. Also, some coatings can be thinned to the point of optical transparency without achieving complete coating removal, thus complicating any attempt to assess adhesion strength.

The second limitation arises from the fact that, as with the indentation test, the scratch test is mechanically complex. The act of pushing the stylus into the coating gives rise to very high stresses and deformations in both the coating and the substrate, thus bringing into play the full range of highly nonlinear viscoplastic material behaviors. The usual type of elastic mechanical calculations do not give quantitatively reliable results and can be used only for a more heuristic analysis of scratch test data. Also, the basic failure modes of the scratch test are only poorly understood, and experience gained on a given coating/substrate system may not be reliably carried over to a different one.

2.6.4 SUMMARY AND RECOMMENDATIONS

The scratch test is one of the most popular adhesion tests currently in use in both industry and academia. This stems largely from the great versatility of this technique for evaluation of a wide range of coating/substrate systems and the ready availability of commercial equipment that can perform a variety of functions, such as surface inspection, surface roughness measurements, and the evaluation of coating mechanical properties. For the case of hard brittle coatings, the scratch test is quite likely the best available technique for most situations. For softer coatings, this method may also be able to give useful results under certain conditions and can be used in a complementary manner with other techniques, such as the pull test.

2.7 BLISTER TEST

2.7.1 INTRODUCTION

One constant complaint about all of the adhesion tests mentioned above is the fact that they involve a mechanically complex process with large deformations and strains giving rise to highly nonlinear viscoplastic response behavior by the coating/substrate system investigated. An immediate consequence of this is the fact that the analysis of these systems in terms of continuum and fracture mechanics concepts is exceedingly difficult, if not impossible. The blister test is an attempt to circumvent these difficulties by developing a blister in the coating in a well-defined manner that will propagate a delamination front between the coating and substrate in a controlled manner using only moderate deformations and strains.

Dannenberg[27] was apparently the first investigator to apply this technique to measure the adhesion of polymer coatings. Lai and Dillard[28] gave an insightful account of the mechanics of several versions of this test. Figure 2.10 is a schematic representation of four different versions of this test mentioned by these authors. Figure 2.10a illustrates the standard blister test configuration. The main limitation of this version occurs when the coating ruptures before the coating can delaminate. This problem limits the standard blister test to coatings with either very high fracture toughness or relatively low adhesion to the underlying substrate. To circumvent this problem, Allen and Senturia[29,30] devised the island blister test shown in Figure 2.10b. Because of the much smaller debond front presented by the inner island, the driving force for delamination is much greater here than at the much larger circumference at the outer radius. Thus, delamination can be made to occur at the inner island at a much lower applied pressure than would be required in the standard blister test.

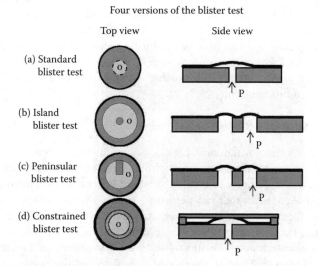

Four versions of the blister test

FIGURE 2.10 Four different configurations of the blister test: (a) The standard test involves applying hydrostatic pressure to the bottom of the coating, thus inducing a blister. The main drawback of this approach is that the coating may burst before significant delamination occurs. (b) The island blister test is one attempt to get around the coating bursting problem. In this configuration, the driving force for delamination is much higher on the central island than on the periphery. Thus, one is more likely to get delamination before the coating fails. (c) The peninsular test is a variation of the island test that also alleviates the coating failure problem while improving on the stability of the test. (d) For the constrained blister test, placing a transparent rigid cover over the blister allows control of the coating deformation while observing the delamination front.

One problem with the island blister test is that it tends to be unstable. To overcome this difficulty, Dillard and Bao* conceived the peninsular blister test depicted in Figure 2.10c. This version of the blister test maintains the high driving force for delamination as in the island blister test while maintaining a more steady delamination front.

Figure 2.10d shows the constrained blister test, which cleverly supplies a simple cover for the standard blister test to prevent the problem of film rupture. The earliest investigations of this technique were apparently carried out by Napolitano et al.[31,32] and nearly simultaneously by Dillard[33-35] and co-workers. Napolitano and co-workers[32] managed to derive, using simple thermodynamic arguments, the following formula for the expanding area of the propagating blister:

$$\frac{A(t)}{A(t_0)} = \exp[\frac{\beta p^2 h}{ph - \gamma}(t - t_0)]$$ (2.2)

where

$A(t)$ = Blister area at time t
p = Applied pressure
h = Spacer height
γ = Interfacial fracture energy
β = Dissipative coefficient
t_0, t = Initial and current time, respectively

Equation (2.2) was used to analyze constrained blister test data taken on a pressure-sensitive adhesive tape. The interfacial fracture energy was determined by noticing at what combination of spacer height h and applied pressure p the onset of delamination occurred. The interfacial fracture energy could then be computed from the following simple formula:

$$\gamma = (hp)_{threshold}$$ (2.3)

where it is to be noted that the pressure p in Equation (2.3) is that which just causes blister delamination to progress. The dissipative coefficient could then be obtained from data on blister area versus time. Plotting the log of both sides of Equation (2.2) yields a straight line with a slope that gives the parameter β, knowing γ from Equation (2.3).

Liang et al.[36] gave a very interesting application of the constrained blister test as applied to electropolymerized polymer coatings on copper substrates. They improved significantly on the work of Napolitano et al. by bringing to bear advanced image analysis methods implemented on a powerful modern workstation. With this advanced hardware and software, they were able to measure in real time the critical blister growth front parameters and thereby analyze their data using a fracture mechanics result for the strain energy release rate derived by Chang et al.[37] given by the following formula:

---

* "The Peninsula Blister Test — A High and Constant Release Rate Fracture Specimen for Adhesives," D.A. Dillard and Y. Bao, *Journal of Adhesion*, 33, 253 (1991).

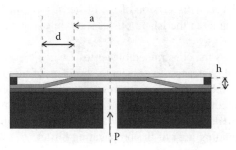

Geometric parameters in constrained blister test
(Refer to Eq. 2.4)

FIGURE 2.11 Illustration of the configurational parameters for the constrained blister test appearing in Equation (2.4).

$$G = ph\left[1 - \left(\frac{d}{2a}\right) + \left[\left(\frac{d}{3a}\right) - 1/2\right]\left(\frac{\partial d}{\partial a}\right)\right] \qquad (2.4)$$

In this equation, G is the strain energy release rate, and p is the applied pressure; the dimensional parameters are explained in Figure 2.11.

An elementary analysis of the island blister test was given by Allen and Senturia.[38] By assuming that the coating can be treated as a membrane (i.e., it is so thin that it has no unsupported stiffness) and that the major component of the driving force is caused by residual stress, they came up with the following simple formula for the surface fracture energy of adhesion:

$$\gamma_a = \frac{(p_c a_1)^2}{32\sigma_0 t}\left[\frac{\beta^2 - 1}{\ln\beta} - 2\right]^2 \qquad (2.5)$$

where
γ_a = Surface fracture energy
p_c = Critical pressure for delamination propagation
σ_0 = Residual film stress
t = Film thickness
$\beta = a_1/a_2$
a_1, a_2 = Geometric parameters shown in Figure 2.12

Allen and Senturia also gave an interesting comparison between the island blister test and the standard version and demonstrated why the former gives a much higher delamination force on the inner island than the latter can achieve on the outer circumference of the suspended membrane.

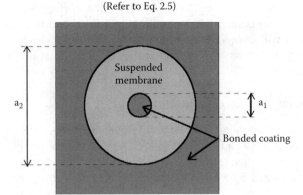

Geometric parameters of island blister test
(Refer to Eq. 2.5)

FIGURE 2.12 Illustration of the configurational parameters for the island blister test appearing in Equation (2.5).

2.7.2 ADVANTAGES OF THE BLISTER TEST

As can be ascertained from the discussion, the blister test has a number of advantages; mainly, it is the first test discussed here that lends itself readily to a fully quantitative analysis based on fracture mechanics methods. This is mainly because this test imposes relatively low strains on the coating material, thus avoiding the complex nonlinear viscoplastic behavior that greatly complicates methods such as the peel test. However, like the peel test the blister test also concentrates the maximum stress at the delamination front, thus constraining the failure crack to be close to the coating/substrate interface of interest. Also, given the fact that a number of different versions are available, as shown in Figure 2.10, the user has great flexibility in testing coatings with varying levels of adhesion.

2.7.3 DISADVANTAGES OF THE BLISTER TEST

Whereas the blister test lends itself fairly readily to quantitative analysis, this advantage is purchased at the price of ease of sample preparation. It seems in the realm of adhesion testing that nothing comes free. If you gain an advantage in one quarter, then you pay for it in another. The main impediment comes with drilling the hole at the center of the blister through which the pressurizing gas enters. This can be accomplished in a number of ways, but the most popular is the use of etchants, which will erode the substrate material and not attack the coating. For coatings on silicon substrates, all the methods of microelectronic lithography are available to etch holes in the silicon and construct the various structures required for tests such as the island blister configuration. However, to use these methods a wafer fabrication facility must be available to carry out the involved series of steps required to construct the desired structures. This work also requires the use of very nasty etchants, such as buffered hydrofluoric acid. In addition, the blister test is limited to fairly flexible coatings, such as polymer-based paints and soft metals. Hard brittle coatings will tend to crack before forming a blister under the influence of the applied pressure. Finally, this test will have severe problems with coatings under high compressive stress because this will cause the coating to buckle as soon as it is lifted from the supporting substrate.

2.7.4 SUMMARY AND RECOMMENDATIONS

The blister test is most suitable when a fully quantitative analysis of the adhesion strength of a coating to a particular substrate is required. Great flexibility is available in the types of samples that can be used, and a number of investigators have provided detailed analyses that can be used to interpret the data in a fully quantitative manner. Also, the fact that delamination is achieved at low strain levels makes the results of the test much more relevant to coating adhesion than large strain-inducing methods such as the peel test. On the downside, however, sample preparation is far more cumbersome than for many other techniques. Thus, the blister test is not recommended when only qualitative or semiquantitative data are required because many other simpler methods are available.

2.8   BEAM-BENDING TESTS

2.8.1 INTRODUCTION

There are a variety of tests for measuring adhesion that rely on the relatively simple mechanics of the elastic beam to simplify the required analysis for the stress intensity factors and the strain energy release rates that drive the delamination process. Precise definitions of these quantities are given in Chapter 4, which treats the methods of fracture mechanics in detail. The most attractive feature of beam-bending tests is that the stress field induced by the bending operation is comparatively simple and can thus be analyzed by elementary methods. Details of this type of analysis are given in Chapter 3, in which the fundamentals of continuum theory are discussed in some detail. Figure 2.13 illustrates several of the most popular adhesion tests that rely on the mechanics of a bending beam. The following sections discuss each in turn.

2.8.2 THREE-POINT BEND TEST

A standard configuration for the three-point bend test is shown in Figure 2.13a. Using the three-point bend test, McDevitt and Baun[39] carried out one of the earliest studies on metal-to-metal adhesive joints. These investigators found the curious result that apparently the three-point bend test was more sensitive to interfacial weaknesses than other tests they were performing such as the T-peel, the wedge test, and the lap shear test. Load-versus-deflection data were gathered on metal/adhesive/metal sandwich samples as depicted schematically in Figure 2.14. The top curve in this figure represents a nonbonded sample in which a cured strip of adhesive was simply laid between two metal layers with no apparent bonding other than simple friction. This load-displacement curve thus serves as a baseline for a completely nonbonded

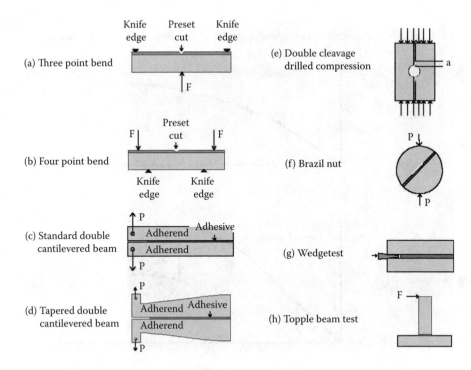

FIGURE 2.13 Eight different adhesion tests that rely on the elastic bending properties of stiff beams.

joint. Figure 2.14b shows the case where the metal/uncured adhesive/metal sandwich was cured in an oven to achieve maximal bonding between all the layers. Note that this sample achieves a much higher load level before yielding. In addition, the authors attributed the break in the curve to failure at the metal/adhesive interface. With basic calibration data in hand for completely bonded and unbonded specimens, the authors were able to test bonded samples subjected to a variety of thermal and environmental stress conditions.

Roche et al.[40] investigated an interesting variation of the three-point bend test that is also sensitive to conditions at the adherate/substrate interface. The basic configuration is depicted in Figure 2.15. At a high enough load, the adherend will detach from the substrate starting from the edge and proceeding to the center, as shown in the figure. With the configuration shown in Figure 2.15, the stress distribution in the adherend can be monitored using the photoelastic setup shown in Figure 2.15b. In a typical experiment, the sample stiffness and stress distribution are monitored as the load is increased. An important piece of information is the load at which delamination of the adherend just begins. This number is sensitive to the detailed substrate preparation procedure before application of the adherend. It is clearly shown that this technique can be a powerful tool for investigating the effect of different surface preparation procedures and adhesive formulations on the adherend/ substrate adhesion strength. Making multiple identical samples also allows one to

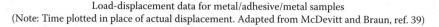

Load-displacement data for metal/adhesive/metal samples
(Note: Time plotted in place of actual displacement. Adapted from McDevitt and Braun, ref. 39)

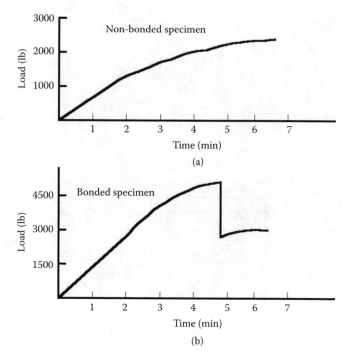

FIGURE 2.14 Bending load versus time for three-layer sandwich samples using the three-point bend test: (a) Poorly bonded sample shows essentially continuous load buildup, with increasing bending basically reflecting the stiffness of the most rigid layer. (b) Strongly bonded sample shows significantly higher load versus deflection behavior than the poorly bonded case, indicating strong coupling and mutual reinforcement between all the layers of the sample. In addition, a sharp drop in the load is detected when the adhesive joint between the layers fails.

study the effect of thermal cycling and other environmental stress factors on the durability of the bond.

2.8.3 FOUR-POINT BEND TEST

The standard configuration for the four-point bend test is shown in Figure 2.13b. Figure 2.16 shows the typical failure modes of the coating/substrate system as the bending load is increased. Note that there are two basic failure modes for this experiment. In the first case, failure is by clean delamination of the coating from the substrate. It can also happen that after propagating a short distance an initial delamination will dive into the substrate, initiating a cohesive failure. Experiments and theory confirm that the crack will dive to some fixed depth below the interface and then propagate indefinitely parallel to the interface. The precise depth of penetration is determined by the elastic properties of the coating and the substrate. This

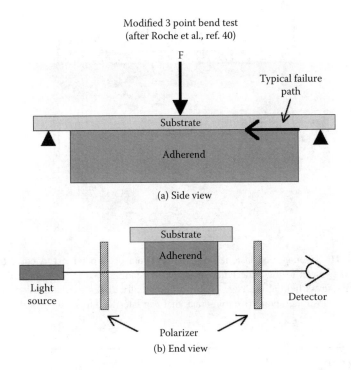

FIGURE 2.15 Variation of the three-point bend test in which the stress in the adherend is monitored by photo elastic measurement: (a) Side view showing most likely delamination path. (b) End view showing schematic of photoelastic measurement setup.

problem is treated in detail in Chapters 3 and 4, which cover the fundamentals of continuum theory and fracture mechanics required to understand this type of behavior.

The four-point bend test is favored largely by those interested in obtaining fully quantitative information on the adhesion strength of coatings by taking advantage of the relatively simple beam mechanics involved in the fracture analysis of this problem. In particular, Evans and co-workers used this test extensively to study the adhesion of a variety of metals to alumina (Al$_2$O$_3$) substrates. This work has been summarized by Evans.[41] A synopsis of their results is given in Table 2.2.

The basic conclusion from this work was that interfaces that were clean and free from contamination were inherently tough and ductile. Failure occurred either by rupture of the ceramic or ductile fracture of the metal. Moisture was observed to cause stress corrosion effects that greatly weakened some of the interfaces.

2.8.4 STANDARD DOUBLE CANTILEVERED BEAM TEST

Figure 2.13c depicts the simplest form of what is generically known as the double cantilevered beam test (the popular acronym is DCB test). One of the earlier reviews of this test was given by Bascom and co-workers[42] as applied to the adhesion of metals to glass and ceramic materials. Bascom et al.[43] also gave an overview of the

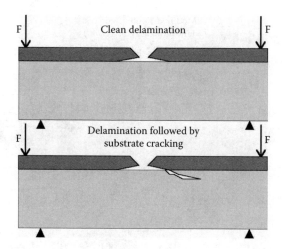

FIGURE 2.16 Schematic of basic failure modes of four-point bend test. A cut is first made in the coating down to the underlying substrate. A bending load is then gradually applied. The singular stress field at the edge of the cut eventually drives a delamination, which can turn into the substrate, giving rise to a crack that can later lead to spalling of the coating.

TABLE 2.2
Summary of Adhesion Strength of Various Metals to Alumina (Al₂O₃) Substrates

Metal	Conditions	Range of Adhesion Strength (J/m²)	Comments
Nickel	High humidity	4–8	Impurity segregation at interface
γ-Nickel (Cr)	Dry air	>100	Failure in metal or Al_2O_3
Gold	High humidity	1.5–2.5	Impurity segregation at interface
Gold	Dry air	>200	Failure in metal or Al_2O_3
Aluminum	All conditions	>50	Al_2O_3 rupture above 300 J/m²
Copper	All conditions	>150	Failure in metal or Al_2O_3

Note: Data summarized from work of A.G. Evans, in *Adhesion Measurement of Films and Coatings*, Vol. 2, K.L. Mittal, Ed., VSP, Utrecht, The Netherlands, 2001, p. 1.

---

application of this technique to the study of structural adhesives. These authors pointed out that one of the primary needs for the development of this technique was the evaluation of the strength and reliability of structural adhesives required to bond the new lightweight, high-performance composite materials required for the next generation of naval vessels, such as hovercraft and hydrofoils. One problem that became apparent right away is the fact that the structural adhesive materials considered showed significantly different strength properties when used as thin adhesive layers as opposed to bulk samples. The cantilevered beam configuration solved this problem with the added benefit of allowing for a relatively simple analysis of the data based on elementary beam mechanics. As shown in Figure 2.13c, the sample is prepared by carefully gluing together two identical beams of the substrate material of interest with a thin layer of the adhesive material under test. Great care must be taken regarding the surface preparation and cleaning of the beams and controlling the thickness and cure conditions of the adhesive. A spacer of some inert material such as Teflon is used at the ends of the sample to set the layer thickness and provide a well-controlled initial crack length. The sample is then mounted in a standard tensile testing apparatus in which the load-displacement behavior can be carefully determined.

A fairly elementary fracture mechanics analysis based on standard beam mechanics and the classic Griffith failure criteria gives the following equation for the apparent fracture toughness of the adhesive bond:

$$\gamma_{Ic} = \frac{4F_c^2}{b^2 E}\left(\frac{2a^2}{h^3} + \frac{1}{h}\right) \tag{2.6}$$

where

F_c = Critical minimum load required to initiate crack propagation
γ_{Ic} = Critical mode I fracture energy for crack propagation
a = crack length
E = Beam modulus
h = Beam height
b = Beam width

To determine γ_{Ic}, the load on the sample is ramped up until the preexisting crack just begins to propagate. This load is recorded as F_c, which then can be inserted into Equation (2.6) along with the other beam parameters to determine the critical fracture energy.

2.8.5 Tapered Double Cantilevered Beam Test

Equation (2.6) is problematic from the experimental point of view in that one has to control both the crack length and the applied load simultaneously to establish a stable constant value of γ_{Ic}. This is because of the involved dependence of γ_{Ic} on the beam height. Someone apparently noticed that if you tapered the beam height h such that the quantity $(a^2/h^3 + 1/h)$ in Equation (2.6) remains constant, then one only has to worry about controlling the applied load. Thus, the equation describing the critical fracture energy for the tapered beam shown in Figure 2.13d can be written as

$$\gamma_{Ic} = \frac{4F_c^2}{b^2 E}C$$

$$C = \left(\frac{2a^2}{h^3} + \frac{1}{h}\right) = \text{constant} \tag{2.7}$$

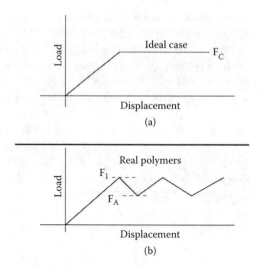

FIGURE 2.17 Typical load-versus-displacement data for the DCB experiment: (a) Hypothetical ideal case in which load increases up to the failure point and then propagates at a constant value determined by the fracture toughness of the adhesive joint; (b) real-world data in which the failure load oscillates between high and low values because of the strain rate sensitivity of most adhesives.

According to Equation (2.7), if we use a tapered beam with the beam height h adjusted so that the factor C in Equation (2.7) is constant, then the critical fracture energy will depend solely on the applied load at fracture. Thus, the output of the tensile test machine that generates load-versus-displacement data as the beam is pulled apart should look like the schematic plot given in Figure 2.17a. In this figure, we see the load ramping up in a linear fashion until the adhesive material gives way or starts to delaminate from either the top or bottom beam. At this point, assuming a uniform failure process, the load will stabilize at a constant level F_c, which can be substituted into Equation (2.7) to determine the critical fracture energy γ_{Ic}. However, even assuming a uniform failure process, actual data tend to look more like Figure 2.17b, with a sequence of peaks and valleys instead of a steady plateau. The reason behind this behavior stems from the fact that the polymer materials that comprise the major part of the adhesive tend to exhibit strain rate-dependent failure behavior. What this amounts to in practice is that γ_{Ic} tends to decrease as the crack propagates. Thus, referring to Figure 2.17b, the load increases to the level F_I, at which point the crack starts to propagate. Now, as just mentioned, γ_{Ic} and thus the resistance to crack propagation decreases, resulting in a lower load level to drive the crack. The lower load level, however, decreases the strain energy in the sample that is driving the crack. Eventually, the crack slows to the point at which it is completely arrested, which occurs at some load F_A. At this point, the fracture resistance of the sample has returned to its zero strain rate value, and the load on the sample can now increase back to the level F_I, at which the whole process can repeat. The upper value of γ_{Ic} is referred to as the *initiation energy* and the lower value as the *arrest energy*.

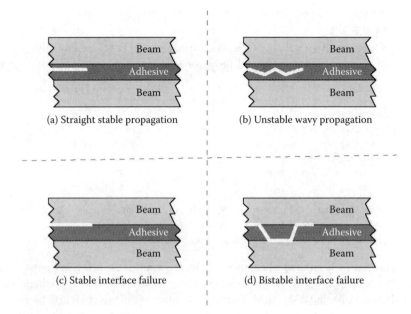

FIGURE 2.18 Known modes of failure in a DCB specimen.

In practice, the range of behavior of DCB samples is even more complex than Figure 2.17 would suggest. Hutchinson and Suo[44] reviewed the analysis of the DCB specimen and pointed out that there are at least four modes of crack propagation that have been identified (shown schematically in Figure 2.18). Each of these modes will have its own characteristic load-displacement behavior. Of particular relevance for adhesion measurement purposes is the stable interface failure mode shown in Figure 2.18c. It turns out that this mode can be selected for by using a nonsymmetric beam configuration in which one beam is stiffer than the other. In this way, the failure mode can be forced to be interfacial as opposed to cohesive in the adhesive.

The DCB technique has seen extensive use in characterizing adhesive systems of all types. An entire volume could be dedicated to this activity alone. For the present purposes, however, just to give some idea of the range of values to be expected, Table 2.3 gives sample critical fracture energy data taken on a variety of polymer systems.

2.8.6 DOUBLE-CLEAVAGE DRILLED COMPRESSION TEST

Figure 2.13e shows the double cleavage drilled compression (DCDC) sample, which is similar to the DCB sample in that an adhesive layer is sandwiched between two stiff beams. The main difference between the two techniques is in the mode of loading. As with the DCB sample, the DCDC sample is fabricated by gluing together two beams using an adhesive material and suitable spacers to achieve a desired uniform thickness of the adhesive layer and convenient initial crack regions. The sandwich sample is then drilled in the center symmetrically with respect to the initial crack region. The purpose of the hole is to provide a region of weakness that will

TABLE 2.3
Representative Values of the Measured Critical Fracture Energies of Common Materials and Polymer Systems

Material	Critical Fracture Energy (J/m²)
Inorganic glasses	$\gamma_{Ic} < 10$
Unmodified epoxies, polyesters, and polyimides	$70 < \gamma_{Ic} < 175$
Poly(methylmethacrylate)	$350 < \gamma_{Ic} < 800$
Rubber-modified epoxies	$1750 < \gamma_{Ic} < 3500$
Aluminum/nylon/epoxy adhesive	$7000 < \gamma_{Ic} < 8800$
Metals	$\gamma_{Ic} > 17,000$

Note: Data retabulated from work of W.D. Bascom, R.L. Cottington, and C.O. Timmons, *Naval Engineers Journal*, 73 (August 1976).

buckle outward when the axial compressive load is applied as shown in Figure 2.13e. The outward buckling of the beam drives the interface crack in a predominantly mode I tensile opening fashion. Michalske and Fuller[45] used this sample to investigate the fracture behavior of silicate glass. These authors pointed out the following advantages of using this test:

1. Use of compressive loading simplifies gripping and loading the sample and allows for more convenient use in corrosive environments, at high temperatures, and at high vacuum conditions.
2. A uniform stress intensity is maintained along the crack front.
3. Stable crack propagation is achieved along the midplane of sample.

Evans[46] pointed out that this sample has the further distinct advantage of decreasing strain energy release rate with increasing crack length. This property greatly simplifies the process of introducing a well-defined precrack before carrying out the adhesion measurement and allows for a more controlled experiment. This can be contrasted, for example, with the pull test, which gives rise to uncontrolled and unbounded crack propagation immediately after the onset of crack initiation.

2.8.7 BRAZIL NUT TEST

Figure 2.13f gives a schematic representation of what has come to be known as the Brazil nut sample. This sample starts out as a section of a solid cylinder that is cut across a diameter and then glued back together with the material to be tested. Appropriate spacers are used to control both the bond spacing and the initial crack length. This rather implausible specimen has the advantage of controlling the mode mixity of the crack propagation by simply adjusting the angle of the bond line with respect to the direction of the applied compressive load. For instance, setting bond line at an angle to the applied compressive load as shown in Figure 2.13f results in a predominantly mode II shearing load. On the other hand, setting the bond line nearly parallel to the applied load gives loading conditions similar to the DCDC test, which is predominantly mode I tensile crack opening. This specimen clearly has a strong advantage when one is interested in testing the bond strength as a function of the mode of loading.

The Brazil nut test has been in use for quite some time to test homogeneous solid specimens[47] and was apparently first adapted to test interfacial toughness by Wang and Suo.[48] An overview of this test has been given by Hutchinson and Suo.[49] One of the interesting and useful features of this test is the fact that residual stresses within the adhesive layer do not contribute to the driving force for crack propagation. Thus, unlike standard tests such as peeling or the DCB method, one does not have to take into account residual stresses within the adhesive layer to compute the true surface fracture energy. This property combined with the ability to adjust the mode mixity of crack propagation easily make this test a favorite when carrying out fundamental studies on interface adhesion.

An interesting study was done by Lee,[50] who used the Brazil nut configuration to study the adhesion of epoxy molding compound to copper lead frame substrates. This study was prompted by the need in the microelectronics industry to understand the adhesion of these materials, which are important for packaging silicon chips. It turns out that although copper is an excellent conductor, it does tend to oxidize easily, and loss of adhesion between oxidized copper and the epoxy compound has been blamed for a loss of adhesion between the two materials that can give rise to a notorious form of blistering known as the popcorn effect. What happens is that moisture can aggregate at the delaminated interface and then form a blister when heated as would happen in a wire bond or die attach procedure. Lee was able to fabricate Brazil nut-type specimens of the epoxy/oxidized copper materials and then investigate the fracture toughness as a function of the mode mixity of the applied load. For mode I (predominantly tensile opening) loading, his data showed that the fracture toughness was very close to that measured for the same materials but using a DCB specimen geometry. However, when loading under conditions carrying a high shear component or mode II-type load, the measured fracture toughness was roughly twice that obtained for mode I conditions. This study clearly indicated the practical usefulness of the Brazil nut specimen because one can clearly be mislead about the adhesion strength when relying on only one mode of loading.

2.8.8 Wedge Test

Figure 2.13g illustrates the wedge test, by which a sandwich laminate is loaded by simply forcing a wedge into one end of the laminate. This test has been used effectively in the aircraft industry to evaluate the durability of sandwich layers of thin aluminum sheets bonded with an adhesive. An interesting study carried out by the Boeing company was informatively reviewed by Pocius.[51] The main advantage of the wedge test is the fact that many samples can be easily and cheaply fabricated and then subjected to any variety of environmental stress conditions to determine the durability of the bond. The success of this method has led to the formulation of

American Society for Testing and Materials (ASTM) Standard D3762, which describes in detail sample preparation and testing procedures.

By loading the sample using the simple insertion of a wedge at one end, a constant load can be applied at a level just below that required for crack propagation. The sample can now be placed in a hostile environment (e.g., high temperature or humidity), and the progress of any preexisting crack can then be monitored as a function of time. For a good bond, the crack should be short and primarily within the adhesive layer itself. A faulty bond is indicated when the crack is relatively long and propagating along the metal/adhesive interface.

The wedge test has found fairly wide application for testing the durability of adhesives. Interesting examples include the work of Sharif et al.,[52] who used the wedge specimen to test the effectiveness of model copolymer adhesives on glass. Davis et al.[53] used the technique to test the effectiveness of plasma sprayed coatings on titanium and aluminum, and Meyler and Brescia[54] used the technique to test a variety of waterborne adhesives on aluminum and aluminum-lithium surfaces. The main advantage of the wedge test lies in the ability to easily create a large number of samples that can be used for long-term environmental stress testing. Also, because the method is basically grounded in the mechanics of bending beams, it can be analyzed quantitatively using standard fracture mechanics analysis.[55-58]

2.8.9 Topple Beam Test

As a final example of an adhesion test based on the mechanics of a bending beam, we mention the topple beam test as shown schematically in Figure 2.13h. This test has much in common with the well-known pull test discussed in Section 2.4, with the crucial difference being the mode of loading the sample. All of the sample preparation procedures prescribed for the pull test apply equally to the topple beam experiment. This experiment in fact shares just about all the advantages and disadvantages encountered in the pull test. One advantage of the topple beam, however, is that the mode of loading is nearly pure bending, allowing for a simplified analysis. Apalak[59] in particular carried out an extensive mechanics analysis of the topple beam geometry, including the effects of geometric nonlinearity caused by large rotation of the beam.

Another distinct advantage of the topple beam over the pull test is the fact that, by applying a lateral load to the top of the beam, the problem of gripping the beam is relatively trivial compared to the pull test, which has to ensure proper gripping to avoid slippage. This can be a significant problem when using beams made of hard brittle materials. Finally, with the pull test one has to be careful to apply the tensile load exactly along the beam axis to avoid applying a bending moment to the beam, a problem that does not arise in the topple beam approach.

An overview of this method was given by Butler,[60] and an application to metal films was provided by Kikuchi et al.[61] It might be added in closing that the topple beam test has much to recommend it for the adhesion of pins on ceramic substrates. As mentioned, the pin pull method is widely used for testing pin adhesion. However, when inserting a pin into a socket there is always a significant bending load present.

From the point of view of reliability testing, one would prefer that the pin should bend as opposed to tearing off the pad. A bent pin can be straightened, but a tear-off failure can render an entire substrate useless. The topple beam experiment is clearly an efficient way to sort out these distinct behaviors in a population of pins attached to a rigid substrate.

2.8.10 ADVANTAGES OF BEAM-BENDING TESTS

It is clear that the main advantage of using any one of the tests described in the previous sections is that they all lend themselves relatively easily to quantitative analysis using the methodology of fracture mechanics, which is described in detail in Chapter 4. The mechanics of bending beams was one of the earliest and quite likely the most studied topic in the history of applied mechanics. An extensive body of literature exists that can be called on to supply stress/strain solutions for nearly any beam configuration that can be imagined. Thus, it is fairly straightforward to convert load-displacement data taken on any of the samples shown in Figure 2.13 into fracture toughness or surface fracture energy results that are directly relevant to the adhesion failure process investigated. This type of data is especially important as input to detailed fracture mechanics-based models, which attempt to predict the failure of structures based on geometry, material properties, and loading conditions.

A second advantage of beam-based tests is the ability to fabricate a large number of samples relatively easily for testing under a variety of conditions.

2.8.11 DISADVANTAGES OF BEAM-BENDING TESTS

As always, we have to recognize that there are no ideal adhesion tests, and the beam-bending methods are no exception. The key advantage of having a simple geometry that is suitable for analysis leads to the main disadvantage of these tests, which is the very limited sample geometries that can be investigated. Ideally, someone manufacturing automobile fenders would like a test that could indicate how well the paint is adhering to the metal and that could be implemented on the manufacturing line to give real-time results on a wide variety of fender shapes. The beam-bending experiments clearly do not fit the bill here and are thus confined to use solely in the testing laboratory.

The second most serious complaint that can be lodged against the beam-bending methods is whether they adequately simulate the stress/strain loading environment that real manufactured parts must endure. In particular, cases for which the main mode of failure is through internal stresses, such as those caused by thermal expansion mismatch strains, the delamination process is driven mainly by the elastic energy generated by the stresses. In such cases as when, for example, a highly stressed coating deadheres spontaneously from its substrate, the delamination process occurs under near-infinitesimal strain conditions. In all of the beam-bending tests, however, the delamination process is driven by external loading at substantial levels of deformation and strain. Although the level of deformation is nowhere near that generated by something like a peel test, nonetheless this calls into question the relevance of beam-bending data for such situations.

2.8.12 Summary and Recommendations

The main conclusion that arises out of all the above discussion is that the bending beam experiments are quite likely the methods of choice when fully quantitative data is required, such as for critical surface fracture energies or mode I and mode II fracture toughness data. The bending beam methods provide the most reasonable way of obtaining fracture and delamination data over the entire range of mode mixity for a wide range of adhesive/adherend combinations. Sample fabrication is demanding but not excessively so. Geometries such as the wedge test allow generation of many samples for testing under a wide range of environmental conditions while giving quantitative data without the need for elaborate test equipment.

On the downside, it must be recognized that the beam-bending tests will find best application in the laboratory environment as opposed to the manufacturing line. This arises primarily from the highly restricted sample geometries involved plus the need for careful sample fabrication and nontrivial data analysis.

2.9 SELF-LOADING TESTS

It is well known that the internal stress in a coating can cause it to delaminate spontaneously from the underlying substrate. This effect is commonly seen in old paint coatings on wood trim and other artifacts in which shrinkage stresses in the coating cause it to peel back from the edges and other internal defects. The idea that naturally suggests itself is to set up an adhesion measurement experiment in which the driving force for delamination is derived solely from well-controlled internal stresses in the coating to be tested. The task of the experimenter is then reduced to monitoring the propagation of the delamination front at a known fixed level of internal stress. A number of experiments along these lines are discussed next. One of the major advantages of this type of experiment is that the loading conditions closely resemble those that initiate failure in real coatings subjected to field conditions.

2.9.1 Circle Cut Test

Figure 2.19 gives a schematic illustration of what is called the circle cut test. In this experiment, one has a coating on a substrate with a uniform biaxial stress σ_0 that can be caused by any number of factors, such as deposition conditions, thermal expansion mismatch strains, shrinkage caused by solvent loss, and so on. At time zero, a circular cut is made somewhere in the interior away from any edge, giving rise to the situation shown in Figure 2.19. As explained in Chapter 4, the act of generating a sharp edge creates a stress singularity at the edge of the cut at the interface between the coating and the substrate. If the adhesion of the coating is relatively weak, then the coating will deadhere from the substrate, forming an annular ring of delaminated material as shown in Figure 2.19. The radius of the cut and the radius of the delaminated annular region are then measured by optical or other means, and this information can then be used along with knowledge of the stress level σ_0 in the coating and its elastic properties to calculate the critical surface fracture energy according to the following formula:

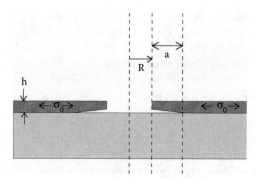

FIGURE 2.19 Schematic of circle cut test. A circular hole of radius R is cut in a coating carrying an internal stress σ_0. If the coating adhesion to the substrate is insufficient, then because of the stress singularity at the edge of the hole, the coating will delaminate out to a radius $R + a$. The driving force for delamination can be easily determined if one knows the elastic constants of the coating, the film stress, and the geometric parameters in the diagram.

$$G_c = \frac{h\sigma_0^2}{E} \left[\frac{2\beta}{\left(1 + \beta \left[\frac{a}{R}\right]^2\right)^2} \right]$$

(2.8)

$$\beta = \frac{1-\nu}{1+\nu}$$

where
$\quad R, h,$ and a = Coating and cut dimensions as shown in Figure 2.19
$\qquad \sigma_0$ = Residual stress in coating
$\qquad E$ = Modulus of coating material
$\qquad \nu$ = Poisson's ratio of coating
$\qquad G_c$ = critical strain energy release rate

A convenient feature of this test is the fact that, according to Equation (2.8), the size of the annular delamination region will always be constrained because the driving force G_c decreases as the size of the delaminated region increases. Thus, at some finite value a, the driving force for deadhesion will be too small to propagate the delamination any further, and the experimenter can make the appropriate measurements at leisure. Farris and Bauer[62] used this method to study the adhesion of polyimide coatings. This approach has a number of distinct advantages, including:

- Close simulation of the mechanism by which coatings tend to delaminate under field conditions of high stress, low strain, and mixed-mode conditions of tension and shear
- Relative ease of sample preparation
- Amenability to quantitative analysis

The main drawback is that because the driving force for delamination relies solely on the internal film stresses, there may not be sufficient strain energy in the coating to initiate a delamination if the level of adhesion is too high. This fact tends to limit this test to coatings with weak adhesion.

2.9.2 Modified Edge Liftoff Test

In view of the limited applicability of the circular cut test to weakly adhering coatings, a related approach called MELT (modified edge liftoff test) has been explored by Hay et al.[63] The main innovation in this approach is the use of a layer of material on top of the coating to be tested that will drive up the total internal stress and thus give sufficient driving force to delaminate any coating. A schematic representation of this test is given in Figure 2.20. As shown in the figure, complications can arise because of crack initiation in the substrate rather than the interface of interest. Hay and co-workers exercised this technique on a multilayer sandwich consisting of epoxy/dielectric layer/silicon nitride/native oxide/silicon. They sidestepped the substrate fracture problem by using a hydrofluoric acid solution to etch out a large initial crack in the silicon nitride layer. In addition, they polished the edge of the sample with fine sandpaper to eliminate any flaws left behind by the wafer dicing operation. A thick epoxy top layer between 150 and 200 μm provided the driving force for crack propagation caused by thermal expansion mismatch with the silicon substrate. Under these conditions, they observed clean delamination of the dielectric layer from the silicon nitride.

These workers noticed an interesting property of their epoxy-loaded samples that seemed to contradict previous work that used chromium as a superlayer instead of epoxy. From the chromium work, it was noticed that the driving force for delamination depended on the size of the initial crack length and did not settle down to a constant value until the delamination length had reached at least 20 times the chromium layer thickness. With the epoxy-loaded samples, there was no apparent

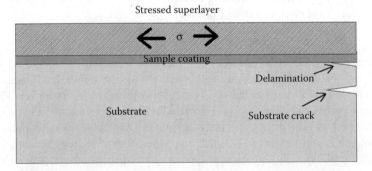

FIGURE 2.20 Schematic of the modified edge liftoff test (MELT), which attempts to circumvent the limitation of the self-loading test to the case of relatively weak adhesion, as in the circular cut test shown in Figure 2.19. A top layer of material with known internal stress behavior is applied to increase the delamination driving force to any desired level. As can be expected, a number of complications can arise, including possible failure of the underlying substrate.

dependence of the driving force for delamination on the initial crack size. A subsequent finite element analysis of the problem comparing the driving force for delamination for both the epoxy- and chromium-loaded systems indicated that indeed the behavior of the epoxy and chromium systems is very different. In the epoxy case, the driving force rises so sharply with crack length it nearly resembles a step function. In the chromium system, the driving force rises much more slowly and essentially follows the rule of 20 times the chromium thickness before leveling off to a steady state.

2.9.3 MICROSTRIP TEST

The microstrip test is closely related to MELT in that the same principle of using a superlayer on top of the layer to be tested is employed to obtain a sufficient driving force for delamination. A schematic of the basic sample configuration is shown in Figure 2.21. The version of the test described here was carried out by Bagchi et al.[64] and reviewed by Evans.[65] What is shown in the figure are the various layers of the adhesion measurement test structure. The substrate may be a block of ceramic or glass or whatever the material of interest is. On top of the substrate, narrow parallel strips of a release layer material are first deposited. A typical release layer strip is visible in the uncoated regions separating the sample strips shown in Figure 2.21. Next, the coating is applied as a blanket layer. On top of that, the superlayer is deposited as a blanket coating. A good candidate for this material when testing thin metal films is chromium, which can develop internal stress levels up to 1 GPa. The blanket coatings are now etched into thin strips perpendicular to the release layer strips using photolithographic methods.

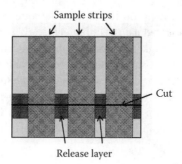

Sample strips

Cut

Release layer

FIGURE 2.21 The microstrip test is yet another self-loading adhesion test that takes advantage of photolithographic methods to create a large number of test specimens on a single substrate. The figure shows a greatly enlarged view of three test strips. The horizontal stripe is a release layer material to which the coating has poor adhesion. After applying blanket coatings of the test film and the loading layer, a large number of test strips are then etched out using lithographic methods. A cut is then created in the strips in the region where they overlay the release layer. Because of the poor adhesion to the release layer, delamination is immediately initiated and will proceed to propagate into the strip. However, the nature of the driving force is such that delamination will always terminate before the strip is entirely peeled off the substrate. The length of remaining adhered ligament can then be used to compute the surface fracture energy.

The adhesion test begins after a thin cut is made down the center of the release layer strip using either etching or milling techniques. As the adhesion at the release layer is very weak by design, a delamination is immediately initiated starting from the cut to the edge of the release layer. Depending on conditions, the delamination front will proceed past the edge of the release layer and into the interface between the sample coating and the substrate. If the delamination stops at the edge of the release layer, then clearly there is not enough strain energy in the chromium superlayer, and a thicker layer of chromium needs to be applied. At a sufficient thickness of the chromium layer, the delamination will proceed at the coating/substrate interface or possibly proceed as a subcrack in the substrate depending on how well adhered the coating is. Assuming that the delamination proceeds at the coating substrate interface, it will go for a certain distance and then arrest. In this test, all of the test strips are finite length, and the driving force for delamination decreases toward zero as the crack front nears the end of the strip. Therefore, the delamination front must arrest, and the length of the remaining ligament is therefore indicative of the adhesion between the coating and the substrate.

2.9.4 ADVANTAGES OF SELF-LOADING TESTS

Self-loading tests come very close to mimicking the conditions under which real coatings delaminate in terms of strain level and mode mixity; this is their main advantage as an adhesion test. This is especially true for microelectronic structures, for which intrinsic stress is one of the major causes of device failure. In addition, the self-loading tests lend themselves readily to quantitative analysis because the strain level at failure is typically quite low, and linear elastic theory suffices to carry out the required fracture mechanics evaluation of the strain energy release rate. When used to evaluate microelectronic devices, the test structures required to carry out the adhesion measurements can be fabricated using equipment nearly identical to that used in the production line. Thus, although sample preparation is fairly involved, it does not require resources beyond what should already be available in the existing manufacturing facility. Finally, the test structures can be integrated into the actual manufacturing process as diagnostic probe sites capable of giving near-real-time quality control and monitoring information on real manufacturing parts. Of all the adhesion tests discussed so far, the self-loading tests come the closest to an ideal adhesion measurement method.

2.9.5 DISADVANTAGES OF SELF-LOADING TESTS

The main disadvantage of the self-loading tests is the need to know the internal stress and mechanical properties of the coating to be tested. The problem of determining the internal stress in a coating is a rather large subject, and an entire chapter in this volume (Chapter 6) is devoted to just this problem. Further complications arise if the internal stress level is insufficient to cause delamination and one of the superlayer methods must be implemented, thus greatly increasing the work required for sample preparation. Finally, the trajectory of the delamination crack front may deviate from the interface into either the substrate or the coating, depending on conditions. This will significantly complicate the fracture mechanics analysis involved.

In essence, the self-loading tests share the common problems of all fully quantitative tests in that they inevitably involve much more work than the simply qualitative or semiquantitative methods.*

2.9.6 SUMMARY AND RECOMMENDATIONS

The self-loading adhesion tests are most suitable when quantitative data is required to support numerical modeling simulations of the structures under fabrication. Multi-level wiring structures from the microelectronics industry are a prime example of an application that can benefit greatly from this type of test. Such structures are replete with interfaces between dissimilar materials that are subject to numerous manufacturing processes that can introduce high levels of internal stress. As a specific example, consider the use of organic insulators as dielectric layers in multilevel wiring schemes. Some of the most popular materials for this application are the polyimides. These materials are typically coated as a viscous liquid using either spin- or spray-coating methods. The resulting film must be cured at an elevated temperature, generally above 300°C to achieve the correct electrical and mechanical properties required. Cooling to room temperature will induce a large internal thermal expansion mismatch strain in the coating if the underlying substrate is either silicon or a ceramic material because most polyimide materials have thermal expansion coefficients 10 to 30 times larger than either silicon or most ceramics. An elementary calculation using the principles of continuum theory covered in Chapter 3 shows that the resulting stress level in the coating can come to nearly 50% of the yield stress of the polyimide itself. In such cases, appropriate stress modeling supported by fully quantitative adhesion data are required to carry out the type of engineering design analysis necessary to support the fabrication and manufacture of a useful device. The large amount of work involved in implementing the self-loading methods will quite likely limit them to this type of application.

2.10 MORE EXOTIC ADHESION MEASUREMENT METHODS

2.10.1 LASER SPALLATION: EARLY WORK

Up to this point, the adhesion tests under discussion have been of a fairly prosaic nature in that the mode of loading has always been some kind of mechanical tension, compression, or shear applied by a mechanical test apparatus, by the residual stress in the coating, or by some applied superlayer material. For the laser spallation test, however, the loading conditions are rather exotic in that they involve highly transient shock waves induced by ultrafast laser pulses. In a typical test, a laser pulse with a fluence of 500,000 J/m^2 is focused onto a 1-mm diameter spot for 30 nsec, which amounts to energy deposition at a power rate of 13 MW.

---

* As a general rule of thumb, when going from a qualitative test to a fully quantitative one, you can expect to work not only twice as hard to obtain acceptable results but more likely 10 times as hard. This almost seems to be some kind of thermodynamic theorem proved through difficult experience if not from any formal axiomatic demonstration.

This test was apparently first discussed in the open literature by Vossen[66] and Stephens and Vossen,[67] who used it to investigate the adhesion of metals to glass. A typical experimental setup is shown in Figure 2.22. In the earlier experiments, an array of islands of the coating to be tested were evaporated onto a glass substrate. Small islands were used to avoid the effect of tearing of the film, which would occur during the spallation of a continuous metal film. The opposite side of the glass substrate was coated with an energy-absorbing layer such as titanium. During a standard experiment, a high-intensity, short-duration laser pulse is directed at the titanium layer on the back side of the sample. The effect on impact is literally explosive vaporizing of the titanium and blasting it off while simultaneously depositing a large amount of heat into the underlying glass in a very short time. The recoil of the blasted titanium combined with the rapid thermal expansion of the glass generates a compressive shock wave that propagates to the front side of the glass, where the islands of deposited metal are waiting. The compression wave is reflected at the front surface of the substrate as a tensile wave caused by the rapid tensile snap back of the substrate. The elastic snap back propagates across the substrate metal interface, providing the rapid tensile loading that ejects the metal islands off the substrate. To determine the adhesion strength of the coating, the energy of the laser pulse is ramped up systematically until coating failure is first observed. Depending on conditions, a variety of collateral damages can occur to both the substrate and the coating tested. Vossen pointed out in particular that brittle substrates tend to shatter. However, one experiment used islands of gold deposited on glass and clearly demonstrated clean delamination.

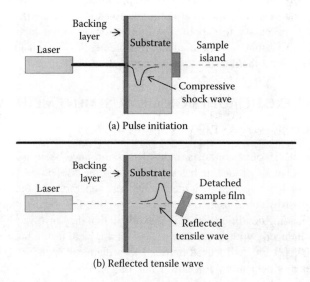

(a) Pulse initiation

(b) Reflected tensile wave

FIGURE 2.22 Laser spallation experiment: (a) A high-intensity laser pulse impinges on the back of the substrate, initiating a strong compressive stress pulse because of the rapid heating at the point of impingement; (b) the compressive pulse reflects off the coating/substrate interface, generating a strong tensile reflective wave, which can cause delamination of the coating.

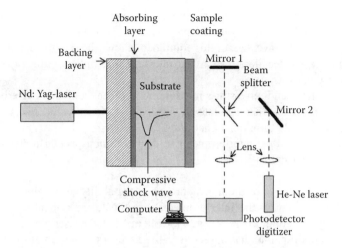

FIGURE 2.23 Improved laser ablation experiment. The basic principle is the same as in Figure 2.22 with the following improvements: (1) A backing layer is added over the energy-absorbing layer to carry away excess thermal energy that would otherwise create a long tail on the induced compressive pulse, which significantly reduces the resolution of the method. (2) A laser Doppler interferometer is employed to monitor the displacement of the sample surface in real time as it deforms because of the impinging stress pulse. This allows for an independent determination of the critical stress level for delamination.

2.10.2 LATER REFINED EXPERIMENTS

The early experiments on laser spallation demonstrated that the technique could be used to test the adhesion of coatings in a unique way. A number of problems hindered progress, however, including the use of expensive finicky equipment that required careful and time-consuming calibration procedures. In addition, a large thermal stress tail followed the main pulse and limited the resolution that could be achieved in stressing the coating substrate interface. Additional problems, such as fracture of brittle materials, excessive stress damping, and potential decomposition of viscoelastic media, essentially limited the method to testing refractory materials.

Advances in technology and techniques, however, have greatly improved the laser spallation test, as discussed by Gupta et al.[68] Figure 2.23 illustrates an improved version of the laser spallation experiment shown in Figure 2.22. The new features in this version of the laser spallation experiment include

1. Inclusion of an absorbing layer between the back side of the substrate and a backing layer. This approach allows the user to nearly eliminate the long thermal stress tail associated with the earlier approach by essentially ablating away the excess thermal energy into the backing layer instead of letting it diffuse into the substrate, where it induces a large thermal stress. This long-tail thermal stress can confound the data analysis, especially when dealing with very thin sample coatings.

2. Use of a laser Doppler interferometer to monitor in real time the deformation and velocity of the sample coating during impact by the reflected tensile shock wave. Use of this equipment allows the user to quantitatively estimate the tensile stress at the interface that initiates film separation from the substrate.
3. Using the quantitative stress measurements taken from the interferometer data, one can further estimate an "intrinsic" surface fracture energy associated with the film delamination.
4. Use of a modified interferometer to deal with nonspecular reflection off rough surfaces.

The main advantage of the new approach is the ability to use the velocity/displacement data taken by the interferometer to estimate the critical stress level at which coating spallation occurs. As with the old method, the laser power is increased incrementally until spallation is observed. Table 2.4 lists sample values for the critical stress for coating liftoff for several coating substrate systems.

The strength values listed in Table 2.4 are similar to those that might be obtained in a pull test or other form of tensile test. The main difference is the dynamic nature of the tensile shock pulse, which hits the interface at a strain rate on the order of 10^7 sec^{-1}. At such an enormous strain rate, we can expect the material response to be significantly different from what would occur in a standard pull test. In particular, viscoelastic effects should be "frozen out," giving rise to nearly pure brittle fracture because most molecular relaxation behavior will be too slow to follow the rapid deformation pulse.

Finally, by using a recently developed scaling relationship for the binding energy-versus-distance curves for metals, Gupta et al.[69] were able to convert their

---

TABLE 2.4

Representative Values of the Interface Strength for Selected Coating Substrate Systems

Substrate	Coating	Interface Strength σ_{max} (GPa)	Comments
Polyimide	Silicon	0.46	5.3-μm polyimide
Copper	Silicon nitride	0.041	0.1-μm copper on nitrided silicon
Copper	Pyrolytic graphite	0.015	
Silicon carbide	Pyrolytic graphite	3.4–7.48	2.1-μm SiC on 1.69mm disc
Alumina/diamond composite	CVD diamond	1.61	5% of 1-μm diamond particles
Alumina/diamond composite	CVD diamond	1.41	5% of 10-μm diamond particles

Note: Data retabulated from V. Gupta, J. Yuan, and A. Pronin, in *Adhesion Measurement of Films and Coatings*, K.L. Mittal, Ed., VSP, Utrecht, The Netherlands, 1995, p. 367.

interface strength data to fracture toughness values similar to that obtained in the bending beam tests. By using physical arguments based on the nature of metal-to-metal interactions at the interface between different metals, these investigators derived the following relationship between their measured critical interface stress and what they called the intrinsic toughness or ideal work of interfacial separation G_{co}:

$$\sigma_{max}^2 = \frac{G_{co}E_o}{e^2 h} \tag{2.9}$$

where

σ_{max} = Critical interface strength as shown in Table 2.4

E_o = Modulus at interface

G_{co} = Intrinsic interface toughness

h = Unstressed separation distance between atoms

e = Mathematical constant epsilon, which is the base of the natural logarithms = 2.71828183

The quantity σ_{max} is measured directly by the laser spallation experiment, and representative values are shown in Table 2.4. The unstressed separation distance h can be determined by direct transmission electron microscopy measurements, and the interface modulus E_o is inferred from delamination experiments and from published data. With this data in hand, Equation (2.9) can be used to estimate the apparent critical surface fracture energy G_{co}. Values for selected systems are summarized in Table 2.5.

TABLE 2.5
Values of Critical Intrinsic Surface Fracture Energy from Laser Spallation Measurements

Interface System	Critical Surface Fracture Energy G_{co} (J/m²)	Comments
Silicon carbide/pyrolytic carbon ribbon	3	$h = 0.9$ nm from TEME_o = 107 GPa (average of two media)
Silicon carbide/silicon	3.31	$h = 0.5$ nm; $E_o = 115$ GPa
Cleavage of (111) planes of silicon	0.84	Stress pulse sent normal to (111) planes; $h = 0.32$ nm; $E_{111} = 185$ GPa
Niobium/Al₂O₃	0.0009	Value considered faulty because of interface flaws; should be close to 0.69

Note: Data tabulated from V. Gupta, J. Yuan, and A. Pronin, in *Adhesion Measurement of Films and Coatings*, K.L. Mittal, Ed., VSP, Utrecht, The Netherlands, 1995, p. 395.

In evaluating the data in Table 2.4, one has to keep in mind that the coatings studied were delaminated from their substrates under extreme loading conditions (strain rates near 10^7 sec^{-1}) for which material properties and viscoplastic deformation behavior in particular can be expected to differ significantly from those in effect for the more standard adhesion tests, such as the bending beam experiments discussed in this chapter. The authors of this work went so far as to suggest that the quantity they called G_{co} be called an intrinsic critical surface fracture energy in distinction to the usual G_c because many of the viscoplastic processes that contribute to G_c will be frozen out in the laser spallation experiment. In other words, the various atomic/molecular deformation processes that can contribute to the measurement of G_c are too slow to play a part in the laser spallation experiment.

2.10.3 LASER-INDUCED DECOHESION SPECTROSCOPY EXPERIMENT

A variant of the laser spallation experiment goes under the acronym LIDS, which stands for laser-induced decohesion spectroscopy* has been discussed by Meth.[70] A schematic of the LIDS experimental setup is shown in Figure 2.24. Strictly speaking, the LIDS experiment is not a spallation test as discussed above because the coating is not removed from the substrate; however, laser energy is used to deadhere the coating, which is sufficient qualification to be included. Referring to Figure 2.24, the basic experiment consists of directing an infrared laser beam onto the sample to be tested, which is a multilayer laminate composed of a substrate, an energy-absorbing layer, and a top coating. The absorption properties of the coating are such that the laser radiation passes through and is soaked up predominantly by the underlying absorbing layer, which rapidly decomposes, giving rise to a gas pressure that initiates blistering of the coating much as one would observe in the standard blister test. At a sufficient level of laser power, the blister will propagate into an annular debonded region that can be monitored by optical means. The curvature of the blister is monitored simultaneously by reflecting a second laser beam from a low-power He-Ne laser off the top of the blister. The reflected laser beam passes through a fixed aperture and then onto a photodetector, which measures the intensity of the light passing through the aperture. When no blister is present, a fixed and calibrated intensity of reflected light is measured. After blistering, the reflected light is spread out in proportion to the blister curvature; because the aperture is fixed, less light reaches the photodetector, and the measured intensity drops. Thus, the blister curvature can be inferred from the light intensity observed at the photodetector.

Meth[70] discussed the details of the calculation used to determine the curvature based on a Gaussian beam shape. With the size and curvature of the blister in hand plus knowledge of the elastic properties of the coating, the gas pressure causing the blister can be inferred through a classical continuum mechanics calculation that treats the blistered coating as a plate subjected to uniform pressure from below. Much of this analysis is already known from standard blister test theory, and full details are covered by Meth.[70] By carefully observing the laser power required to

---

* A more appropriate designation would be laser-induced deadhesion spectroscopy. Another problem is that this experiment has nothing to do with spectroscopy as it is commonly understood.

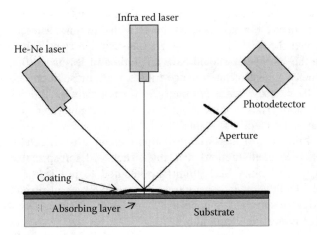

FIGURE 2.24 LIDS experiment. Strictly speaking, this is not a laser ablation experiment but is close enough to warrant consideration here. A strong laser pulse is directed at the coating substrate interfaces. Gases created during the intense heat buildup create a blister in the coating, which can further delaminate as in the standard blister test. The curvature, and thus the size, of the blister is monitored by reflection of a weak laser beam off the top of the blister into a photodetector.

just initiate annular debonding of the blister, one can further apply standard blister test methods to determine the critical surface fracture energy for delamination.

2.10.4 ADVANTAGES OF LASER SPALLATION TESTS

The main advantage of the laser spallation test derives from the unique way in which the delamination load is applied to the coating under investigation. No gripping scheme is required, and the ultrahigh loading rate precludes many viscoelastic/viscoplastic mechanisms, which tend to cloud the analysis of more conventional adhesion tests, such as the bending beam methods discussed in Section 2.8. Laser spallation may in fact be the only effective method for analyzing refractory systems such as the diamond-on-alumina samples. This technique is also amenable to full quantitative analysis because of the availability of advanced instrumentation such as the laser Doppler interferometer. The LIDS version of this test has the potential of application to a wide variety of surfaces, including actual parts from the manufacturing line, because the laser pulse and surface displacement measurement are applied directly to the top surface of the coating under study. The fact that this is a noncontact method that can be fully automated makes it possible to implement the LIDS experiment directly on the manufacturing line as a real-time quality control tool.

2.10.5 DISADVANTAGES OF LASER SPALLATION TEST

The same properties that give the laser spallation test its unique advantages also impart a number of its most important disadvantages. The most significant disadvantage derives from the ultrahigh strain rates involved. To perform the laser Doppler measurements required to gain quantitative information on the adhesion strength of the coating, one needs to measure stress pulses with rise times shorter than 5 nsec, which necessitates the use of 10- to 100-GHz transient digitizers. The photodetector for measuring the laser pulse should have a rise time of less than 200 psec, which will allow a measurement of interference fringes with a resolution on the order of 0.2 nsec. Needless to say, this is not the type of equipment likely to be found at a local hardware store. Thus, it is clear that this method is going to require expensive equipment that is not readily available.

A second limitation of the laser spallation technique is the relatively difficult and limited sample configurations available. The need for special backing and absorbing layers will quite likely limit this method to use in the development laboratory only. Finally, the fact that ultrahigh strain rates are involved in removing the coating from the substrate complicates the problem of interpreting the resulting data in terms of typical field conditions.

2.10.6 SUMMARY AND RECOMMENDATIONS

The laser spallation adhesion test is one of the more exotic adhesion measurement methods and has some unique advantages to recommend it in certain situations. In particular, it may be the only method available to reasonably test highly refractory coatings that are either too brittle or too hard to be tested by more conventional methods. The instrumentation required for fully quantitative measurements is not commonly available but can be purchased from specialty vendors. This equipment is expensive, so the need for the required data must be sufficiently compelling.

The LIDS version of this technique has a number of potentially useful applications given the fact that it is a noncontact method that can be fully automated. There is a definite prospect for use in real-time manufacturing quality control applications because it can be applied to very small test sites and requires no special sample preparation whatever. The only special requirement is the need to tailor the spot-forming laser pulse to be absorbed preferentially by an existing underlayer of the coating to be tested.

2.11 ELECTROMAGNETIC TEST

Another fascinating adhesion test method is the electromagnetic test, which utilizes the surprisingly large forces that can be developed by currents in a magnetic field. Figure 2.25 gives a schematic illustration of the basic configuration for lifting a metal line off of an insulating substrate. This figure shows a cross section of the metal line that is directed out of the page and carries a current I also directed out of the page. A magnetic field B is applied at a right angle to the flow of the current giving rise to a Lorentz force F acting perpendicular to the metal/substrate interface. For this configuration, the Lorentz force is given by the following formula from basic electromagnetic theory:

$$F = Il \times B \qquad (2.10)$$

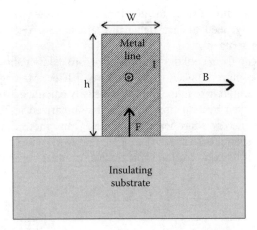

FIGURE 2.25 In the electromagnetic test, a current I flows in a metal test line in the presence of a perpendicular magnetic field. If the current and the magnetic field are sufficiently strong, then the line can be lifted off the substrate because of the induced Lorentz force.

where

F = Lorentz force
I = Current flowing in line
l = Unit length vector in direction of current flow
B = Magnetic field vector

The right-hand side of Equation (2.10) involves the vector cross product between the length vector $\mathbf{l}$ and the magnetic field $\mathbf{B}$. For the geometry in Figure 2.25 with $\mathbf{l}$ and $\mathbf{B}$ mutually perpendicular, the cross product evaluates to lB, and the Lorentz force becomes $F = IlB$ pointing normal to the substrate surface as shown. What is of primary interest, however, is the stress on the line, which is evaluated by dividing the force by the area $A = lw$, which gives the following very simple formula for the normal stress tending to lift the line off the substrate:

$$\sigma_n = \frac{IB}{w} \qquad (2.11)$$

Equation (2.11) gives the normal stress on the metal line caused by the Lorentz force induced by the interaction of the current I with the magnetic field B where w is the width of the line. If we assume a magnetic field of 2 T (Tesla) and a current of 30 A (ampere) with a line width of 1 μm Equation (2.11) gives the substantial normal stress of 60 MPa. Referring to Table 2.4, we see that this stress level is more than sufficient to lift a copper coating off of a silicon nitride substrate as evaluated by the laser spallation method. However, those familiar with electromagnetics will immediately detect a potential problem in that 30 A is a substantial current to pass through a 1-μm wide line, or any metal wire for that matter, without vaporizing it. Therefore, we really need to investigate the temperature rise we can expect by performing this experiment.

One way to keep from cooking the sample before delamination can be observed is to use very short current pulses on the order of 1 μsec. Making this assumption, the expected temperature rise in the line can be easily calculated from Fourier's heat transfer equation given by Equation (3.36) and summarized in Table 3.1. Because we are talking about very short heating times, we can ignore diffusion effects and thus leave out the spatial derivatives (i.e., $\nabla^2 T$) and work with the following simple relation, which will be valid for the short timescale under consideration:

$$\frac{dT}{dt} = \frac{\gamma}{\rho c} \qquad (2.12)$$

Because all quantities on the right-hand side are constants, this equation readily integrates to give the following relation for the expected temperature rise:

$$\Delta T = T - T_0 = \frac{\gamma t}{\rho c} \qquad (2.13)$$

where

ΔT = Temperature rise above ambient in degrees centigrade
γ = Heating rate per unit volume caused by joule heating (J/sec m³)
ρ = Density of metal (kg/m³)
c = Specific heat of metal [J/(kg degrees centigrade)]

The heating rate can be estimated from the joule heating caused by the electric current I. Elementary circuit theory tell us that this is given by the product $I^2 R$, where I is the current, and R is the line resistance. The line resistance is further given in terms of the metal resistivity r by the simple formula $R = rl/A$, where r is the resistivity, l is the length of the line segment, and A is the cross-sectional area. Dividing the input power $I^2 R$ by the total volume of the line segment (lA) gives the heating rate per unit volume as $\gamma = I^2 r/A^2$. Inserting this expression into Equation (2.13) gives our final expression for the expected temperature rise in the line segment on application of a current I for a time duration t:

$$\Delta T = \frac{I^2 rt}{\rho c A^2} \qquad (2.14)$$

where I is the applied current, A is the line cross-sectional area, and all other quantities are as defined in Equation (2.13). It is illustrative to see what this equation predicts for our 30-A, 1-μsec pulse if applied to a copper line 1 μm wide and 10 μm high. For this case, the quantities in Equation (2.14) are

$I = 30$ A line current
$t = 10^{-6}$ sec current pulse duration
$r = 1.7 \times 10^{-8}$ ohm-m
$c = 385$ J/(kg degrees centigrade)
$\rho = 8960$ kg/m^3
$A = 10^{-11}$ m^2

Plugging these numbers into Equation (2.14) gives the rather sobering estimate of a 4400 C temperature rise in the line. Thus, we are far more likely to vaporize the line rather than delaminate it. Early workers in the field facing this problem noticed that the killer factor in Equation (2.14) is the very small cross-sectional area that appears in the denominator squared. Krongelb[71] in particular got around this problem by fabricating top-heavy lines, as shown in Figure 2.26. This line has close to 10 times the cross-sectional area of the one shown in Figure 2.19 while maintaining the same common interfacial area with the substrate. Thus, all quantities remain the same in Equation (2.14) except that the A^2 term is 100 times larger than previously, which lowers the calculated temperature rise in the line to 44 C, which brings the whole experiment back down to Earth and within the realm of possibility. In actual samples fabricated by Krongelb, the top portion of the test line is not a nice rectangle as shown in Figure 2.26 but looks more like a semicircle. The principle remains the same, however; the top part of the line carries the heavy current load and prevents the metal line from vaporization before it can be lifted off by the Lorentz force.

In addition to the work of Krongelb, Baranski and Nevin[72] used a version of this test to measure the adhesion of aluminum to polyimide coatings, demonstrating the usefulness of this method for evaluating the adhesion of metal lines to organic as well as inorganic substrates.

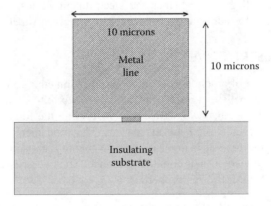

FIGURE 2.26 Modified line structure for the electromagnetic adhesion test. A much larger cross-sectional area is created to avoid vaporizing the metal line before delamination can occur.

2.11.1 Advantages of the Electromagnetic Test

Like the laser spallation test, the electromagnetic test has distinct advantages for certain special applications. The electromagnetic test is perhaps the only test available for investigating the adhesion of 1-µm wide metal lines under conditions closely resembling those of actual microelectronic devices. As with the laser spallation test, there is no problem with gripping the sample. In addition, the very high strain rates on the order of 10^6 sec^{-1} make it likely that viscoplastic effects will also be suppressed in this test.

Finally, with the electromagnetic test there is the possibility of performing chip-scale device testing on the manufacturing line because all of the equipment can be automated, and the sample preparation uses only well-developed lithography techniques that are compatible with standard device fabrication equipment.

2.11.2 Disadvantages of the Electromagnetic Test

The most obvious disadvantage of the electromagnetic test is that it can be applied to conducting materials only, which essentially limits it to metals such as copper and aluminum. Even considering this limitation, the test still suffers from the constraint of requiring large currents and very high magnetic fields to achieve sufficient lifting force to induce delamination. Constraints of sample fabrication essentially limit the usefulness of this method to electronic devices, for which it has certain natural advantages. Finally, the instrumentation required for this test, although not particularly exotic, is not standard, and every setup is likely to be of the one-of-a-kind variety.

2.11.3 Summary and Recommendations

The overall evaluation of the electromagnetic test runs pretty much parallel to that for laser spallation. This is a test that has clear advantages in certain applications but is unlikely to find wide application because of the constraints and limitations mentioned. Availability of the required resources and the specific needs of the intended project will govern whether the electromagnetic test will be the best-available procedure for adhesion measurement in any given situation.

2.12 NONDESTRUCTIVE TESTS

Whereas the concept of a nondestructive adhesion is something of an oxymoron in that it is difficult to believe a result that is supposed to tell you how strongly two materials are stuck together without actually measuring the force required to pull them apart, it is nonetheless a useful concept. The fundamental concept behind nondestructive adhesion tests is to measure some quantity that depends on how well two materials are joined at their common interface. The two tests discussed in this section depend on the following properties:

- Ability to transmit strain/deformation across an interface
- Ability to transmit vibrations along an interface

The dynamic modulus test relies on the ability to transmit strain across an interface. For two perfectly bonded adherends, there is a boundary condition at their common interface that the local displacement field be continuous. If the adhesion is not perfect, then there should be some slippage, which can be detected by some type of measurement. The surface wave propagation test relies on the fact that the velocity of surface waves in an interfacial region will depend on how well adhered the separate layers are. These two measurement techniques are investigated in more detail next.

2.12.1 Dynamic Modulus Test

The dynamic modulus test was used by Su et al.[73] to investigate the adhesion of metals to glass. The basic experimental setup is based on an internal friction apparatus originally developed by Berry and Pritchet[74] to study atomic relaxation phenomena in thin films. A schematic diagram of the apparatus involved is given in Figure 2.27. In a typical experiment, the apparatus in Figure 2.27 can be used to measure the internal friction and dynamic modulus of thin, reedlike specimens depicted in Figure 2.28. The reed is a thin, cantilevered beam, and usual dimensions are roughly 4.5 cm long by 0.5 cm wide by 50 μm thick. By applying a harmonic driving force to the tip of the reed using a frequency oscillator, the reed can be made to vibrate in one of its normal modes of vibration, the frequencies of which are given by the following formula:

$$f_n = \frac{\alpha_n^2 d}{l^2 \sqrt{3\pi}} \sqrt{\frac{E}{\rho}} \qquad n = 1,2,3\ldots\infty \qquad (2.15)$$

where

f_n = Frequency of the nth normal mode of vibration
α_n = Mode number associated with the nth normal mode of vibration
E = Young's modulus of reed
ρ = Reed density
d = Reed thickness
l = Reed length

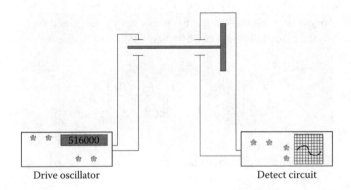

Drive oscillator   Detect circuit

FIGURE 2.27 Vibrating reed experiment for measuring internal friction and dynamic modulus. A cantilevered beam is set into vibration by an oscillatory signal applied at the beam tip. On removing the applied field, the resulting damped vibration is monitored by a second set of electrodes near the base of the sample.

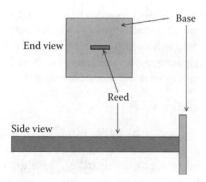

FIGURE 2.28 Typical configuration of the reed sample used in the vibration reed experiment of Figure 2.27.

If the dimensions and density of the reed material are accurately known, then Equation (2.15) provides a precise way of determining the modulus of the reed material from the normal mode frequencies. For the experimental setup shown in Figure 2.27, the frequency of the drive oscillator is varied until it comes close to one of the normal mode frequencies. Because the output of the drive oscillator is capacitively coupled to the tip of the reed through two drive electrodes, the reed will start vibrating in the mode corresponding to the applied frequency, as given by Equation (2.15). The drive signal is then turned off, and the frequency of the now-decaying signal is measured by the detection circuitry, which is capacitively coupled through detect electrodes near the base of the reed. The rate of damping is determined by the internal friction behavior of the reed material and is measured by observing the log decrement, which essentially is calculated from the number of vibration cycles before the amplitude of vibration reaches half of its initial value.

A variation of the above experiment can be used to measure the modulus of a thin film deposited on top of an initially bare reed. This technique was used by Lacombe and Greenblatt[75] to measure the modulus of 0.5-μm polystyrene films on top of a vitreous silica reed. The basis of this measurement is the fact that a thin coating on top of a bare reed will cause the frequency of the coated reed to be shifted with respect to the bare reed frequency by an amount given by the following formula:

$$\left(\frac{f_n^c}{f_n^0}\right)^2 = 1 + \frac{2t}{d}\left[\frac{3E_f}{E_s} - \frac{\rho_f}{\rho_s}\right] \tag{2.16}$$

where

f_n^c = Vibration frequency of the nth mode of coated reed
f_n^0 = Vibration frequency of the nth mode of bare reed
t = Coating thickness
d = Bare reed thickness
E_f = Modulus of coating
E_s = Modulus of bare reed

ρ_f = Coating density

ρ_s = Bare reed density

From Equation (2.16), we see that there are two effects causing the frequency shift of the coated reed. The modulus of the coating adds a stiffening term that causes the frequency to increase. The mass of the coating adds extra inertia, which decreases the vibration frequency as given by the negative term in the ratio of the densities.

The derivation of Equation (2.16) assumes that the coating is perfectly bonded to the underlying reed substrate. Su et al.[73] wondered how Equation (2.16) would be changed if the bonding between the coating and the substrate was not perfect. To investigate, they introduced a coupling parameter γ, which can vary between 0 and 1, with 0 indicating no adhesion and 1 perfect adhesion. In this case, Equation (2.16) becomes

$$\left(\frac{f_n^c}{f_n^0}\right)^2 = 1 + \frac{2t}{d}\left[\frac{3\gamma^2 E_f}{E_s} - \frac{\rho_f}{\rho_s}\right] \tag{2.17}$$

All of the quantities in Equation (2.17) are the same as those in Equation (2.16) with the exception of the coupling parameter. Su et al. used Equation (2.17) to investigate the adhesion of NiTi films to silicon substrates. A replot of their data is shown in Figure 2.29. Note that because of imperfect adhesion, the frequency of the coated reed shown in the bottom curve was less than that of the bare reed and the predicted behavior for perfect adhesion (middle and top curves, respectively). If the adhesion were perfect $\gamma = 1$, Equation (2.17) would predict that the modulus term on the right-hand side of Equation (2.17) should dominate, giving rise to an increased frequency, whereas the observed decrease makes it clear that imperfect coupling was the cause for the frequency decrease. The value of γ in this case was 0.7. Thus, γ^2 is 0.14, therefore reducing the modulus term in Equation (2.17) by nearly an order of magnitude, which accounts for the dominance of the negative density terms. The dynamic modulus test in effect is measuring the efficiency with which bending strains are transferred across the substrate coating interface. Quite interestingly, Su et al. were able to measure γ as a function of coating thickness and annealing treatments, thereby performing a unique set of nondestructive adhesion measurements for the NiTi/Si system.

2.12.2 Advantages of the Dynamic Modulus Test

The main advantages of the dynamic modulus test are its nondestructive nature and the fact that it can be integrated into microelectronic devices as shown in Figure 2.30. Using standard lithography techniques, cantilevered reedlike structures can be fabricated for direct use in testing a variety of material coatings and thin films used in electronic fabrication processes. Further, using micromechanical electronic machine systems technology, the reed structures can be made very small, to the point at which they can be integrated seamlessly into production wafers for purposes of real-time monitoring of adhesion under manufacturing conditions.

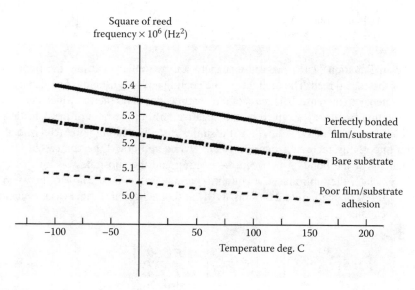

FIGURE 2.29 Square of resonant frequency versus temperature for a silicon reed. Middle curve: bare silicon reed. Bottom curve: reed coated with 2-µm thick NiTi film. Top curve: predicted frequency for coated reed according to Equation (2.17) assuming perfect adhesion, $\gamma = 1$. The fact that the data for the coated reed are lower than that of the bare substrate indicates poor adhesion for this coating. (Data adapted and replotted from Q. Su, S.Z. Hua, and M. Wuttig, in *Adhesion Measurement of Films and Coatings*, K.L. Mittal, Ed., VSP, Utrecht, The Netherlands, 1995, p. 357.)

FIGURE 2.30 Cantilevered silicon reed etched out of a chip. Lithographic methods can be used to miniaturize this structure, making it possible to use it as a diagnostic test site on manufacturing wafers.

A second advantage of the dynamic modulus test arises from the fact that the sample specimen lends itself naturally to internal friction measurements as well as dynamic modulus measurements. Internal friction data can be very useful in monitoring defect structures and impurity migration along grain boundaries in metal films and is also a sensitive way to detect the glass transition and secondary relaxation processes in polymer coatings. Thus, the dynamic modulus test can be a very versatile technique not only for measuring the adhesion on thin films, but also as a potential diagnostic tool for monitoring a wide range of material relaxation phenomena in both metal and insulator films.

2.12.3 DISADVANTAGES OF THE DYNAMIC MODULUS TEST

The clearest disadvantage of the dynamic modulus test lies in the limited sample configuration and the difficulty of sample preparation. The reed samples must be fabricated to precise tolerances to yield accurate data. Also, the nondestructive nature of the test makes it difficult to relate to more standard tests. In particular, there is at present no known relationship between the coupling parameter γ and more conventional measures of adhesion strength, such as the critical fracture energy or the critical fracture stress intensity factor.

2.12.4 SUMMARY AND RECOMMENDATIONS

At this stage of its development, the dynamic modulus test will be most useful within the research-and-development laboratory for specialized applications of the adhesion of thin films to microelectronic substrate materials such as silicon and vitreous silica. The main advantages of the test stem from its nondestructive nature and its versatile capabilities as a method for evaluating both adhesion strength and the thermal-mechanical properties of thin films through internal friction measurements. A significant drawback of the method is that there is as yet no way of relating the measured coupling parameter γ to more conventional adhesion strength quantities such as the critical surface fracture energy. It should be possible to establish an empirical correlation between these quantities through experiments by which the same coating substrate system is tested by both the dynamic modulus test and by one of the more conventional methods discussed in this chapter. This test has significant potential for use in the semiconductor industry, in which the performance of thin metal films is of great concern.

2.13 SURFACE ACOUSTIC WAVES TEST

If you bang on the surface of a block of solid material, then you will send a variety of compression and transverse waves propagating throughout the medium. These waves will typically propagate throughout the entire structure, reflecting off of any boundaries or internal discontinuities. The entire field of ultrasonic inspection is based on a much more controlled version of this primitive experiment. Piezoelectric coupling or another controllable method is used to send very high frequency waves through the structure under carefully controlled conditions. For purposes of adhesion testing, there is a special class of waves that propagate only on the surface of the material; these are known as surface acoustic waves (SAWs). These waves can be induced by a number of methods, including piezoelectric coupling, and are sensitive to the surface conditions of the material.

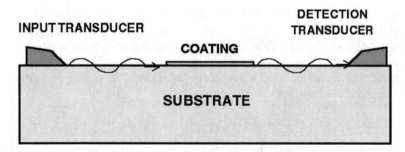

FIGURE 2.31 Schematic of a typical surface acoustic wave (SAW) experiment. The input transducer sends out a surface wave that travels through the coating/substrate interface before it is received by the detection transducer. Changes in the amplitude and velocity of the surface wave on passing through the coating/substrate interface can reveal useful information concerning coating/substrate interactions.

Figure 2.31 illustrates an experiment in which a SAW is generated by a transducer, made to propagate along an interface between a coating and substrate and is subsequently detected by another transducer on the opposite side of the coating. As the surface wave travels through the interface between the coating and the substrate, both its velocity and its amplitude are affected. The shift in velocity gives rise to dispersion effects that are detectable when several different frequencies are present, as would be the case for a wave pulse. Amplitude effects can be used to detect loss mechanisms that might be indicative of poor coupling between the coating and substrate.

The SAW method was reviewed by Ollendorf et al.,[76] who studied TiN films on steel and compared the SAW method to several other more standard techniques, including the scratch test and the four-point bend test. These investigators used the SAW method to determine the modulus of the TiN coatings and found the value so determined correlated well with adhesion measurements on the same films as determined by the four-point bend and scratch tests. The basic effect is quite similar to the dynamic modulus test. As was seen in the dynamic modulus test, propagation of the SAW through the interface depends on how efficiently transverse strains are transmitted from the substrate to the coating. Voids, regions of loose packing, and other surface imperfections clearly interfere with efficient transmission of strain at the interface and thus affect the apparent modulus measurement of the film as inferred by the SAW method. Because these defects also reduce the film adhesion, it is clear why there should be a correlation between the modulus measurement and the film adhesion. An extensive review of the use of SAWs to study the performance of adhesive joints was given by Rokhlin.[77]

In another application of the SAW method, Galipeau et al.[78] studied the adhesion of polyimide films on quartz substrates. Their basic setup is shown schematically in Figure 2.32. Because the substrate used is quartz, which is a natural piezoelectric material, the input and detection transducers consist simply of interdigitated lines that are patterned on the substrate using standard photolithographic methods. An oscillating input applied to the electrodes of the transducer couples directly into the substrate, setting up an analogous mechanical surface wave. This wave travels as an

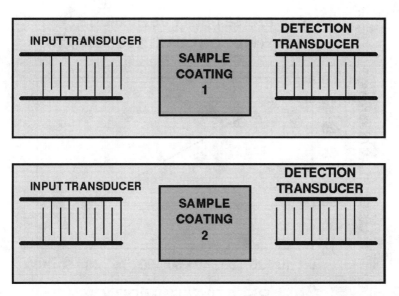

FIGURE 2.32 Surface acoustic wave experiment for polyimide on quartz. Because quartz is a natural piezoelectric material, transducers can simply be fabricated by evaporating metal lines onto the quartz as shown. Electrical signals in the metal lines are then directly converted into mechanical vibrations in the quartz. In the single delay line experiment using one substrate, the propagated surface wave is compared to a reference signal generated by the input driver. The dual delay line experiment employs two samples in which everything is the same except for the preparation of the coating/substrate interface. The two signals are compared at the detection transducers.

SAW into the coating/substrate interface and out the other end, where it is detected by an identical transducer, which converts the mechanical SAW back into an electrical signal. Note that this differs from the method used by Ollendorf et al. in that these authors used a laser pulse to generate a pulsed wave with many harmonics, which would then show dispersion effects going through the coating/substrate interface. The setup in Figure 2.32 will generate a single harmonic that will not show dispersion. The velocity of this wave will however be affected by passing through the interface, and this is detected in one of two ways. In the first method, the input signal is compared to that of the detector transducer in what is called a single delay line experiment. The second method uses two delay lines to compare the signals from two coatings prepared differently and therefore with different levels of adhesion.

Figure 2.33 illustrates some sample data replotted from results presented by Galipeau et al. The curve shows the phase delay incurred when the surface wave passed through the interface between a polyimide film and the quartz substrate as a function of ambient humidity. The negative phase delay is indicative of slowing of the wave. Thus, as the humidity increases, it seems to have a softening effect on the interface, slowing the wave more and more as the relative humidity increases. Because it is well known from standard adhesion tests that high humidity reduces the adhesion of polyimide, we can assume that the apparent softening of the interface region is also partly responsible for this effect. One can perform the same type of

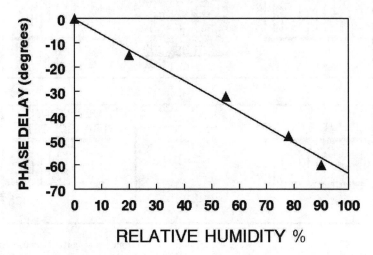

FIGURE 2.33 Typical data showing the effect of relative humidity on the passage of a surface acoustic wave through a polyimide/quartz interface. The negative phase delay is indicative of slowing of the wave, indicating that the moisture accumulated at the interface is having a "softening" effect, thereby slowing the signal.

experiment using the dual delay line method in which two polyimide films, one with an adhesion promoter such as a silane material and one without, are compared. The dual delay line test clearly shows that the SAW travels faster through the coating with the adhesion promoter, thus accounting for the higher adhesion observed when the adhesion promoter is present.

2.13.1 Advantages of the Surface Acoustic Waves Test

The key advantages of the SAW method lie in its nondestructive character and the possibility that it can be applied in a manufacturing environment for online quality control purposes. In addition, the SAW method can be calibrated against standard destructive adhesion measurement experiments to provide quantitative results related to surface fracture energies.

2.13.2 Disadvantages of the Surface Acoustic Waves Test

The main disadvantage of the SAW approach is that it does not yield direct adhesion strength data and must be calibrated against one of the more standard tests. In addition, sample preparation is an important problem with this method because it is not always a trivial matter to couple the input and output transducers to the substrate. Unlike the quartz substrates, most materials do not have natural piezo-electric behavior; therefore, some other method of coupling the electrical signal to the substrate must be devised. Many times, this involves the use of a coupling fluid, which works but also complicates the experiment.

2.13.3 SUMMARY AND RECOMMENDATIONS

From the discussion, it is fairly clear that the SAW adhesion measurement has its greatest potential for use as a real-time quality control tool. Using the laser pulse method to generate the surface waves coupled with interferometry as a detector, this method could be used to inspect coatings rapidly and without contact.

Notes

1. For further details on this most fascinating story, a first-rate account is given in "Stern and Gerlach: How a Bad Cigar Helped Reorient Atomic Physics," B. Friedrich and D. Herschbach, *Physics Today*, 53 (December 2003).
2. "Adhesion Measurement of Films and Coatings: A Commentary," K.L. Mittal, in *Adhesion Measurement of Films and Coatings*, K.L. Mittal, Ed. (VSP, Utrecht, The Netherlands, 1995), p. 1.
3. "An Experimental Partitioning of the Mechanical Energy Expended During Peel Testing," R.J. Farris and J.L. Goldfarb, in *Adhesion Measurement of Films and Coatings*, K.L. Mittal, Ed. (VSP, Utrecht, The Netherlands, 1995), p. 265.
4. "Peel Adhesion and Pressure Sensitive Tapes," D. Satas and F. Egan, *Adhesives Age*, 22 (August 1966).
5. "Failure Mechanisms in Peeling of Pressure-Sensitive Adhesive Tape," D.W. Aubrey, G.N. Welding, and T. Wong, *Journal of Applied Polymer Science*, 13, 2193 (1969).
6. "Quantifying the Tape Adhesion Test," G.V. Calder, F.C. Hansen, and A. Parra, in *Adhesion Aspects of Polymeric Coatings*, K.L. Mittal, Ed. (Plenum Press, New York, 1983), p. 569.
7. Ibid., p. 570.
8. "Salt Bath Test for Assessing the Adhesion of Silver to Poly(ethylene terephthalate) Web," J.M. Grace, V. Botticelle, D. R. Freeman, W. Kosel, and R.G. Spahn, in *Adhesion Measurement of Films and Coatings*, K.L. Mittal, Ed. (VSP, Utrecht, The Netherlands, 1995), p. 423.
9. "Pull-off Test, an Internationally Standardized Method for Adhesion Testing — Assessment of the Relevance of Test Results," J. Sickfeld, in *Adhesion Aspects of Polymeric Coatings,* K.L. Mittal, Ed. (Plenum Press, New York, 1983), p. 543.
10. "Analysis of Pull Tests for Determining the Effects of Ion Implantation on the Adhesion of Iron Films to Sapphire Substrates," J.E. Pawel and C.J. McHargue, in *Adhesion Measurement of Films and Coatings*, K.L. Mittal, Ed. (VSP, Utrecht, The Netherlands, 1995), p. 323.
11. "Measurement of the Adhesion of Diamond Films on Tungsten and Correlations with Processing Parameters," M. Alam, D.E. Peebles, and J.A. Ohlhausen, in *Adhesion Measurement of Films and Coatings*, K.L. Mittal, Ed. (VSP, Utrecht, The Netherlands, 1995), p. 331.
12. "Indentation-Debonding Test for Adhered Thin Polymer Layers," P.A. Engel and G.C. Pedroza, in *Adhesion Aspects of Polymeric Coatings,* K.L. Mittal, Ed. (Plenum Press, New York, 1983), p. 583.
13. Ibid., p. 588.

14. "Mechanics of the Indentation Test and Its Use to Assess the Adhesion of Polymeric Coatings," R. Jayachandran, M.C. Boyce, and A.S. Argon, in *Adhesion Measurement of Films and Coatings*, K.L. Mittal, Ed. (VSP, Utrecht, The Netherlands, 1995), p. 189.

15. "Observations and Simple Fracture Mechanics Analysis of Indentation Fracture Delamination of TiN Films on Silicon," E.R. Weppelmann, X.-Z. Hu, and M.V. Swain, in *Adhesion Measurement of Films and Coatings*, K.L. Mittal, Ed. (VSP, Utrecht, The Netherlands, 1995), p. 217.

16. "Hardness and Adhesion of Filmed Structures as Determined by the Scratch Technique," J. Ahn, K.L. Mittal and R.H. MacQueen, in *Adhesion Measurement of Thin Films, Thick Films and Bulk Coatings*, ASTM STP 640, K.L. Mittal, Ed. (American Society for Testing and Materials, West Conshohocken, PA, 1978), p. 134.

17. "Some Factors Influencing the Adhesion of Films Produced by Vacuum Evaporation," O.S. Heavens, *Journal of the Physics of Radium*, 11, 355 (1950).

18. "L'epitaxie dans les Lames Polycrystallines," O.S. Heavens and L.E. Collins, *Journal of the Physics of Radium*, 13, 658 (1952).

19. "Measurement of Adhesion of Thin Film," P. Benjamin and C. Weaver, *Proceedings of the Royal Society* (London), 254A, 163 (1960).

20. "The Stylus or Scratch Method for Thin Film Adhesion Measurement: Some Observations and Comments," D.W. Butler, C.J.H. Stoddard, and P.R. Stuart, *Journal of Physics (D): Applied Physics*, 3, 887 (1970).

21. Note 16, p. 150.

22. "Threshold Adhesion Failure: An Approach to Aluminum Thin-Film Adhesion Measurement Using the Stylus Method," J. Oroshnik and W.K. Croll, in *Adhesion Measurement of Thin Films, Thick Films and Bulk Coatings*, ASTM STP 640, K.L. Mittal, Ed. (American Society for Testing and Materials, West Conshohocken, PA, 1978), p. 158.

23. "Scratch Indentation, a Simple Adhesion Test Method for Thin Films on Polymeric Supports," G.D. Vaughn, B.G. Frushour, and W.C. Dale, in *Adhesion Measurement of Films and Coatings*, K.L. Mittal, Ed. (VSP, Utrecht, The Netherlands, 1995), p. 127.

24. "Micro-scratch Test for Adhesion Evaluation of Thin Films," V.K. Sarin, in *Adhesion Measurement of Films and Coatings*, K.L. Mittal, Ed. (VSP, Utrecht, The Netherlands, 1995), p. 175.

25. "Can the Scratch Adhesion Test Ever Be Quantitative?" S.J. Bull, in *Adhesion Measurement of Films and Coatings*, Vol. 2, K.L. Mittal, Ed. (VSP, Utrecht, The Netherlands, 2001), p. 107.

26. "Scratch Test Failure Modes and Performance of Organic Coatings for Marine Applications," S.J. Bull, K. Horvathova, L.P. Gilbert, D. Mitchell, R.I. Davidson, and J.R. White, in *Adhesion Measurement of Films and Coatings*, K.L. Mittal, Ed. (VSP, Utrecht, The Netherlands, 1995), p. 175.

27. "Measurement of Adhesion by a Blister Method," H. Dannenberg, *Journal of Applied Polymer Science*, 5, 125 (1961).

28. "A Study of the Fracture Efficiency Parameter of Blister Tests for Films and Coatings," Y.-H. Lai and D.A. Dillard, in *Adhesion Measurement of Films and Coatings*, K.L. Mittal, Ed. (VSP, Utrecht, The Netherlands, 1995), p. 231.

29. "Analysis of Critical Debonding Pressures of Stressed Thin Films in the Blister Test," M.G. Allen and S.D. Senturia, *Journal of Adhesion*, 25, 303 (1988).

30. M.G. Allen and S.D. Senturia, *Journal of Adhesion*, 29, 219 (1989).

31. M.J. Napolitano, A. Chudnovsky, and A. Moet, *Proceedings of ACS, Division of Polymer Material Science and Engineering*, 57, 755 (1987).

32. "The Constrained Blister Test for the Energy of Interfacial Adhesion," M.J. Napolitano, A. Chudnovsky, and A. Moet, *Journal of Adhesion Science and Technology*, 2, 311 (1988).

33. "The Constrained Blister — A Nearly Constant Strain Energy Release Rate Test for Adhesives," Y.S. Chang, Y.H. Lai, and D.A. Dillard, *Journal of Adhesion*, 27, 197 (1989).

34. "An Elementary Plate Theory Prediction for Strain Energy Release Rate of the Constrained Blister Test," Y.H. Lai and D.A. Dillard, *Journal of Adhesion*, 31, 177 (1990).

35. "Numerical Analysis of the Constrained Blister Test," Y.H. Lai and D.A. Dillard, *Journal of Adhesion*, 33, 63 (1990).

36. "Adhesion (Fracture Energy) of Electropolymerized Poly(*n*-octyl maleimide-*co*-styrene) on Copper Substrates Using a Constrained Blister Test," J.-L. Liang, J.P. Bell, and A. Mehta, in *Adhesion Measurement of Films and Coatings*, K.L. Mittal, Ed. (VSP, Utrecht, The Netherlands, 1995), p. 249.

37. Note 33.

38. Note 29, p. 311.

39. "The Three Point Bend Test for Adhesive Joints," N.T. McDevitt and W.L. Braun, in *Adhesive Joints: Formation, Characteristics and Testing*, K.L. Mittal, Ed. (Plenum Press, New York, 1984), p. 381.

40. "Practical Adhesion Measurement in Adhering Systems: A Phase Boundary Sensitive Test," A.A. Roche, M.J. Romand, and F. Sidoroff, in *Adhesive Joints: Formation, Characteristics and Testing*, K.L. Mittal, Ed. (Plenum Press, New York, 1984), p. 19.

41. "Interface Adhesion: Measurement and Analysis," A.G. Evans, in *Adhesion Measurement of Films and Coatings*, Vol. 2, K.L. Mittal, Ed. (VSP, Utrecht, The Netherlands, 2001), p. 1.

42. "Use of Fracture Mechanics Concepts in Testing of Film Adhesion," W.D. Bascom, P.F. Becher, J.L. Bitner, and J.S. Murday, in *Adhesion Measurement of Thin Films, Thick Films and Bulk Coatings*, ASTM STP 640, K.L. Mittal, Ed. (American Society for Testing and Materials, West Conshohocken, PA, 1978), p. 63.

43. "Fracture Design Criteria for Structural Adhesive Bonding — Promise and Problems," W.D. Bascom, R.L. Cottington, and C.O. Timmons, *Naval Engineers Journal*, 73 (August 1976).

44. "Mixed Mode Cracking in Layered Materials," J.W. Hutchinson and Z. Suo, *Advances in Applied Mechanics*, 29, 63 (1991).

45. "Closure and Repropagation of Healed Cracks in Silicate Glass," T.A. Michalske and E.R. Fuller Jr., *Journal of the American Ceramics Society*, 68, 586 (1985).

46. Note 41, p. 3.

47. "Combined Mode Fracture via the Cracked Brazilian Disk Test," C. Atkinson, R.E. Smelser, and J. Sanchez, *International Journal of Fracture*, 18, 279 (1982).

48. "Experimental Determination of Interfacial Toughness Using Brazil-Nut-Sandwich," J.S. Wang and Z. Suo, *Acta Met.*, 38, 1279 (1990).

49. Note 44, p. 120.

50. "Fracture Toughness Curves for Epoxy Molding Compound/Leadframe Interfaces," H.-Y. Lee, *Journal of Adhesion Science and Technology*, 16, 565 (2002).

51. "Adhesion and Adhesives Technology," A.V. Pocius (Carl Hanser Verlag, Munich, 2002), p. 186.

52. "The Role of Interfacial Interactions and Loss Function of Model Adhesives on Their Adhesion to Glass," A Sharif, N. Mohammadi, M. Nekoomanesh, and Y. Jahani, *Journal of Adhesion Science and Technology*, 16, 33 (2002).

53. "Plasma Sprayed Coatings as Surface Treatments of Aluminum and Titanium Adherends," G.D. Davis, P.L. Whisnant, D.K. Shaffer, G.B. Groff, and J.D. Venables, *Journal of Adhesion Science and Technology*, 9, 527 (1995).

54. "Effectiveness of Water Borne Primers for Structural Adhesive Bonding of Aluminum and Aluminum-Lithium Surfaces," K.L. Meyler and J.A. Brescia, *Journal of Adhesion Science and Technology*, 9, 81 (1995).

55. J. Cognard, *Journal of Adhesion*, 22, 97 (1987).

56. M.F. Kauninen, *International Journal of Fracture*, 9, 83 (1973).

57. H.R. Brown, *Journal of Material Science*, 25, 2791 (1990).

58. "Strength of Polystyrene Poly(Methyl Methacrylate) Interfaces," K.L. Foster and R.P. Wool, *Macromolecules*, 24(6), 1397 (1991).

59. "On the Non-linear Elastic Stresses in an Adhesively Bonded T-Joint With Double Support," M.K. Apalak, *Journal of Adhesion Science and Technology*, 16, 459 (2002).

60. "A Simple Film Adhesion Comparator," D.W. Butler, *Journal of Physics (E): Science of Instrumentation*, 3, 979 (1970).

61. "Measurement of the Adhesion of Silver Films to Glass Substrates," A. Kikuchi, S. Baba, and A. Kinbara, *Thin Solid Films*, 124, 343 (1985).

62. "A Self-Delamination Method of Measuring the Surface Energy of Adhesion of Coatings," R.J. Farris and C.L. Bauer, *Journal of Adhesion*, 26, 293 (1988).

63. "Measurement of Interfacial Fracture Energy in Microelectronic Multifilm Applications," J.C. Hay, E. Lininger, and X.H. Liu, in *Adhesion Measurement of films and Coatings,* Vol. 2, K.L. Mittal, Ed. (VSP, Utrecht, The Netherlands, 2001), p. 205.

64. The relevant references are: "A New Procedure for Measuring the Decohesion Energy for Thin Ductile Films on Substrates," A. Bagchi, G.E. Lucas, Z. Suo, and A.G. Evans, *J. Mater. Res.,* 9(7), 1734 (1994); "Thin Film Decohesion and its Measurement," A. Bagchi and A.G. Evans, *Mat. Res. Soc. Symp. Proc.,* 383, 183 (1995); "The Mechanics and Physics of Thin Film Decohesion and its Measurement," A. Bagchi and A.G. Evans, *Interface Science,* 3, 169 (1996).

65. "Interface Adhesion: Measurement and Analysis," A.G. Evans, in *Adhesion Measurement of Films and Coatings,* Vol. 2, K.L. Mittal, Ed. (VSP, Utrecht, The Netherlands, 2001), p. 9.

66. "Measurement of Film-Substrate Bond Strength by Laser Spallation," J.L. Vossen, in *Adhesion Measurement of Thin Films, Thick Films and Bulk Coatings*, ASTM STP 640, K.L. Mittal, Ed. (American Society for Testing and Materials, West Conshohocken, PA, 1978), p. 122.

67. "Measurement of Interfacial Bond Strength by Laser Spallation," A.W. Stephens and J.L. Vossen, *Journal of Vacuum Science and Technology*, 13, 38 (1976).

68. "Recent Developments in the Laser Spallation Technique to Measure the Interface Strength and Its Relationship to Interface Toughness With Applications to Metal/Ceramic, Ceramic/Ceramic and Ceramic/Polymer Interfaces," V. Gupta, J. Yuan, and A. Pronin, in *Adhesion Measurement of Films and Coatings,* K.L. Mittal, Ed. (VSP, Utrecht, The Netherlands, 1995), p. 367.

69. Ibid., p. 395.

70. "Effect of Primer Curing Conditions on Basecoat-Primer Adhesion — A LIDS Study," J.S. Meth, in *Adhesion Measurement of Films and Coatings*, Vol. 2, K.L. Mittal, Ed. (VSP, Utrecht, The Netherlands, 2001), p. 255.

71. "Electromagnetic Tensile Adhesion Test Method," S. Krongelb, in *Adhesion Measurement of Thin Films, Thick Films and Bulk Coatings*, ASTM STP 640, K.L. Mittal, Ed. (American Society for Testing and Materials, West Conshohocken, PA, 1978), p. 107.

72. "Adhesion of Aluminum Films on Polyimide by the Electromagnetic Tensile Test," B.P. Baranski and J.H. Nevin, *Journal of Electronic Material,* 16, 39 (1987).
73. "Nondestructive Dynamic Evaluation of Thin NiTi Film Adhesion," Q. Su, S.Z. Hua, and M. Wuttig, in *Adhesion Measurement of Films and Coatings*, K.L. Mittal, Ed. (VSP, Utrecht, The Netherlands, 1995), p. 357.
74. "Vibrating Reed Internal Friction Apparatus for Films and Foils," B.S. Berry and W.C. Pritchet, *IBM Journal of Research and Development*, 19, 334 (1975).
75. "Mechanical Properties of Thin Polyimide Films," R.H. Lacombe and J. Greenblatt, in *Polyimides,* Vol. 2, K.L. Mittal, Ed. (Plenum, New York, 1984), p. 647
76. "Testing the Adhesion of Hard Coatings Including the Non-destructive Technique of Surface Acoustic Waves," H. Ollendorf, T. Schülke, and D. Schneider, in *Adhesion Measurement of Films and Coatings*, Vol. 2, K.L. Mittal, Ed. (VSP, Utrecht, The Netherlands, 2001), p. 49.
77. "Adhesive Joint Characterization by Ultrasonic Surface and Interface Waves," S.I. Rokhlin, in *Adhesive Joints: Formation, Characterization and Testing,* K.L. Mittal, Ed. (Plenum Press, New York, 1984), p. 307.
78. "Adhesion Studies of Polyimide Films Using a Surface Acoustic Wave Sensor," D.W. Galipeau, J.F. Vetelino, and C. Feger, in *Adhesion Measurement of Films and Coatings*, K.L. Mittal, Ed. (VSP, Utrecht, The Netherlands, 1995), p. 411.

3 Theoretical Foundations of Quantitative Adhesion Measurement Methods

Since the subject became popular with mathematicians about 150 years ago and I am afraid that a really formidable number of unreadable, incomprehensible books have been written about elasticity, and generations of students have endured agonies of boredom in lectures about materials and structures. In my opinion the mystique and mumbo-jumbo is overdone and often beside the point. It is true that the higher flights of elasticity are mathematical and very difficult—but then this sort of theory is probably only rarely used by successful engineering designers. What we find difficult about mathematics is the symbolic presentation of the subject by pedagogues with a taste for dogma, sadism and incomprehensible squiggles.

J. E. Gordon, *Structures or Why Things Don't Fall Down*

3.1 INTRODUCTION TO CONTINUUM THEORY

Because every method for performing an adhesion measurement involves applying a force or load of some type to the coating/substrate system, considerations of stress, strain, and deformation must inevitably be confronted. Notwithstanding the comments of Prof. Gordon quoted above, this chapter attempts to provide a brief introduction to the subject of the continuum theory of solids and to the greatest extent possible tries to avoid technical jargon while at the same time bringing into sharper focus the essential mechanics that need to be understood to grasp the role stress plays in adhesion measurement. The main focus is on developing a sufficient level of understanding of the fundamental theory so that the reader will be able to follow subsequent analytical developments, have sufficient background to confront the technical literature, and be able to perform simple but useful calculations in a competent manner. The development of necessity will be limited in scope, but a number of references to the more lucid and useful literature are provided to give ample means for the interested reader to delve deeper into the subject.

Continuum theory is a subject with foundations that were essentially completed by the end of the 19th century by such famous physicists and mathematicians as Navier, Poisson, Green, Cauchy, Hooke, Young, and a number of others. By the early 20th century, the fashion in physics had shifted dramatically to the then-emerging theory of quantum mechanics, which was required to understand the behavior of matter at the atomic level. Continuum theory was essentially relegated to the engineers and a small coterie of mathematicians who found significant challenges in the mathematics of the subject. Indeed, by the time I was carrying out my undergraduate and graduate studies in physics, the subject of the continuum theory of matter had all but dropped out of the curriculum in deference to more current topics, such as particle mechanics, electromagnetism, optics, and of course quantum theory. Part of the problem stems from the fact that continuum theory was viewed as only an approximation to the reality of matter and was in fact clearly wrong at the atomic level. However, even though it may only be an approximate description of material behavior it proved to be an extremely useful one and is in fact indispensable as a modern engineering design tool for everything from safety pins to the space shuttle. For present purposes, it is assumed that matter at the scale of tenths of micrometers to tens of meters is accurately described by the field equations of continuum theory. The presentation is carried out in four parts.

Sections 3.1.1 through 3.1.3 define the concept of stress in a solid. Stress is something of a mystery to most people, especially those who are more comfortable thinking about forces. A force is a vector quantity requiring three components to be given for a complete specification, one for each of the three mutually orthogonal directions in space. Stress, however, is a tensor quantity requiring in general nine components for a full description. Why all the complication? Why mess around with something three times more complicated than simple forces that most people can understand and that should, in principle, be sufficient to describe the loads applied to any solid? The discussion in Part 1 will try to motivate and explain the rationale behind the concept of stress.

Section 3.1.4 introduces the concepts of deformation and strain. This is a relatively straightforward development because nearly everyone has a fairly intuitive sense of deformation. If you apply a load to an object, it will deform by bending, twisting, or stretching in some manner. However, there are a number of subtle and nonintuitive aspects to this subject that need careful attention and are appropriately highlighted.

Section 3.1.5 introduces constitutive relations. At this point, all of the kinematical relations for defining the stresses and deformations in solids are in place but a further essential ingredient is missing, what is called the constitutive relations. The basic role of the constitutive relations is to connect the stresses to the strains. They describe a fundamental material property of the solid under investigation. Different materials require different constitutive relations. Without a constitutive relation, you really do not know the material with which you are dealing, whether it is a piece of metal, a slab of glass, or an item of ceramic cookware. The simplest such relation is Hooke's law for elastic solids, which in essence says that the stress is linearly proportional to the strain, with the constant of proportionality the elastic modulus, often referred to as Young's modulus. All of this is covered in detail in Section 3.1.5.

Finally, in Section 3.2 and following, all of these ingredients are put together to form what are known as the field equations of thermoelasticity. As an extra benefit, Fourier's equation of heat transfer is thrown into the pot so that problems of thermal stress loading can be addressed. These are very important loads for many problems. The final confection of field equations forms one of the crowning achievements of 19th century theoretical physics, and these equations are rivaled only by Maxwell's equations, which form the foundation of electromagnetic theory. Maxwell's theory is truly phenomenal. It is valid over spatial and timescales of at least 15 orders of magnitude and was relativistically correct at its inception. The only correction that had to be made was to allow for the quantum nature of matter. This gave rise to quantum electrodynamics as developed by Feynman, Schwinger, and Tomonaga. The field equations of thermoelasticity are not relativistically correct or valid at the atomic level. However, for a vast range of practical problems, they are as valid in the 21st century as they were in the 19th century. The errors committed by ignoring relativistic and quantum mechanical effects are well below the sensitivity of most practical measurements. The largest source of error in fact arises from not having truly correct constitutive relations for the material substances under investigation.

The general field equations of thermoelasticity are, to put it mildly, exceedingly difficult to solve by paper-and-pencil methods. The common practice in engineering applications is to resort to powerful numerical methods such as finite element analysis or boundary element analysis. Given the enormous computational power of modern workstations, these methods are highly effective at generating useful numerical solutions to the field equations. Surprisingly, however, there are a number of nontrivial special cases that can be solved "on the back of an envelope." These cases are discussed in some detail as they provide a number of useful computational formulae for estimating stresses in situations of practical interest and supply a handy set of tools for the user.

3.1.1 CONCEPT OF STRESS IN SOLIDS

As pointed out, nine numbers are required in general to specify the state of stress at a point in a solid. In most cases, certain symmetries apply, and only six numbers are needed, but this is still twice as many as are required to determine a force vector, which should amount to the same thing. The problem arises because, in a continuous medium such as a solid, the relevant quantity is not the force vector alone but the force acting on some element of area. The simplest example is the pressure in a pneumatic tire, which is always specified in pounds per square inch (psi) or Newtons per square meter (Pascal or Pa). The point is that, when speaking of a continuous lump of matter, forces are generally applied to some element of area that gives rise to the concept of a stress that is a force per unit area. The pressure in a tire is a very special stress state known as a hydrostatic stress. It is the same in all directions and thus specified by only one number. This comes about from the fluid nature of the compressed gas in the tire, which cannot support anything but a hydrostatic load. Fluids in general cannot support either tensile or shear loads and thus must be confined to some type of sealed container in which they can be loaded hydrostatically. Solids, on the other hand, support both types of loads, and the state of stress in a solid is sensitive to how it is supported and the details of the applied force field that will typically act on the surface. For large objects, there is also what is called body forces, which strictly speaking act only on a specific point in the body. Gravitational loading is the most important example. Each and every element of a solid has a mass, which is computed as the density times the volume of the element. Each such mass element is under a force loading because of the Earth's gravitational field. This gravitational loading will give rise to stresses in the solid at the surfaces where it is supported. Thus, again forces per unit area must be considered.

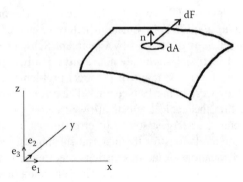

FIGURE 3.1 An element of surface dA subjected to a localized force dF. The complete description of dA requires specification of the unit normal vector **n**, which defines the orientation of dA in three dimensions. The vector element d**A** is given by the product (dA)**n**.

Figure 3.1 depicts a hypothetical solid surface with a small element of area shown under the influence of some applied force vector **dF**. The area element also has a vector character because it is characterized by both a magnitude dA and a unit vector **n** that is perpendicular to the tangent plane that contains the area element. The vector **n** is commonly referred to as the unit normal that characterizes the direction of the area element dA. Although one does not normally associate a direction with the notion of area, it is clearly important in continuum theory because every surface in three-dimensional space is oriented in some way, and specifying the unit normal is a precise and convenient way of describing this property. The force **dF** arises because of the action of the material in the immediate neighborhood of the area element dA. It can be of an intrinsic nature, such as forces arising from material defects such as voids, inclusions, or dislocations. Alternately, it can arise from loads applied to the surface of the material that are transmitted to the interior in a chainlike fashion from one volume element to another.

In general, the genesis of the local force vector is quite complex and not obtainable by simple means. Fortunately, for present purposes only the existence of this force is required. At equilibrium, an equal and opposite force **−dF** operates on the other side of the area element dA. Together, these forces create a state of stress acting on dA. Because each separate element of surface will require a different unit normal, it is clear that to come up with a general definition of stress, which is force per unit area, two vector quantities will be involved; this is the reason behind the tensor nature of the stress field. The area element vector is defined in the obvious way as follows:

$$d\mathbf{A} = dA\ \mathbf{n} \tag{3.1}$$

More useful for our purposes, however, is the inverse area element, which also has an obvious definition:

$$d\mathbf{A}^{-1} = \frac{\mathbf{n}}{dA} \tag{3.2}$$

The vector dA^{-1} represents the surface element by assigning it a direction $\mathbf{n}$ and a magnitude (1/dA). The stress acting at the area element dA can now be defined in a straightforward manner:

$$\mathbf{T} = d\mathbf{F}dA^{-1} \qquad (3.3)$$

Equation (3.3) now fits the bill for the required tensor quantity that has the magnitude of a force per unit area but points in two directions at the same time. One direction is that of $d\mathbf{F}$; the other is the direction of the area unit normal $\mathbf{n}$. Thus, Equation (3.3) reads that $\underline{\mathbf{T}}$ is the stress at the location of the area element dA caused by the vector force $d\mathbf{F}$ acting on the area element dA^{-1} pointing in the direction $\mathbf{n}$. $\underline{\mathbf{T}}$ is a dyadic quantity because it is the composition of two vectors that generally act in separate directions and thus exist in a higher dimensional space than simple vectors. Tensors and dyadics are roughly mathematical equivalents and are often used interchangeably as a matter of convenience. One can imagine even more complex quantities such as triadics, tetradics, and so on that can act in three or more directions. Refer to Appendix A on mathematical foundations for a more detailed treatment of this subject.

A more useful description of $\underline{\mathbf{T}}$ is provided by referring both $d\mathbf{F}$ and dA^{-1} to some appropriate Cartesian coordinate system as depicted in Figure 3.1:

$$d\mathbf{F} = dF_1\mathbf{e}_1 + dF_2\mathbf{e}_2 + dF_3\mathbf{e}_3$$

$$dA^{-1} = \left(\frac{\mathbf{e}_1}{dA_1} + \frac{\mathbf{e}_2}{dA_2} + \frac{\mathbf{e}_3}{dA_3} \right) \qquad (3.4)$$

The area elements dA_1, dA_2, and dA_3 are projections of the element dA onto the coordinate planes perpendicular to the respective unit vectors $\mathbf{e}_1$, $\mathbf{e}_2$, and $\mathbf{e}_3$ depicted in Figure 3.1. The stress tensor $\underline{\mathbf{T}}$ now becomes:

$$\underline{\mathbf{T}} = d\mathbf{F}\,dA^{-1} = \left(dF_1\mathbf{e}_1 + dF_2\mathbf{e}_2 + dF_3\mathbf{e}_3 \right)\left(dA_1^{-1}\mathbf{e}_1 + dA_2^{-1}\mathbf{e}_2 + dA_3^{-1}\mathbf{e}_3 \right)$$

If we carry out all the implied multiplications and collect terms appropriately, we see that the stress tensor can be written as

$$\underline{\mathbf{T}} = \begin{array}{l} \dfrac{dF_1}{dA_1}\mathbf{e}_1\mathbf{e}_1 + \dfrac{dF_2}{dA_1}\mathbf{e}_2\mathbf{e}_1 + \dfrac{dF_3}{dA_1}\mathbf{e}_3\mathbf{e}_1 \\[3mm] \dfrac{dF_1}{dA_2}\mathbf{e}_1\mathbf{e}_2 + \dfrac{dF_2}{dA_2}\mathbf{e}_2\mathbf{e}_2 + \dfrac{dF_3}{dA_2}\mathbf{e}_3\mathbf{e}_2 \\[3mm] \dfrac{dF_1}{dA_3}\mathbf{e}_1\mathbf{e}_3 + \dfrac{dF_2}{dA_3}\mathbf{e}_2\mathbf{e}_3 + \dfrac{dF_3}{dA_3}\mathbf{e}_3\mathbf{e}_3 \end{array} \qquad (3.5)$$

Following standard convention, we write the stress tensor $\mathbf{T}$ in terms of three traction vectors $\mathbf{T}_1$, $\mathbf{T}_2$, and $\mathbf{T}_3$:

$$\underline{\mathbf{T}} = \mathbf{e}_1\mathbf{T}_1 + \mathbf{e}_2\mathbf{T}_2 + \mathbf{e}_3\mathbf{T}_3 \tag{3.6}$$

where

$$\mathbf{T}_1 = \frac{dF_1}{dA_1}\mathbf{e}_1 + \frac{dF_2}{dA_1}\mathbf{e}_2 + \frac{dF_3}{dA_1}\mathbf{e}_3$$

$$\mathbf{T}_2 = \frac{dF_1}{dA_2}\mathbf{e}_1 + \frac{dF_2}{dA_2}\mathbf{e}_2 + \frac{dF_3}{dA_2}\mathbf{e}_3$$

$$\mathbf{T}_3 = \frac{dF_1}{dA_3}\mathbf{e}_1 + \frac{dF_2}{dA_3}\mathbf{e}_2 + \frac{dF_3}{dA_3}\mathbf{e}_3$$

The main purpose in defining the traction vectors is that they are the most convenient quantities for defining surface loads on solid bodies. They are convenient vector quantities defined directly in terms of the applied forces while at the same time are genuine stress quantities with the dimension of force per unit area. Thus, if one solid body impinges on another, there is always some finite area of contact involved, and the applied load is distributed over this area. Therefore, the actual load is a force per some area or a traction vector as defined above. The equations can be greatly simplified by introducing the standard notation for the components of the stress tensor σ_{ij}:

$$\mathbf{T}_1 = \sigma_{11}\mathbf{e}_1 + \sigma_{12}\mathbf{e}_2 + \sigma_{13}\mathbf{e}_3$$

$$\mathbf{T}_2 = \sigma_{21}\mathbf{e}_1 + \sigma_{22}\mathbf{e}_2 + \sigma_{23}\mathbf{e}_3 \tag{3.7}$$

$$\mathbf{T}_3 = \sigma_{31}\mathbf{e}_1 + \sigma_{32}\mathbf{e}_2 + \sigma_{33}\mathbf{e}_3$$

By comparing Equations (3.6) and (3.7), it is immediately clear what the definitions of the stress components σ_{ij} are. For example σ_{11} is the stress generated by $\mathbf{dF}_1$ acting on area element dA_1, which points in the direction $\mathbf{e}_1$. Likewise, σ_{12} is the stress generated by $\mathbf{dF}_2$ acting on area element dA_1 pointing in the direction $\mathbf{e}_2$. Note that σ_{12} is a shear component because the force acts perpendicular to the area unit normal as opposed to σ_{11}, which is called a tensile component because it acts parallel to the area unit normal. Similar observations hold for the remaining components.

Working with quantities such as the stress tensor is awkward, to say the least. Because of this, a number of different notations have been invented to try to make the situation more manageable. Those who like matrices have come up with the obvious matrix definition:

$$\underline{\sigma} = \begin{bmatrix} \sigma_{11} & \sigma_{12} & \sigma_{13} \\ \sigma_{21} & \sigma_{22} & \sigma_{23} \\ \sigma_{31} & \sigma_{32} & \sigma_{33} \end{bmatrix} \tag{3.8}$$

This notation is handy when dealing with matrix operations on vectors and is also easily programmed on computers.

A second approach is to use what is called indicial notation, by which one always deals with the tensor components as indexed scalar quantities. This notation has a number of advantages and is a favorite of the theoreticians. Thus, the stress tensor is simply referred to as σ_{ij} where it is understood that the indices i and j run over the values 1, 2, and 3. The main advantage of the indicial notation is the easy ability to generalize and expand the definitions to tensor quantities of higher dimension and rank. Thus, the indices can run from 1 to 4 if the problem requires working in a four-dimensional space time as is the case in relativity theory. In addition, higher rank quantities are easily generated by simply adding more indices. In elasticity theory, the elastic constants of the most general class of crystals are specified by the fourth-order elasticity tensor C_{ijkl}, which has 81 elements. The matrix version of this quantity is most useful simply to display the elements on a page as opposed to performing formal manipulations. Einstein's field equations from his general relativity theory involve tensor quantities that are truly extravagant to the point of being mind boggling and have no matrix counterparts. For the purposes of this section, a hybrid vector/tensor notation is adopted to provide a compact and general notation for introducing the field equations of elasticity. Refer to Appendix A for a more detailed treatment of vector and tensor notation and manipulation.

The primary thrust of this discussion, however, is to develop the basic field equations of continuum theory in a reasonably transparent manner and show that they arise in a straightforward way from the simpler equations of Newtonian particle mechanics. To this point, we have established that the main quantities of interest are the components of the stress tensor as opposed to the simpler vector forces that act on the body in question. This arises because the vector forces inevitably act on surface elements when a solid is considered as a continuum as opposed to a collection of point particles; as Equation (3.3) demonstrates, this leads to the formation of a higher dyadic quantity called the stress tensor.

The usefulness of the stress tensor formulation becomes apparent when the stress on some arbitrary surface is required and not just the special orthogonal surfaces of the coordinate planes. The stress on any surface whatever can be related to the components of the stress tensor given in some fixed reference system in the following manner. Figure 3.2 shows the traction vector $\mathbf{T}_n$ acting on some arbitrary area element dA with surface normal vector $\mathbf{n}$ that is not parallel to any of the surfaces dA_1, dA_2, or dA_3 associated with the coordinate planes and therefore must intersect each in some fashion, forming a tetrahedron figure shown in Figure 3.2. Equilibrium requires that the vector sum of the forces acting on all the faces of the tetrahedron should add to zero; thus,

$$F_n - F_1 - F_2 - F_3 = 0 \qquad (3.9)$$

Using Equations (3.6) and (3.7), Equation (3.9) can be satisfied on a component-by-component basis by collecting all terms acting in the direction $\mathbf{e}_1$ and setting them to zero and proceeding similarly for the directions $\mathbf{e}_2$ and $\mathbf{e}_3$. The result is

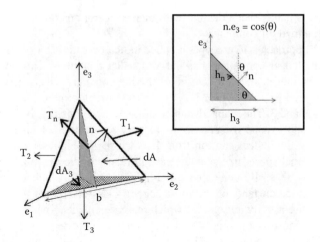

FIGURE 3.2 A surface element within a solid (dA) having some arbitrary orientation **n** with respect to a fixed coordinate system characterized by the unit vectors $\mathbf{e_1}$, $\mathbf{e_2}$, and $\mathbf{e_3}$. The area element dA is subjected to a surface traction $\mathbf{T_n}$ that is not parallel to the surface normal vector **n**. T1, T2, and T3 are the components of $\mathbf{T_n}$ along the coordinate axes. The shaded area labeled dA_3 is the projection of dA onto the $\mathbf{e_1}$, $\mathbf{e_2}$ plane and has unit normal $\mathbf{e_3}$. The inset shows that the element dA_3 is related to dA by the simple formula $dA_3 = dA\mathbf{n}.\mathbf{e_3}$. A similar relation holds for the remaining projections dA_1 and dA_2 (not shown).

$$\left(T_n\right)_1 dA - \sigma_{11}dA_1 - \sigma_{21}dA_2 - \sigma_{31}dA_3 = 0$$

$$\left(T_n\right)_2 dA - \sigma_{12}dA_1 - \sigma_{22}dA_2 - \sigma_{32}dA_3 = 0 \qquad (3.10)$$

$$\left(T_n\right)_3 dA - \sigma_{13}dA_1 - \sigma_{23}dA_2 - \sigma_{33}dA_3 = 0$$

Equations (3.10) can be greatly simplified by noting that the area elements dA_1, dA_2, and dA_3 are just the projections of the area element dA onto the main coordinate planes. This fact is demonstrated by the inset in Figure 3.2 for the case of area element dA_3. The primary result is the following useful set of relations:

$$dA_1 = dA(\mathbf{n}.\mathbf{e_1}) = dA\cos\theta_1 = n_1 dA$$

$$dA_2 = dA(\mathbf{n}.\mathbf{e_2}) = dA\cos\theta_2 = n_2 dA \qquad (3.11)$$

$$dA_3 = dA(\mathbf{n}.\mathbf{e_3}) = dA\cos\theta_3 = n_3 dA$$

where $n_1 = \cos\theta_1$, $n_2 = \cos\theta_2$, and $n_3 = \cos\theta_3$ are the common direction cosines for the normal vector n.

By substituting Equations (3.11) into Equations (3.10), the common factor dA can be seen to drop out, and with some obvious rearrangement, the following result is achieved:

$$\left(T_n\right)_1 = \sigma_{11}n_1 + \sigma_{21}n_2 + \sigma_{31}n_3$$

$$\left(T_n\right)_2 = \sigma_{12}n_1 + \sigma_{22}n_2 + \sigma_{32}n_3 \tag{3.12}$$

$$\left(T_n\right)_3 = \sigma_{13}n_1 + \sigma_{23}n_2 + \sigma_{33}n_3$$

These equations have a very convenient matrix representation:

$$\left[\left(T_n\right)_1, \left(T_n\right)_2, \left(T_n\right)_3\right] = \left[n_1, n_2, n_3\right]\begin{bmatrix} \sigma_{11} & \sigma_{12} & \sigma_{13} \\ \sigma_{21} & \sigma_{22} & \sigma_{23} \\ \sigma_{31} & \sigma_{32} & \sigma_{33} \end{bmatrix} \tag{3.13}$$

Using compact matrix notation, Equation (3.13) can be written very succinctly as $\mathbf{T} = \mathbf{n} \cdot \underline{\sigma}$. Equation (3.13) demonstrates the utility of the stress tensor $\underline{\sigma}$ because once it is established in any particular convenient coordinate system, the tractions across any desired area element can be obtained by simple matrix operations knowing only $\underline{\sigma}$ and the components of the area normal $\mathbf{n}$.

3.1.2 SPECIAL STRESS STATES AND STRESS CONDITIONS

3.1.2.1 Principal Stresses

On contemplating Equation (3.13), the alert reader may wonder, given the fact that the stress tensor can be defined in any convenient coordinate system, whether there are particular coordinates in which the stress tensor assumes a particularly simple form. This question has an affirmative answer. Quite generally, at any particular point in any elastic solid there is a coordinate system in which the stress tensor has a simple diagonal form:

$$\underline{\sigma} = \begin{bmatrix} \sigma_I & 0 & 0 \\ 0 & \sigma_{II} & 0 \\ 0 & 0 & \sigma_{III} \end{bmatrix} \tag{3.14}$$

Equation (3.14) is very nice indeed because diagonal matrices are much simpler to manipulate than the general case. The components σ_I, σ_{II}, and σ_{III} are called the principal stress components and are typically arranged so that $\sigma_I > \sigma_{II} > \sigma_{III}$. However, there is a very important string attached in that the special coordinate system in which the stress tensor assumes a diagonal form will in general vary from point to point in the solid. This takes much of the fun out of using the principal stress coordinates, but they are still quite useful nonetheless. Indeed, the largest component σ_I, called the maximum principal stress, is useful in fracture mechanics work because it represents the largest stress component that can exist at a particular point. Furthermore, the direction in which σ_I acts can give an important clue to the direction

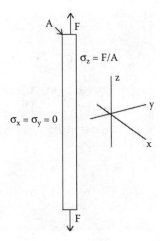

FIGURE 3.3 Simplest uniaxial stress state for a slender tensile sample.

in which a potential crack will propagate. All of this is valid for homogeneous isotropic materials and is discussed in more detail in Chapter 4, dealing with the fracture mechanics of delamination processes.

By far the simplest example of a principal stress state is the case of uniaxial tension shown in Figure 3.3. The principal axes are intuitively obvious for this case, and the stress tensor assumes the simplest possible form as follows:

$$\underline{\sigma} = \begin{bmatrix} 0 & 0 & 0 \\ 0 & 0 & 0 \\ 0 & 0 & \sigma_z \end{bmatrix} \tag{3.15}$$

This is the form of the stress tensor that is assumed for all tensile testing experiments.

3.1.2.2 St. Venant's Principle

The uniaxial tension case provides a clear and simple example of a useful and general rule of stress analysis. The basic idea is most simply illustrated by Figure 3.4, which exhibits three entirely different ways of clamping a uniaxial tensile sample. The fact that the net load is identical for each sample means that they are "statically identical" in the parlance of engineering mechanics. Because of the different clamping configurations, however, the stress state near the top end will be widely different for each sample. But, because each sample is of identical cross section and each is loaded in a statically identical manner (i.e., the net load F is identical in each case), the state of stress far from the clamped region near the center of the sample will be identical in each case and is in fact given by F/A, the net force divided by the cross-sectional area. This is a fundamental fact underlying all tensile testing experiments. The general principle highlights the fact that the action of surface tractions is

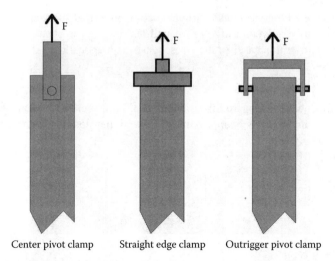

Center pivot clamp   Straight edge clamp   Outrigger pivot clamp

FIGURE 3.4 St. Venant's principle for three equally loaded but differently clamped tensile samples. The details of the stress field near the clamped edge will differ significantly for each sample; however, because each sample is equally loaded, the stress state far from the clamped edge will be identical for each sample. This is a consequence of the localized nature of the surface tractions imposed by the different clamping arrangements.

essentially local in that the fine details of the local stress state decay away as one proceeds away from the point of application.

The stress state in the "far field" depends only on the net loading and the boundary conditions. Although this principle is most clearly illustrated by the uni-axial tension case, it holds for a very general class of solids in one, two, and three dimensions. The precise meaning of *far away* depends on the details of the problem, but for each case a characteristic length can be defined, and the definition of far away is typically a small multiple of this length. An example of particular relevance to this work is a coating on a rigid substrate. The characteristic length in this case can be taken as the coating thickness and far away typically turns out to be about five times this thickness.

3.1.2.3 Two-Dimensional Stress States

3.1.2.3.1 Plane Stress

The case of plane stress is of particular interest in adhesion work because nearly all coatings are in a state of plane stress in which the stress component perpendicular to the coating surface is zero. This is typically taken to be the z direction, so the stress tensor assumes the following form:

$$\underline{\sigma} = \begin{bmatrix} \sigma_{xx} & \sigma_{xy} & 0 \\ \sigma_{yx} & \sigma_{yy} & 0 \\ 0 & 0 & 0 \end{bmatrix} \qquad (3.16)$$

The case of a homogeneous isotropic coating on a rigid substrate is particularly simple and can be solved easily with pencil and paper. The shear components σ_{xy} and σ_{yx} are zero in this case, so the stress tensor is diagonal, and σ_{xx} and σ_{yy} are in fact principal stresses.

3.1.2.3.2 Plane Strain

When dealing with a long rodlike sample of arbitrary cross section, the state of stress away from the ends is approximately two dimensional, and the stress tensor has the following form:

$$\underline{\sigma} = \begin{bmatrix} \sigma_{xx} & \sigma_{xy} & 0 \\ \sigma_{yx} & \sigma_{yy} & 0 \\ 0 & 0 & \sigma_{zz} \end{bmatrix} \tag{3.17}$$

where $\sigma_{zz} = \nu(\sigma_{xx} + \sigma_{yy})$.

The plane strain case is also very important from the point of view of adhesion testing because nearly all fracture mechanics tests on crack propagation are performed in a state of plane strain. More is said on this matter in Chapter 4.

3.1.2.3.3 Hydrostatic Stress

A final simplified stress state of interest is that of hydrostatic compression or tension. The stress tensor in this case is as simple as it can be:

$$\underline{\sigma} = \sigma \begin{bmatrix} 1 & 0 & 0 \\ 0 & 1 & 0 \\ 0 & 0 & 1 \end{bmatrix} \tag{3.18}$$

The stress tensor is simply a scalar times the identity matrix and is thus effectively a scalar quantity. The pressure in a pneumatic tire is the simplest example of this stress state. The pressure in most fluids is similarly characterized. In solids, the scalar factor σ can be positive, indicating a state of hydrostatic tension. This can be achieved when the sample is constrained along each of the three coordinate axes and can lead to void formation at high stress levels. This is an important failure mode in some rubbers.

3.1.3 EQUATION OF MOTION IN SOLIDS

The whole purpose of the presentation up to this point has been to set the stage for the development of the basic equation of motion for an elastic body. Figure 3.5 shows a small element of material embedded within a larger object. The question is how this object responds to the forces acting on it. The answer is the same as the one given by Sir Isaac Newton in the 17th century in his second law of particle mechanics; that is, the mass of the object times the acceleration is equal to the net vector sum of all the forces acting on it. This is typically written as follows:

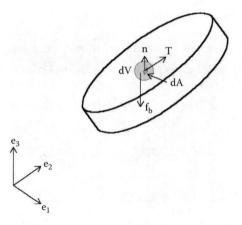

FIGURE 3.5 An element of volume dV embedded in a larger solid. The surface element dA with unit normal **n** is acted on by a traction **T**. A body force $\mathbf{f_b}$ acts on the entire volume element.

$$\sum_i F_i = m\frac{d^2\mathbf{u}}{dt^2} \tag{3.19}$$

where

m = mass of volume element

$\mathbf{u}$ = displacement vector

Equation (3.19) is the typical expression given in texts on classical particle mechanics, which treat all objects as point masses, ignoring their internal structure as continuous bodies. This treatment is quite adequate for problems in ballistics, for which the projectile can be treated as a point mass acted on by the Earth's gravitational field and various frictional forces due to the Earth's atmosphere. This treatment calculates the projectile trajectory and where will it land. If it is desired to know what will happen when the projectile hits another object, then the continuum nature of the missile cannot be ignored, and continuum theory must be applied. Equation (3.19) must be modified to reflect that the projectile is in fact an extended body. The discrete mass m is replaced by the density times a volume element, and the list of forces must now include surface tractions caused by the impacting surfaces as well as the usual external forces, such as gravity. The following relations must now be used:

$$dm = \rho dV$$

$$F = \int_V f_b dV + F_{int} \tag{3.20}$$

where

ρ = material density

f_b = body force/unit volume (gravitational, electromagnetic, etc.)

F_{int} = total internal traction forces = $\int_A T dA = \int_A \underline{\sigma}.n dA$

In light of these new definitions, Newton's second law as expressed in Equation (3.19) now becomes

$$\int_V \mathbf{f_b} dV + \int_A \underline{\sigma}.\mathbf{n} dA = \int_V \rho \frac{d^2 \mathbf{u}}{dt^2} dV \qquad (3.21)$$

At this point, one of the truly marvelous bits of legerdemain from vector calculus, known as the divergence theorem, is invoked to transform the surface integral over the stress tensor into a volume integral. The theorem, called the Divergence Theorem, is general and essentially states that

$$\int_A \underline{\sigma}.n \, dA = \int_V \nabla \bullet \underline{\sigma} \, dV \ .$$

Readers who need to brush up on vector calculus or would like more detailed references on the subject should refer to Appendix A. The main point in doing this is as follows: All the terms can now be gathered within the same volume integral, which is set equal to zero. Because the volume is essentially arbitrary (i.e., the equation must hold for all shapes), the only way the integral can be zero is if the integrand itself is identically zero, which then leads to the required differential relation. Applying these steps to Equation (3.21) gives

$$\int_V f_b dV + \int_A \underline{\sigma}.n dA = \int_V \rho \frac{d^2 \mathbf{u}}{dt^2} dV$$

(rearrange + divergence theorem)

$$\int_V f_b dV + \int_V \nabla \cdot \underline{\sigma} dV - \int_V \rho \frac{d^2 \mathbf{u}}{dt^2} = 0$$

(collect terms under common volume integral)

$$\int_V \left(f_b dV + \nabla \cdot \underline{\sigma} - \rho \frac{d^2 \mathbf{u}}{dt^2} \right) dV = 0$$

Because the last volume integral must hold for objects of arbitrary shape, size, and loading condition, the integrand itself must be zero, which gives

$$\boxed{\mathbf{f_b} + \nabla \cdot \underline{\sigma} - \rho \frac{d^2 u}{dt^2} = 0} \qquad (3.22)$$

where

ρ = material density (kg/m^3)

$\mathbf{f_b}$ = body force density (N/m^3)

$\underline{\sigma}$ = stress tensor (N/m^2)

u = displacement field vector (m)

Equation (3.22) represents Newton's second law as applied to a continuous solid. The main feature that differs from the standard particle mechanics version given by Equation (3.19) is the divergence of the stress tensor, which accounts for the action of the internal stresses acting within the body.

3.1.4 DEFORMATION AND STRAIN

With the basic equation for stresses now in hand, the discussion proceeds to the complementary topic of deformation and strain. When a stress is applied to any material object, there is a resulting deformation, so the key questions now become how deformation is defined and how it is related to the applied stress. To most readers, the idea of a deformation should seem quite obvious because nearly everyone has an intuitive feel for what deformation is. If you run your car into a rigid post, the resulting deformation of the fender is obvious, so it is somewhat surprising that the mathematical description arising from continuum theory is fairly recondite, giving rise to dense topics such as deformation gradients and invariant strain measures.

The basic difficulties arise from the need for mathematical precision and the fact that the world is three dimensional. In one dimension, everything falls out quite naturally. The stress is proportional to the strain (Hooke's law), and the strain is just the increment in length divided by the original length. However, when describing general three-dimensional deformations, these concepts simply do not pass muster, and one must resort to something more heavy duty.

The basic approach of continuum theory, therefore, is to treat any object what-soever as a vector field as shown in Figure 3.6. In this figure, the object labeled B$_u$ (subscript u for undeformed) is described as a vector field that is a region of space labeled by a collection of vectors $\mathbf{X} = (X_1, X_2, X_3)$. It is clear then that just about any object can be so defined in terms of its geometric shape. Other properties are also required but that comes later in Section 3.1.5. At this point, to be able to define deformations in three dimensions the geometric configuration must first be defined, and the vector field $\mathbf{X}$ does the job. The precise shape of the object is defined by also defining the boundary of B$_u$ (commonly designated as ∂B_u), but that is not needed to obtain a general definition of the deformation of an object.

Now, the mathematicians are perfect wizards at performing deformations through entities called mappings, which turn out to be vector fields themselves and that describe transformations between different vector spaces. The vector $\mathbf{K(X)}$ shown in Figure 3.6 is such a mapping. The figure shows that $\mathbf{K(X)}$ is just the original

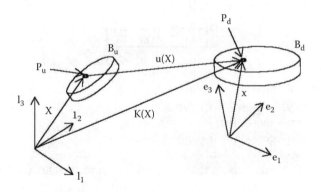

FIGURE 3.6 Continuum description of the deformation of a body B_u.

position vector **X** plus a deformation vector **u**, which in general will depend on **X**. It should be noted at this point that the mapping **K** will take the entire space to which B_u belongs (described by basis vectors l_1, l_2, l_3) into an entirely new space described by basis vectors e_1, e_2, and e_3. Thus, the mapping **K(X)** can be thought of as taking the point **X** in the space described by l_1, l_2, and l_3 into the point **x** in the space with basis vectors e_1, e_2, and e_3 or **x = K(X)**.

It is also clear that the configuration of the body can be equally well described using either coordinate system because the mapping **K(X)** is assumed to be continuous with continuous derivatives and suitably well behaved so that its inverse $K^{-1}(X)$ exists and is equally well behaved. This is both a powerful and convenient feature of the continuum description of the deformation process and, unfortunately, also a source of much confusion, so a few words must be said regarding nomenclature for the description used. When using the l_1, l_2, l_3 basis vectors, one is said to be employing the Lagrangian formulation, and if using the e_1, e_2, e_3 basis, then the Eulerian description is in effect. The main reason for noting this is for those who consult the continuum mechanics literature. For present purposes, the deformations and strains will be relatively small, and in this case the Lagrangian and Eulerian descriptions turn out to be nearly identical.

Getting back now to the general description of deformation and strain, the basic relation for the deformation **u** can be written by referring to Figure 3.6 as

$$u(X) = K(X) - X \qquad (3.23)$$

Equation (3.23) describes the overall deformation process, but for purposes of describing the internal strain in the body something must be said about the gradient of the deformation or how steeply the deformation separates contiguous points in the object under consideration. Thus, taking the gradient of both sides of Equation (3.23) leads to

$$\nabla u(X) = \nabla K(X) - \underline{I}$$

or   (3.24)

$$\nabla u(X) = F - \underline{I}$$

The new tensor quantity $\underline{F} = \nabla K$ is called the deformation gradient, and the tensor ∇u is called the displacement gradient. Equations (3.23) and (3.24) lay a foundation for defining general deformations, but what is needed now is a proper definition of strain. This is somewhat trickier than just defining deformations because a true strain will always give rise to a concomitant stress. The general deformation given by Equation (3.23) contains translations and rotations as well as stretches and twists. The pure translations and rotations clearly will not induce any kind of stress in an object as they are rigid deformations in which every part of the object moves in unison. Simply moving a glass from the table and inverting it on the dresser will not induce any kind of stress. It is the stretches and twists that actually cause material particles within the object to separate relative to each other and that give rise to strains that induce a level of stress in the object. Take a glass rod and either twist it or bend it, and you will induce a strain that will cause a stress.

There is no unique way of defining the strain. However, one approach is to take the deformation gradient $\underline{F}$ and cobble together a formula that will flush out those parts of the deformation that are essentially rigid rotations or translations and leave only the bona fide stretches, compressions, and twists. One such definition is as follows:

$$\underline{D} = 1/2 \left[\underline{F}\,\underline{F}^{T} - \underline{I} \right] \tag{3.25}$$

It can be shown using the polar decomposition theorem that the finite strain tensor $\underline{D}$ defined in Equation (3.25) fits the bill in terms of eliminating all rigid body motions from the general deformation formula given by Equation (3.23) and thus includes only those stretches, compressions, and twists that will give rise to a stress. Unfortunately, $\underline{D}$ turns out to be a rather difficult entity to deal with in general because of what are commonly known as nonlinear effects. These typically arise when relatively large deformations occur, such as when one pulls a bar of taffy to two times its original length. However, if the deformations are relatively small, as in the case of shrinkage strains in a paint coating, then another, simpler strain measure can be used, which is defined as follows:

$$\underline{\varepsilon} = 1/2[(\nabla u) + (\nabla u)^{T}] \tag{3.26}$$

The strain tensor $\underline{\varepsilon}$ is closely related to $\underline{D}$, as can be shown by the following simple manipulations:

$$\underline{D} = 1/2\left[\underline{F}\underline{F}^T - \underline{I}\right]$$

where

$$\underline{F} = \nabla \mathbf{u} + \underline{I}$$
$$\underline{F}^T = (\nabla \mathbf{u})^T + \underline{I}$$

Substituting the expressions for $\underline{F}$ and $\underline{F}^T$ into the expression for $\underline{D}$ gives the following result:

$$\underline{D} = 1/2[\underline{F}\ \underline{F}^T - \underline{I}] = 1/2[(\nabla \mathbf{u} + \underline{I})((\nabla \mathbf{u})^T + \underline{I}) - \underline{I}]$$

$$= 1/2[\nabla \mathbf{u} + (\nabla \mathbf{u})^T] + 1/2(\nabla u)(\nabla \mathbf{u})^T$$

$$= \underline{\varepsilon} + 1/2(\nabla \mathbf{u})(\nabla \mathbf{u})^T$$

The final definitions for strain can be summarized as follows:

$$\underline{D} = 1/2\left[\underline{F}\underline{F}^T - \underline{I}\right] \qquad \text{Finite deformations}$$

$$\underline{\varepsilon} = 1/2\left[\nabla \mathbf{u} + (\nabla \mathbf{u})^T\right] \qquad \text{Small strains only} \qquad (3.27)$$

$$\underline{D} = \underline{\varepsilon} + 1/2(\nabla \mathbf{u})(\nabla \mathbf{u})^T$$

Note that $\underline{D}$ and $\underline{\varepsilon}$ are nearly equal when $\nabla \mathbf{u}$ is small, which is always the case for small strain problems in the theory of linear elasticity.

One further issue has to be confronted before the development of a suitable strain measure can be considered completed. The basic problem comes down to this: If we know the displacement field $\mathbf{u}$, then Equations (3.27) give perfectly good recipes for conjuring up the required strain fields. However, in many problems of practical interest one first computes the strains, and the question arises as to whether Equations (3.27) will be sufficient to determine the corresponding displacements. The question is a subtle and tricky one, and the answer is generally "No." Equations (3.27) alone do not guarantee that the derived strains will correspond to acceptable displacements. It turns out that one further relation is required to ensure compatibility between a derived set of strains and the corresponding displacements as follows:

$$\nabla \times \nabla \times \underline{\varepsilon} = 0 \qquad (3.28)$$

Equation (3.28) is known as the equation of compatibility for the strain field $\underline{\varepsilon}$. A clear and convincing development of Equation (3.28) has been presented by Gurtin,[1] who also gives an excellent, if somewhat high-powered, development of the general deformation/strain equations. The now-complete continuum description of the deformation and strain fields can be summarized as follows:

$$
\begin{array}{lll}
\text{a) } x = \mathbf{K}(\mathbf{X}) & \text{(Deformation mapping)} \\[4pt]
\text{b) } \mathbf{u} = \mathbf{K}(\mathbf{X}) - \mathbf{X} & \text{(Displacement vector)} \\[4pt]
\text{c) } \underline{\mathbf{F}} = \nabla \mathbf{K} & \text{(Deformation gradient)} \\[4pt]
\text{d) } \underline{\mathbf{J}} = \nabla \mathbf{u} - \underline{\mathbf{I}} & \text{(Displacement gradient)} \\[4pt]
\text{e) } \underline{\mathbf{D}} = 1/2\left[\underline{\mathbf{F}}\,\underline{\mathbf{F}}^{T} - \underline{\mathbf{I}}\right] & \text{(Finite strain tensor)} \\[4pt]
\text{f) } \boldsymbol{\varepsilon} = 1/2\left[\nabla \mathbf{u} + (\nabla \mathbf{u})^{T}\right] & \text{(Infinitesimal strain tensor)} \\[4pt]
\text{g) } \nabla \times \nabla \times \boldsymbol{\varepsilon} = 0 & \text{(Equation of compatibility)} \\[4pt]
\text{h) } \underline{\mathbf{D}} = \boldsymbol{\varepsilon} + 1/2\,(\nabla \mathbf{u})(\nabla \mathbf{u})^{T} & \text{(relation between } \underline{\mathbf{D}},\ \boldsymbol{\varepsilon},\ \text{and } \mathbf{u})
\end{array}
\tag{3.29}
$$

Equations (3.29) essentially complete the second step in the continuum mechanics program outlined on page 76, to describe and compute deformations and stresses in solid bodies. A few words on nomenclature are in order for those who will consult the general literature on the subject. The tensor quantity $\underline{\mathbf{F}}\,\underline{\mathbf{F}}^{T}$ is called the Green's strain after George Green, the famous 19th century mathematician. All of what has been developed to this point has been worked out in the Lagrangian description as defined here (meaning in the l_1, l_2, l_3 coordinate system). One can easily imagine that by using the Euler description a somewhat different tensor quantity can be conjured up that will be essentially equivalent to $\underline{\mathbf{F}}\,\underline{\mathbf{F}}^{T}$ but referred to different coordinates. Names such as Cauchy's strain tensor or Euler's (Almansi's) strain tensor will pop up in various places, which tends to give rise to much confusion regarding which is which. The reader can take heart from the fact that all descriptions give the same result for the small strain tensor $\boldsymbol{\varepsilon}$, and that this is the most commonly used strain measure in the practical engineering literature. This arises from the fact that when the strains are small, the quantity $\nabla \mathbf{u}$ is small, and thus the term $\nabla \mathbf{u}(\nabla \mathbf{u})^{T}$ in Equation (3.27) is of second order in a small quantity and can be neglected. In this case, $\boldsymbol{\varepsilon}$ and $\underline{\mathbf{D}}$ become identical.

Caution must be exercised, however. A common problem arises when someone using a finite element code, for instance, does not understand which strain measure is used. In particular, if $\boldsymbol{\varepsilon}$ is used and the problem involves a significant degree of rotation, major errors can creep into the calculation even though the actual strains may be low. This comes about because $\boldsymbol{\varepsilon}$ does not accurately account for large rotations as the general deformation measure $\underline{\mathbf{D}}$ does.

3.1.5 CONSTITUTIVE RELATIONS OR CONNECTING THE STRESS TO THE STRAIN

3.1.5.1 General Behavior

At this point, Steps 1 and 2 of the continuum mechanics presentation have been accomplished. General definitions of stress, deformation, and strain are now in place.

What is clearly lacking, however, is a description of how these quantities are related in a real material. What is required is referred to as the *constitutive relation* or *material law* for substances of interest. In general, this is a very difficult topic and goes to the very heart of the thermodynamic nature of materials and how they respond to the myriad array of loads and deformations to which they can be subjected. Everything from the bending of a pencil to the gravitational collapse of a star requires a fundamental understanding of the constitutive behavior of the material in question. Many common materials show a complex range of behaviors when subjected to relatively modest loading conditions. Materials such as mayonnaise tend to deform more easily as the deformation rate increases (shear thinning). Contrary materials such as Silly Putty do just the opposite. You can easily deform this material with your fingers like a putty, but if you drop it on the floor, it will bounce like a rubber ball (shear thickening). This type of behavior is commonly called thixotropic, which is basically a way of saying we really do not know what is going on, so let us just give it a fancy name and move on.

There exists a vast body of literature dealing with the constitutive behavior of a wide range of materials, ranging from rubber bands to desert soils. The topic can easily require an entire encyclopedia for anything approaching a definitive treatment. Fortunately, however, a wide range of useful applications deal with materials undergoing relatively small deformations* for which the constitutive relation between the stress and the strain is essentially linear. In this circumstance, there exists a general formulation of the constitutive relation that is useful and is dealt with reasonably easily. It should also be added that focusing on the linear behavior of materials has other merits in addition relative tractability. When attacking the general nonlinear behavior of a material, for instance, it is useful to look at a linear approximation first to see the nature of the error incurred. In addition, nearly all numerical algorithms for solving full nonlinear problems proceed by a sequence of linear approximations that, if handled correctly, will converge to the true nonlinear solution.

The problem now is as follows: What is the most general linear relationship that can exist between the stress tensor σ_{ij} and the strain tensor ε_{ij}? Given the complex nature of the tensor quantities involved, it is no surprise that the answer is also complex, involving the elasticity tensor C_{ijkl}, which was mentioned earlier, thus:

$$\sigma_{ij} = \sum_{k,l=1}^{3} C_{ijkl}\varepsilon_{kl} \tag{3.30}$$

Equation (3.30) expresses the most general linear relationship between two second-order tensors that can exist. The fact that a fourth-order tensor is required to make the connection is not surprising because after summing over the two indices k and l, which describe the components of the strain tensor $\boldsymbol{\varepsilon}$, there have to be two indices left over (*i* and *j* in this case) to designate the corresponding component of

---

* Beware that small deformations and strains do not necessarily imply small stresses. Most engineering materials, for instance, have a fairly high modulus, so that relatively small strains can give rise to stresses that cause permanent deformations or in some cases lead to fracture.

the stress tensor σ. Because the indices i, j, k, l run over the values 1, 2, 3, the elasticity tensor C_{ijkl} can have a maximum of 81 (3^4) independent components. Fortunately, a number of physical symmetries come to the rescue, which cuts down this horrendous number considerably. In particular, because the tensor quantities σ_{ij} and ε_{ij} are symmetric*, they have at most 6 independent components. This means that C_{ijkl} will have at most 36 (6^2) independent components. This is still an outlandish number, but luckily real materials have further symmetries that cut this number down still further. Cubic** crystals, for instance, which have many rotational and reflection symmetries, require only 3 independent elastic constants.

Ultimately, for homogeneous isotropic materials with full spherical symmetry (amorphous glasses, for instance), the final number comes down to only two elastic constants. It is quite surprising*** that in most applications the assumption of a homogeneous isotropic material seems to work out reasonably well. Even materials such as metals, which are nominally crystalline, can be successfully treated as homogeneous isotropic materials. This stems in large part from the fact that in their final use condition most metals are polycrystalline in nature, which means that they are made up of zillions of very small crystals oriented in random directions. The net effect is that, when viewed over a macroscopically small but microscopically large**** region, the material appears to be homogeneous isotropic.

The main types of situations in which one can get into trouble by assuming homogeneous isotropic behavior is when dealing with materials that are large single crystals, composites containing highly oriented fibers, or nonisotropic coatings. The most ubiquitous example of large single crystals is the silicon wafers used in the microelectronics industry. These are near-perfect cubic crystals, and it is important to know how the parent crystal was cut in order to know the mechanical behavior of the product wafer. For example, some laboratories perform wafer-bending experiments to assess the stress state of the wafer. Not knowing which crystal plane the wafer was cut along can lead to very serious errors in such experiments. In particular, it is highly desirable in such experiments to have wafers that have been cut along the (111) plane and thereby have in-plane isotropy; otherwise, one has to deal with a significant level of anisotropy to obtain accurate results. Circuit boards are another example of objects that exhibit a high level of anisotropy because of their composite nature. The common FR4 board material, for example, is made up of a mat of glass fibers embedded in an epoxy matrix. All the fibers lie in the plane of the board, thus giving rise to a large difference between in-plane and out-of-plane mechanical properties.

A final example of anisotropy that is of significant concern to the coatings industry is polymer systems made up of stiff chains that tend to orient parallel to the surface on which they are deposited after drying or during the curing process.

---

* The symmetry of the stress tensor arises from the constraint that conservation of angular momentum must hold as well as linear momentum, which is expressed in Equation (3.22). For a detailed derivation, refer to page 20 in the work of Gould.[2] Also beware that there are formulations of continuum theory in which the stress and strain tensors are not symmetric.

** A derivation of how this comes about was given by Feynman, Note 8, Chapter 39.

*** Or perhaps this is not so surprising given the general human proclivity toward indolence.

**** This is what is commonly referred to as the coarse graining approximation.

Coatings made from these polymers tend to be highly anisotropic with far different mechanical properties in the plane compared to out-of-plane behavior. This is commonly referred to as orthotropic behavior, and an example is given in Table 3.1.

3.1.5.2 Homogeneous Isotropic Materials

It is no accident that the most popular constitutive relation is the one that deals with homogeneous isotropic materials that require only two elastic constants, namely, the Young's modulus and Poisson's ratio. It seems that people already regard it as a burden to deal with 2 elastic constants, not to mention 36. If you are working with a newly synthesized polymer, for example, and you browse the manufacturer's data sheet, you will most likely find only two useful constitutive properties mentioned: the Young's modulus and the thermal expansion coefficient. In fact, if you inquire about it, the people who made the material, primarily chemists, are most likely unaware of what a Poisson's ratio is. However, the main reason why it is difficult to find good data on Poisson's ratio is that it is much more difficult to measure than the Young's modulus.

This problem is discussed in more detail in Section 3.3.1, but now it is important to establish the constitutive relation for homogeneous isotropic materials. This can be done by essentially applying the constraints of spherical symmetry to Equation (3.30). An excellent discussion of the procedure has been given by Gould,[2] and the result comes down to the following:

$$\varepsilon_{11} = \frac{1}{E}\left[\sigma_{11} - v\left(\sigma_{22} + \sigma_{33}\right)\right] + \alpha\Delta T$$

$$\varepsilon_{22} = \frac{1}{E}\left[\sigma_{22} - v\left(\sigma_{11} + \sigma_{33}\right)\right] + \alpha\Delta T$$

$$\varepsilon_{33} = \frac{1}{E}\left[\sigma_{33} - v\left(\sigma_{11} + \sigma_{22}\right)\right] + \alpha\Delta T \qquad (3.31)$$

$$\varepsilon_{12} = \frac{\sigma_{12}}{2G}; \ \varepsilon_{23} = \frac{\sigma_{23}}{2G}; \ \varepsilon_{31} = \frac{\sigma_{31}}{2G}$$

Equations (3.31) are the simplest three-dimensional expression of Hooke's law. In fact, it is easy to see that by collapsing down to one dimension the familiar one-dimensional version pops out. Thus, for a simple tensile specimen $\sigma_{22} = \sigma_{33} = \sigma_{12} = \sigma_{23} = \sigma_{31} = 0$, and the remaining equation for σ_{11} becomes $\sigma_{11} = E\,\varepsilon_{11}$, which is the familiar standard form taught in elementary texts.

All of this is very satisfying, but yet another essential ingredient is missing. Not only is the real world three dimensional, but also it is governed very strictly by the laws of thermodynamics, which state that everything exists at some finite temperature. Not only that, but also the temperature tends to vary substantially from moment to moment and place to place, giving rise to thermal gradients; these gradients can give rise to strains, which also induce stresses in ordinary objects. In fact, in many practical situations thermal stresses can entirely dominate the stress distribution in an object.

The equation governing the temperature distribution in a body under most conditions is derived from three separate equations, the first of which is in essence a statement of the conservation of energy. Let **h** be defined as the heat flux vector, which describes the flow of heat (disordered energy) across some small surface area dA in a body per unit time. The quantity **h** is a vector that will generally vary from point to point in the body. In any particular small volume within the body, heat may be passing through from neighboring volume elements, and there may also be some local heat source within the element generating heat at some rate γ (joules per cubic meter per unit time). Letting q denote the total amount of heat energy per unit volume within the volume at any given time, the conservation of energy requires that

$$\nabla \cdot \mathbf{h} + \gamma = \frac{\partial q}{\partial t} \tag{3.32}$$

Equation (3.32) is commonly referred to as an equation of continuity for the heat flow within a material body. It says in essence that the rate at which heat builds up or drains out of any particular volume is the sum of two terms. The first $\nabla \cdot \mathbf{h}$ accounts for the net flow in from or out to neighboring volume elements. The second term γ accounts for local heat generation, which could be caused by radioactive decay processes or embedded wires. Equation (3.32) states that the time rate of change of heat buildup or loss in any particular volume element is the net sum of energy flow through the bounding surface plus any local contribution from embedded sources.

The second equation comes directly from thermodynamics and the definition of the specific heat, that is,

$$dq = \rho c_v dT \tag{3.33}$$

or

$$\frac{dq}{dt} = \rho c_v \frac{dT}{dt}$$

In this equation, ρ is the density, and c_v is the specific heat (or heat capacity per unit volume) under conditions of constant volume.* Combining Equations (3.32) and (3.33) leads to

$$\nabla \cdot \mathbf{h} + \gamma = \rho c_v \frac{\partial T}{\partial t} \tag{3.34}$$

---

* Note that there is also a c_p or specific heat under conditions of constant pressure.

The final equation is known as Fourier's law of heat conduction. It is basically the simplest linear constitutive relation for heat flow in solids, relating the heat flux to the gradient of the temperature:

$$\mathbf{h} = -\kappa \nabla T \tag{3.35}$$

Equation (3.35) is simply the statement that when there are spatial variations in the temperature field described by ∇T there will be a concomitant heat flow designated by $\mathbf{h}$, with the constant of proportionality κ known as the thermal conductivity. Combining Equation (3.35) with Equation (3.34) leads to

$$\kappa \nabla^2 T + \gamma = \rho c_v \frac{\partial T}{\partial t} \tag{3.36}$$

Using Equation (3.36) along with appropriate initial and boundary conditions, the temperature field in a body can be calculated. If thermal gradients are present or if the body is a composite of materials with different thermal expansion coefficients and is either heated or cooled uniformly, then thermal strains will arise, given by the following formula:

$$\varepsilon_T = \alpha \Delta T \tag{3.37}$$

with

α = (thermal expansion coefficient
$\Delta T = (T - T_0)$ (temperature difference)

The parameter α is the thermal expansion coefficient characterizing the tendency of the body to expand or contract on heating. ΔT is the temperature excursion from the reference temperature T_0, where the body is in a reference state generally taken to be a condition of zero stress. The thermal strain given by Equation (3.37) must be added to the constitutive relations defined in Equations (3.31) to give the following relations for a thermoelastic solid:

$$
\begin{aligned}
\varepsilon_{11} &= \frac{1}{E}\left[\sigma_{11} - \nu\left(\sigma_{22} + \sigma_{33}\right)\right] + \alpha\Delta T \\[2mm]
\varepsilon_{22} &= \frac{1}{E}\left[\sigma_{22} - \nu\left(\sigma_{11} + \sigma_{33}\right)\right] + \alpha\Delta T \\[2mm]
\varepsilon_{33} &= \frac{1}{E}\left[\sigma_{33} - \nu\left(\sigma_{11} + \sigma_{22}\right)\right] + \alpha\Delta T \\[2mm]
\varepsilon_{12} &= \frac{\sigma_{12}}{2G}; \quad \varepsilon_{23} = \frac{\sigma_{23}}{2G}; \quad \varepsilon_{31} = \frac{\sigma_{31}}{2G}
\end{aligned}
\tag{3.38}
$$

Equations (3.38) represent the constitutive relations for a homogeneous isotropic material and thereby complete the continuum mechanics program by providing the essential link between the stresses and the strains in the body. It should be noted that these equations are by no means the most general that can be formulated and are essentially valid in the realm of small strains and temperature excursions. However, they are valid for a wide variety of practical problems, and they form a foundation for considering more elaborate situations in which the assumptions of isotropy, small strains, and the like are not valid. The complete set of field equations for the thermoelastic behavior of a homogeneous isotropic solid are summarized in Table 3.1.

3.2 EXAMPLES

Table 3.1 essentially summarizes what was known about the thermoelastic behavior of solids by the end of the 19th century. As such, they represent a cumulative intellectual triumph of that century. However, in addition to representing a purely transcendent accomplishment of the human intellect, the field equations catalogued in Table 3.1 are also amazingly useful tools for analyzing the behavior of solid materials that have found constant application in engineering problems since their derivation. The advent of nearly unlimited computing power represented by the now ubiquitous personal computer workstation has made these equations even more powerful as tools of analysis because previously unusable numerical algorithms can now be fruitfully brought to bear to provide accurate solutions to these equations in support of myriad design and development projects. This section presents a number of "back-of-the-envelope" type calculations designed to give some insight into the utility of the field equations listed in Table 3.1 and provide some feel for how they can be put to work in a number of practical situations. The examples presented, however, only scratch the surface of what can be done with pencil and paper. Indeed, all serious work involving these equations is done on powerful workstations using the exceedingly effective finite element and boundary element analysis schemes.

3.2.1 SIMPLE DEFORMATIONS

Some feel for the meaning of the equations for deformation and strain given by Equations (3.29) can be gained by looking at a few simple examples. The simplest possible deformation is a rigid translation. The required deformation mapping comes directly from Equation (3.29a): Consider as a simplest case a rigid translation along the x_1 axis, where the displacement vector is simply given by $\mathbf{u} = \mathbf{b} = (b_1, 0, 0)$. Another way to write this is to combine Equations (3.29a) and (3.29b) to obtain the relation $\mathbf{x} = \mathbf{X} + \mathbf{b}$. Yet another way of looking at it is to write Equation (3.29a) out explicitly; that is, $\mathbf{x} = \mathbf{K}(\mathbf{X})$ where $\mathbf{x} = (x_1, x_2, x_3)$, $\mathbf{X} = (X_1, X_2, X_3)$, and $\mathbf{K}(\mathbf{X}) = (X_1 + b_1, X_2, X_3)$. To make things clearer, this can be set out componentwise as follows:

$$x_1 = K_1(X) = X_1 + b_1$$

$$x_2 = K_2(X) = X_2 \qquad (3.39)$$

$$x_3 = K_3(X) = X_3$$

TABLE 3.1

Collected Field Equations for Homogeneous Isotropic Thermoelastic Material

Formula	Variable/Parameter Descriptions	Comments
Momentum balance $$f_b + \nabla \cdot \underline{\sigma} - \rho \frac{d^2 u}{dt^2} = 0$$	$\underline{\sigma}$ = Stress tensor (N/m²) u = Displacement vector (m) f_b = Body force vector (N/m³) ρ = Material density (kg/m³)	Equation (3.22): Newton's second law generalized for a continuum body
Deformation and strain fields a) *Deformation mapping* $\quad x = K(X)$ b) *Displacement vector* $\quad u = K(X) - X$ c) *Deformation gradient* $\quad \underline{F} = \nabla K$ d) *Displacement gradient* $\quad \underline{J} = \nabla u - \underline{I}$ e) *Finite strain tensor* $\quad \underline{D} = 1/2[\underline{F}\,\underline{F}^T - \underline{I}]$ f) *Infinitesimal strain tensor* $\quad \underline{\varepsilon} = 1/2[\nabla u + (\nabla u)^T]$ g) *Equation of compatibility* $\quad \nabla \times \nabla \times \underline{\varepsilon} = 0$ h) *Relation between* $\underline{D}$, $\underline{\varepsilon}$ *and* u $\quad \underline{D} = \underline{\varepsilon} + 1/2(\nabla u)(\nabla u)^T$	u = Displacement vector (m) $\underline{I}$ = Identity tensor $\underline{F}$, $\underline{D}$, $\underline{J}$, and $\underline{\varepsilon}$ dimensionless	Equations (3.29): Basic definitions of deformation and strain fields; note that $\underline{\varepsilon}$ is reliable for small strains only
Constitutive relations for thermoelastic homogeneous isotropic material *Stresses in terms of strains* $$\underline{\sigma} = \frac{E}{1+v}\left(\underline{\varepsilon} + \frac{v\underline{I}}{1-2v}\,tr\underline{\varepsilon}\right) + E\alpha\Delta T\underline{I}$$ *Strains in terms of stresses* $$\underline{\varepsilon} = \frac{1+v}{E}\underline{\sigma} - \frac{v\,tr\underline{\sigma}}{E}\underline{I} + \alpha\Delta T\underline{I}$$ *Componentwise strains in terms of stresses* $$\varepsilon_{11} = \frac{1}{E}\left[\sigma_{11} - v(\sigma_{22} + \sigma_{33})\right] + \alpha\Delta T$$ $$\varepsilon_{22} = \frac{1}{E}\left[\sigma_{22} - v(\sigma_{11} + \sigma_{33})\right] + \alpha\Delta T$$ $$\varepsilon_{33} = \frac{1}{E}\left[\sigma_{33} - v(\sigma_{11} + \sigma_{22})\right] + \alpha\Delta T$$ $$\varepsilon_{12} = \frac{\sigma_{12}}{2G}; \quad \varepsilon_{23} = \frac{\sigma_{23}}{2G}; \quad \varepsilon_{13} = \frac{\sigma_{13}}{2G}$$	ε_{ij} = Components of strain tensor $\underline{\varepsilon}$ $\sigma_{i,j}$ = Components of stress tensor $\underline{\sigma}$ (N/m³) $\underline{I}$ = Identity matrix $tr\,\underline{\varepsilon} = \varepsilon_{11} + \varepsilon_{22} + \varepsilon_{33}$ $tr\,\underline{\sigma} = \sigma_{11} + \sigma_{22} + \sigma_{33}$ $\Delta T = (T - T_0)$ thermal excursion (K) E = Young's modulus (N/m³) G = Shear modulus (N/m³) v = Poisson's ratio α = Thermal expansion coefficient (K⁻¹) $G = E/[2(1 + v)]$	Equations (3.38): Basic equations connecting the stresses to the strains; define the solid as a material entity with recognizable physical properties; the material is assumed to be homogeneous and isotropic

TABLE 3.1 (continued)
Collected Field Equations for Homogeneous Isotropic Thermoelastic Material

Formula	Variable/Parameter Descriptions	Comments
Constitutive relations for thermoelastic homogeneous orthotropic material	ε_{ij} = Components of strain tensor $\boldsymbol{\varepsilon}$	Basic stress-strain relations for a nonisotropic coating for which the elastic behavior is dependent on direction; these relations form the next level of complexity above Equations (3.38); for the case of in-plane isotropy, the following constraints hold:
Componentwise strains in terms of stresses	σ_{ij} = Components of stress tensor $\boldsymbol{\sigma}$ (N/m^3)	
	$\Delta T = (T - T_0)$ thermal excursion (K)	
$\varepsilon_{11} = \dfrac{1}{E_{11}}\left[\sigma_{11} - \nu_{12}\sigma_{22} - \nu_{13}\sigma_{33}\right] + \alpha_1\Delta T$	E_1, E_1, E_1 = Nonisotropic Young's moduli (N/m^3)	
$\varepsilon_{22} = \dfrac{1}{E_{22}}\left[\sigma_{22} - \nu_{21}\sigma_{11} - \nu_{23}\sigma_{33}\right] + \alpha_2\Delta T$	ν_{ij}, i, j = 1, 2, 3 = Nonisotropic Poisson's ratios	
$\varepsilon_{33} = \dfrac{1}{E_{33}}\left[\sigma_{33} - \nu_{31}\sigma_{11} - \nu_{32}\sigma_{22}\right] + \alpha_3\Delta T$	G_{12}, G_{13}, G_{23} = Nonisotropic shear moduli (N/m^3)	
$\dfrac{\nu_{12}}{E_{11}} = \dfrac{\nu_{21}}{E_{22}}$	$\alpha_1, \alpha_2, \alpha_3$ = Nonisotropic thermal expansion coefficients (K^{-1})	$E_{11} = E_{22} = E;\ E_{33} = E_z$
$\varepsilon_{12} = \dfrac{\sigma_{12}}{2G_{12}};\ \varepsilon_{13} = \dfrac{\sigma_{13}}{2G_{13}};\ \varepsilon_{23} = \dfrac{\sigma_{23}}{2G_{23}}$		$\nu_{12} = \nu_{21} = \nu$
		$\nu_{31} = \nu_{13} = \nu_{23} = \nu_{32} = \nu_z$
		$G_{12} = \dfrac{E}{2(1+\nu)}$
		$G_{13} = G_{23}$
		$= \dfrac{EE_z}{(E + E_z + 2E\nu_z)}$
		$\nu^2 < 1$ and $\nu_z^2 < \dfrac{E_z}{2E}(1-\nu)$
		$\alpha_1 = \alpha_2 = \alpha;\ \alpha_3 = \alpha_z$
		$\alpha_1 + \alpha_2 + \alpha_3 = \alpha_V$
		α_V = Volume thermal expansion
Fourier's equation for temperature field	T = Temperature field (K)	Equation (3.36): Equation governing temperature distribution
$\kappa\nabla^2 T + \gamma = \rho c_v \dfrac{\partial T}{\partial t}$	ρ = Material density (kg/m^3)	
	c_v = Specific heat [J/(kg K)]	
	γ = Heat source (W/m^3)	
	κ = Thermal conductivity (W/m K)	

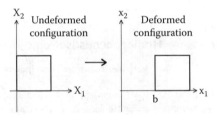

FIGURE 3.7 Example of the simplest possible deformation, which is a translation described mathematically by Equations (3.39).

Figure 3.7 shows this mapping as a simple two-dimensional graphic. From Equations (3.39), the deformation gradient $\underline{\mathbf{F}}$ may be calculated using the definition given by Equation (3.29c). This can be written in matrix form in two separate ways as follows:

$$\underline{\mathbf{F}} = \begin{bmatrix} \dfrac{\partial K_1}{\partial X_1} & \dfrac{\partial K_1}{\partial X_2} & \dfrac{\partial K_1}{\partial X_3} \\[2mm] \dfrac{\partial K_2}{\partial X_1} & \dfrac{\partial K_2}{\partial X_2} & \dfrac{\partial K_2}{\partial X_3} \\[2mm] \dfrac{\partial K_3}{\partial X_1} & \dfrac{\partial K_3}{\partial X_2} & \dfrac{\partial K_3}{\partial X_3} \end{bmatrix} \quad \text{or} \quad \underline{\mathbf{F}} = \begin{bmatrix} \dfrac{\partial x_1}{\partial X_1} & \dfrac{\partial x_1}{\partial X_2} & \dfrac{\partial x_1}{\partial X_3} \\[2mm] \dfrac{\partial x_2}{\partial X_1} & \dfrac{\partial x_2}{\partial X_2} & \dfrac{\partial x_2}{\partial X_3} \\[2mm] \dfrac{\partial x_3}{\partial X_1} & \dfrac{\partial x_3}{\partial X_2} & \dfrac{\partial x_3}{\partial X_3} \end{bmatrix} \tag{3.40}$$

The right-hand version uses a notation commonly found in the continuum mechanics literature. Unfortunately, it is the cause of confusion because the symbol x_i is used to indicate a functional relationship in the right-hand expression of Equation (3.40), whereas the x_i in Equation (3.39) represents coordinate variables. The right-hand notation is a favorite with theorists and those who prefer indicial notation because it allows for certain economies of notation; however, because this work is more concerned with clarity, the left-hand notation is preferred. Those who explore the literature should be forewarned, however. Using Equations (3.39) and (3.40), it is a simple matter to perform the required differentiations and fill in the matrix entries for the deformation gradient tensor $\underline{\mathbf{F}}$:

$$\underline{\mathbf{F}} = \begin{bmatrix} 1 & 0 & 0 \\ 0 & 1 & 0 \\ 0 & 0 & 1 \end{bmatrix} \tag{3.41}$$

As expected for the very simple deformation represented by Equations (3.39), the deformation gradient tensor is also very simple. Referring to Equations (3.29), the following further results are obtained:

$$\nabla \mathbf{u} = \nabla \mathbf{K}(\mathbf{X}) - \nabla \mathbf{X} = \mathbf{I} - \mathbf{I} = 0$$

$$\underline{\mathbf{D}} = 1/2\left[\mathbf{FF}^{T} - \mathbf{I}\right] = 1/2\left[\mathbf{I} - \mathbf{I}\right] = 0$$

$$\underline{\varepsilon} = 1/2\left[\nabla \mathbf{u} + (\nabla \mathbf{u})^{T}\right] = 1/2\left[0 + 0\right] = 0$$

These results are comforting in that they demonstrate that, for the simple translation deformation given by Equations (3.39), both the finite and infinitesimal strain tensors are exactly zero, so there can be no stress associated with such a deformation, as expected. In fact, it is quite straightforward to show using the same procedures as demonstrated above that any simple translation in three dimensions will also give zero for both strain tensors.

The case of a simple rotation is somewhat more involved but can be handled without much difficulty by the same machinery. The relevant mapping is as follows:

$$x_1 = K_1(X) = X_1 \cos\theta - X_2 \sin\theta$$

$$x_2 = K_2(X) = X_1 \cos\theta - X_2 \sin\theta \qquad (3.42)$$

$$x_3 = K_3(X) = X_3$$

The reader can easily be convinced that the mapping given by Equations (3.42) essentially performs the simple rotation depicted in Figure 3.8. Further details on the matrix algebra of simple rotations can be found in Appendix A. The deformation gradient for this mapping follows from the same recipe as used for the simple translation, giving

$$\underline{\mathbf{F}} = \begin{bmatrix} \cos\theta & -\sin\theta & 0 \\ \sin\theta & \cos\theta & 0 \\ 0 & 0 & 1 \end{bmatrix} \qquad (3.43)$$

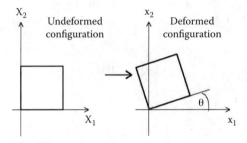

FIGURE 3.8 Example of a rigid rotation as given by Equations (3.42).

From Equation (3.29e), the matrix product $\underline{\mathbf{F}}\,\underline{\mathbf{F}}^T$ is required to compute the finite strain tensor $\underline{\mathbf{D}}$. This is readily carried out as follows:

$$\underline{\mathbf{F}}\,\underline{\mathbf{F}}^T = \begin{bmatrix} \cos\theta & -\sin\theta & 0 \\ \sin\theta & \cos\theta & 0 \\ 0 & 0 & 1 \end{bmatrix} \times \begin{bmatrix} \cos\theta & \sin\theta & 0 \\ -\sin\theta & \cos\theta & 0 \\ 0 & 0 & 1 \end{bmatrix} =$$

$$\begin{bmatrix} \cos^2\theta + \sin^2\theta & 0 & 0 \\ 0 & \cos^2\theta + \sin^2\theta & 0 \\ 0 & 0 & 1 \end{bmatrix} = \begin{bmatrix} 1 & 0 & 0 \\ 0 & 1 & 0 \\ 0 & 0 & 1 \end{bmatrix} = \mathbf{I}$$

$$(3.44)$$

The finite strain tensor can now be readily calculated from Equation (3.29e) to give $\underline{\mathbf{D}} = 1/2[\underline{\mathbf{F}}\,\underline{\mathbf{F}}^T - \underline{\mathbf{I}}] = 1/2[\,\underline{\mathbf{I}} - \underline{\mathbf{I}}\,] = 0$, as would be expected. Thus, a rigid rotation does not induce any strain and therefore no stress in a solid body, which gives us further confidence that the finite strain formula is giving proper results. The situation for the infinitesimal strain formula Equation (3.29f) is a little trickier, however. Noting from Equation (3.29b) that $\nabla\mathbf{u} = \nabla[\mathbf{K}(\mathbf{X}) - \mathbf{X}] = [\nabla\mathbf{K}(\mathbf{X}) - \nabla\mathbf{X}] = \underline{\mathbf{F}} - \underline{\mathbf{I}}$, the infinitesimal strain can be written directly from Equation (3.29f) as $\underline{\boldsymbol{\varepsilon}} = 1/2[\nabla\mathbf{u} + (\nabla\mathbf{u})^T] = 1/2[\underline{\mathbf{F}} + \underline{\mathbf{F}}^T - 2\underline{\mathbf{I}}]$, which can be written explicitly using Equation (3.43) as

$$\underline{\boldsymbol{\varepsilon}} = 1/2\begin{bmatrix} \cos\theta & -\sin\theta & 0 \\ \sin\theta & \cos\theta & 0 \\ 0 & 0 & 1 \end{bmatrix} + 1/2\begin{bmatrix} \cos\theta & \sin\theta & 0 \\ -\sin\theta & \cos\theta & 0 \\ 0 & 0 & 1 \end{bmatrix} - \mathbf{I}$$

$$= \begin{bmatrix} \cos\theta - 1 & 0 & 0 \\ 0 & \cos\theta - 1 & 0 \\ 0 & 0 & 0 \end{bmatrix}$$

The final result can be simplified further by assuming that the rotation angle θ is small and using the small-angle expansion for $\cos\theta = 1 + \theta^2/2 + \text{Order }\theta^4 + \dots$ to give the following result for the infinitesimal strain tensor after a small rotation:

$$\underline{\boldsymbol{\varepsilon}} \cong \begin{bmatrix} \dfrac{\theta^2}{2} & 0 & 0 \\ 0 & \dfrac{\theta^2}{2} & 0 \\ 0 & 0 & 0 \end{bmatrix}$$

Thus, for rigid rotations the infinitesimal strain tensor does not give a zero result and will predict a fictitious stress. However, for small rotations the error will be on

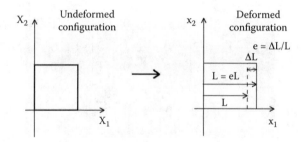

FIGURE 3.9 Example of a simple stretching deformation as given by Equations (3.45).

the order of $\theta^2/2$, which for an angle of $1°$ ($= 0.017$ radians) comes to about 0.00015 or slightly more than 1 part in 10,000. Such a small error is of no consequence in most practical engineering applications, so the infinitesimal strain measure can be used with confidence, but caution must be exercised nonetheless. The most treacherous situation occurs when individuals not fully versed in the arcana of continuum theory come to apply a commercial finite element code to a problem for which they are not aware of which strain measure is in use. Most commercial codes use the infinitesimal strain measure as the default option because it gives rise to a linear problem that the numerical routines quickly and efficiently solve. However, as demonstrated, problems can arise when the physical strains involved are very small, but relatively large rotations can occur. In this case, use of the infinitesimal strain measure can introduce very significant errors. Unfortunately, when using computer codes there is an unwise tendency to believe that whatever the computer puts out is correct by definition. This can have distinctly undesirable consequences.

As a final example of a deformation mapping, consider the case of a simple stretching elongation as shown in Figure 3.9:

$$x_1 = K_1(X) = eX_1 \quad \left(e = \frac{\Delta L}{L} \right)$$

$$x_2 = K_2(X) = X_2 \tag{3.45}$$

$$x_3 = K_3(X) = X_3$$

The deformation gradient for this mapping is easily computed using the explicit matrix formulation given in Equation (3.40):

$$\underline{F} = \begin{bmatrix} 1+e & 0 & 0 \\ 0 & 1 & 0 \\ 0 & 0 & 1 \end{bmatrix} \tag{3.46}$$

The finite strain tensor follows immediately as

$$\underline{D} = 1/2\left[\underline{F}\,\underline{F}^{T} - \underline{I}\right] = 1/2\begin{bmatrix} (1+e)^2 & 0 & 0 \\ 0 & 1 & 0 \\ 0 & 0 & 1 \end{bmatrix} - 1/2\begin{bmatrix} 1 & 0 & 0 \\ 0 & 1 & 0 \\ 0 & 0 & 1 \end{bmatrix}$$

(3.47)

$$= 1/2\begin{bmatrix} (1+e)^2 - 1 & 0 & 0 \\ 0 & 0 & 0 \\ 0 & 0 & 0 \end{bmatrix}$$

For this simple deformation, the finite strain tensor turns out to have only one component and is thus effectively a scalar quantity given by $D = 1/2[(1 + e)^2 − 1] = e + e^2/2$. The infinitesimal strain tensor is even simpler. First, compute $\nabla\mathbf{u}$ as follows:

$$\nabla\mathbf{u} = \underline{F} - \underline{I} = \begin{bmatrix} e & 0 & 0 \\ 0 & 0 & 0 \\ 0 & 0 & 0 \end{bmatrix}$$

The infinitesimal strain tensor is also effectively a scalar quantity given by the usual formula $\underline{\varepsilon} = 1/2[\nabla\mathbf{u} + (\nabla\mathbf{u})^{T}] = 1/2[e + e] = e$. Knowing that $e = \Delta L/L$, the standard Hooke's law definition of the strain is recovered from the formula for the infinitesimal strain tensor. Note that $\underline{D}$ and $\underline{\varepsilon}$ are essentially the same when the strain e is small.

Thus, all of the simple examples presented above underscore the point that the general formulas for deformation and strain given by Equations (3.29) reduce to the simpler concepts of strain and deformation given by standard texts on strength of materials. All of the complications introduced by these equations is simply to account for the three-dimensional nature of real deformations and strains and to further adjust for rigid body motions such as translations and rotations that do not give rise to any strain in the object moved.

3.3 SOLVING THE FIELD EQUATIONS

3.3.1 Uniaxial Tension

It is no secret that it is difficult to deal with the field equations displayed in Table 3.1. They comprise a set of nonlinear partial differential equations notwithstanding the fact that the constitutive relations given by Equations (3.34) are linear and of the simplest form possible for a three-dimensional solid. As demonstrated, if large strains or rotations are present, then what are known as geometric nonlinearities arise to complicate what is already a difficult linear problem. Even the fully linear problem

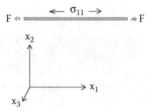

FIGURE 3.10 Sample in uniaxial tension.

becomes involved because of the large number of variables,* but what really causes a problem in most situations is satisfying the boundary conditions, which can be quite complex for useful items such as gears and aircraft fuselages. Finally, if all of this is not enough, the solutions to the field equations can harbor devious singularities in which one or more of the components of the stress tensor go to infinity. Dealing with this last problem requires the use of fracture mechanics concepts and is one of the major issues concerning the adhesion of coatings. This topic alone requires an entire chapter.

What is not generally well known is that there are a number of nontrivial solutions to the field equations that can be solved on the back on an envelope and surprisingly enough also find useful application. The simplest is the case of uniaxial tension exhibited in Figure 3.10. In this case, the body force f_b is generally negligible, and the stress is constant throughout the sample, so the first two terms of Equation (3.22) are zero. Also, because at equilibrium nothing is changing in time, the third term is also zero, so Equation (3.22) is satisfied identically. Because the temperature is assumed constant and there are no heat sources, Equation (3.36) is also satisfied identically. In addition, because the strains are also constant and uniform throughout the sample, all of Equations (3.29) are satisfied, assuming the case of small strains. Finally, the constitutive relations given by Equations (3.38) must be satisfied. Referring to Figure 3.10, it is clear that the following conditions hold from the apparent loading condition and intuitive symmetry relations: $\sigma_{22} = \sigma_{33} = \sigma_{12} = \sigma_{23} = \sigma_{31} = 0$. This leaves the following relations still standing:

$$\varepsilon_{11} = \frac{1}{E}\sigma_{11}$$

$$\varepsilon_{22} = -\frac{v}{E}\sigma_{11} = -v\varepsilon_{11} \tag{3.48}$$

$$\varepsilon_{33} = -\frac{v}{E}\sigma_{11} = -v\varepsilon_{11}$$

---

* There are 10 variables to be solved for: 6 stresses, 3 components of displacement, plus the temperature. Equations (3.22) give 3 relations among the stresses and displacements, Equations (3.38) give 6 more relations, and Equation (3.36) pins down the temperature field, for a total of 10 relations, making the entire system determinate.

Equations (3.48) give the complete solution to the field equations in Table 3.1 for the case of a uniaxial specimen under tension. The reader will quickly note that the first is simply the well-known Hooke's law, albeit in an inverted form. The other relations for the strains ε_{22} and ε_{33} are less well known. These last two relations state what is commonly referred to as a Poisson effect; that is, the sample will contract in the directions perpendicular to the primary strain direction. In most cases involving uniaxial stress-strain measurements, the last two strains are effectively invisible and therefore commonly ignored. A simple example demonstrates why. Say a straightforward tensile test is performed on a sample 10 cm long by 0.1 cm wide. These are good dimensions that ensure that the bulk of the sample is in uniaxial tension, as assumed by the above analysis. Assume further a small strain of 1% is impressed on the sample to ensure linear behavior. For a 10-cm sample, imposing a strain of 1% involves measuring a displacement of 0.1 cm or 1.0 mm (remember $\varepsilon_{11} = \Delta L/L = 0.1/10 = 0.01$). The concomitant perpendicular strains ε_{22} and ε_{33} will be given by the last two equations of Equations (3.48). Assuming a common value for the Poisson ratio of 1/3, the perpendicular strains come to roughly 0.003. However, the displacement that must be measured to detect these strains will be given by the formula $\Delta L = L \times 0.003 = 0.1 \times 0.003 = 0.0003$ mm, which is nearly 100 times smaller than the primary axial displacement. Thus, it is clear why the perpendicular strains in most uniaxial measurements are unnoticed. Furthermore, this also demonstrates why it can be very tricky to try to measure the Poisson ratio by performing uniaxial measurements. The perpendicular deformations must be measured very accurately compared to the primary axial deformations to achieve reasonably accurate results.

3.3.2 BIAXIAL TENSION

A second elementary solution to the field equations can be found for the case of a thin coating on a thick rigid substrate, as illustrated in Figure 3.11. In this case, Equations (3.22) and (3.29) are satisfied identically as in the uniaxial case because nothing is varying with time, and the strains and stresses are constant and uniform

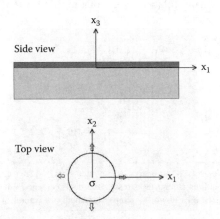

FIGURE 3.11 Sample in biaxial tension derived from a thin coating on a large, rigid substrate.

everywhere. For this case, the loading is assumed to arise from thermal expansion mismatch between the thin coating material and the substrate. An example might be a high-temperature insulator such as one of the polyimide materials deposited on a silicon wafer, cured at an elevated temperature, and subsequently cooled to room temperature. It can be assumed that at the cure temperature there is essentially no stress in the system, so this can be taken as a reference state. On cooling to room temperature, the polyimide material wants to contract much faster than the silicon substrate, which rigidly constrains it; thus, at room temperature the coating is in a state of biaxial tension, as illustrated in Figure 3.11. The situation is much the same as for the uniaxial tension case because the stress and temperature are uniform throughout the sample; all of the field equations listed in Table 3.1 are trivially satisfied except for Equations (3.38).

In this case the thermal expansion coefficient α can be taken as the difference between the coefficients of expansion of the substrate and the coating. For typical polyimide materials coated onto a silicon wafer, this difference can amount to about $30 \times 10^{-6} \, C^{-1}$. The temperature difference ΔT is the difference between the reference temperature (commonly about 300°C for polyimides) and room temperature (about 20 C). For this problem, because of the constraint imposed by the rigid substrate, there are no in-plane mechanical strains and no shear strains. Thus, $\varepsilon_{11} = \varepsilon_{22} = \varepsilon_{12} = \varepsilon_{23} = \varepsilon_{31} = 0$. Also, because there is no constraint on the coating perpendicular to the substrate surface, there is no stress in this direction either; therefore, $\sigma_{33} = 0$. Applying these conditions to Equations (3.38) yields the following set of relations:

$$\frac{1}{E}\left[\sigma_{11} - \nu\left(\sigma_{22}\right)\right] = -\alpha\Delta T$$

$$\frac{1}{E}\left[\sigma_{22} - \nu\left(\sigma_{11}\right)\right] = -\alpha\Delta T \qquad (3.49)$$

$$\varepsilon_{33} = -\frac{\nu}{E}\left[\left(\sigma_{11} + \sigma_{22}\right)\right] + \alpha\Delta T$$

Equations (3.49) are a set of simultaneous linear relations among the quantities σ_{11}, σ_{22}, and ε_{33} and can be solved almost by inspection. The result is

$$\sigma_{11} = \sigma_{22} = \frac{-E\alpha\Delta T}{\left(1 - \nu\right)}$$

$$\varepsilon_{33} = \frac{\alpha\Delta T}{\left(1 - \nu\right)} \qquad (3.50)$$

Equations (3.50) give the solution for the nonzero components of stress and strain in a thin coating in equibiaxial tension caused by a thermal expansion mismatch between it and the underlying substrate. The equation for the in-plane stresses is commonly known as the membrane equation because it applies to thin coatings and can be useful for estimating the stress induced by thermal expansion mismatch strains in a coating on a rigid substrate. For the polyimide material coated on a silicon wafer, the result is quite interesting and useful. For this case, we have the following typical values for the parameters:

$$E = 3 \text{ GPa}$$

$$v = 1/3$$

$$\alpha = \alpha_{(silicon)} - \alpha_{(polyinide)} = (2 - 32) \times 10^{-6} = -30 \times 10^{-6} \text{ C}^{-1}$$

$$\Delta T = T_0 - T = 300 - 20 = 280 \text{ C}$$

Substituting the above data into Equation (3.50) gives the following result for the biaxial stress in the coating after cooling from the cure temperature $\sigma = (3/2) \times 3 \times 30 \times 280 \times 10^{-6}$ GPa = 0.038 GPa.* This result becomes more impressive when one realizes that the ultimate tensile strength of the majority of polyimide materials is in the range of 0.07 to 0.1 GPa. Thus, this simple calculation indicates that these coatings are very highly stressed as they come out of the cure oven because the biaxial tension is nearly 50% of the ultimate strength, as determined by simple tensile measurements.

This lesson was learned the hard way in the microelectronics industry in the early 1980s when polyimide materials were explored as alternative insulators to sputtered glass for wafer- and packaging-scale applications. The initial coatings did not crack right away because they were still below the ultimate strength of the material; however, for an insulator to be useful it must be patterned with holes to allow electrical connection to the underlying substrate. It is well known that such a hole can introduce a stress concentration factor amounting to about a factor of 3, which was more than enough to push these materials over the edge. Much time and energy was wasted in sorting out this problem, which was eventually solved by using only those polyimides with an exceptionally high tensile strength or a special class of materials with expansion coefficients close to that of silicon. Thus, the above simple calculation of the estimated biaxial stresses could have served as a warning of trouble to come had anyone bothered to carry it out.

3.3.3 TRIAXIAL STRESS CASE

As one might expect, there is yet one more very simple case that can be solved immediately with pencil and paper: the case of triaxial tension. Triaxial tension is precisely the negative of hydrostatic compression, in which every element of the body in question is stretched equally in all three orthogonal directions, as depicted in Figure 3.12. One way to generate such a stress state would be to take a cube of

---

* Here, GPa is the SI unit gigapascal or 10^9 pascals. For those more familiar with pounds per square inch (psi), a handy approximate conversion is to divide the result in pascals by 7000 to obtain pounds per square inch. In this case, the result is $0.038 \times 10^9/7000 = 5430$ psi, which is a fairly hefty stress for a polymer material.

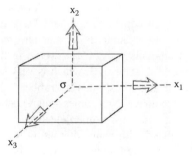

FIGURE 3.12 Sample in a state of triaxial tension.

material and arrange by some means to constrain all six faces of the cube from moving while the material is at some elevated temperature. Subsequent cooling of the entire arrangement while simultaneously maintaining the imposed constraints will leave the material in the cube in a state of equitriaxial tension. The field equations in Table 3.1 are satisfied in a similar manner as the cases of uniaxial and biaxial tension except that now every component of the strain tensor ε_{ij} is set equal to zero, giving rise to the following set of simultaneous linear equations:

$$\frac{1}{E}\left[\sigma_{11} - \nu\left(\sigma_{22} + \sigma_{33}\right)\right] = -\alpha\Delta T$$

$$\frac{1}{E}\left[\sigma_{22} - \nu\left(\sigma_{11} + \sigma_{33}\right)\right] = -\alpha\Delta T \qquad (3.51)$$

$$\frac{1}{E}\left[\sigma_{33} - \nu\left(\sigma_{11} + \sigma_{22}\right)\right] = -\alpha\Delta T$$

From the apparent symmetry of Equations (3.51), they are seen to have the following solution:

$$\sigma_{11} = \sigma_{22} = \sigma_{33} = \frac{-E\alpha\Delta T}{\left(1 - 2\nu\right)} \qquad (3.52)$$

Equation (3.52) is remarkably like Equation (3.50), which is the solution for the biaxial case except for the seemingly harmless factor of 2 multiplying the Poisson ratio in the denominator of Equation (3.52). The reader may be beginning to perceive that continuum theory is somehow riddled with little booby traps that lie in wait to waylay the unwary. Equation (3.52) is one such case in that if one happens to be working with rubbery materials for which the Poisson ratio is remarkably close to 1/2, with values of 0.499 not uncommon. Inserting this value for Poisson's ratio into Equation (3.52) gives a multiplying factor of $1/(1 - 2 \times 0.499) = 500$. This factor is close to 250 times the equivalent multiplier for the same type of loading in biaxial tension and of course 500 times that for the uniaxial case.

What is exhibited here is another case of the Poisson effect mentioned above regarding the lateral shrinkage accompanying uniaxial tension. For the case of uniaxial tension, this effect is all but invisible. For a standard material in biaxial tension with Poisson ratio 1/3, it amounts to a 50% increase in the stress, but in triaxial tension the stress is increased a factor of 3, which is very substantial indeed. When dealing with incompressible materials, such as rubbers, the effect can be totally catastrophic, as demonstrated above. Now, unless the reader should think that all this is entirely academic, be aware that in the microelectronics industry multilevel structures are commonly constructed with many alternating layers of metal and polymer insulator in which the polymer may find itself sandwiched on all sides by metal that has a substantially lower thermal expansion behavior. Consider that on top of this these same structures tend to be subjected to extensive temperature excursions as part of soldering, curing, and other manufacturing operations, then one can begin to appreciate that the consequences of Equation (3.52) can cause substantial mischief.

3.4 APPLICATION TO SIMPLE BEAMS

One of the most important applications of continuum theory is the bending of simple beams. In the case of adhesion measurement, many techniques such as the lap shear test, the double cantilevered beam, the three- and four-point bend tests and a variety of variations of these simple tests rely on knowing the stress and deformation behavior of simple beam samples. The simplest treatment of this problem arises from the theory of strength of materials (SOM). This is basically a collection of ad hoc and "hook-or-by-crook" methods for estimating the deformation and stress buildup in beams subjected to bending, shear, or torsional loads.

A vast body of literature exists on this subject. One of the classic works is the book by Den Hartog.[3] Timoshenko[4] wrote an engaging and informative history of the subject dating from the earliest work on beams by Galileo in the 17th century to the modern theories of the 20th century. A near-encyclopedic collection of results from SOM theory is contained in the classic volume by Roark and Young.[5] Most of the results of this theory are approximations developed in the precomputer age and were suitable for calculations by pencil and slide rule. It was always incumbent on the user to know the limitations of the formulae and under what conditions they could be used without serious error. Needless to say, not everyone using the simple formulae was so conscientious; thus, the literature is riddled with inaccurate results. For the purposes of this work, the results of the primitive theory are rederived both to provide a uniform treatment based on the more accurate formalism of continuum theory and to point out the limitations of the simpler theory.

The basic ideas of SOM theory are best illustrated by examining the classic problem of bending of a cantilevered beam by some concentrated load located at the free end. The configuration is drawn schematically in Figure 3.13 for the case of a beam with rectangular cross section. It is important to understand the following assumptions:

1. The left-hand edge is cantilevered into an immovable wall, meaning that not only is it rigidly fixed in the x, y, and z directions, but also the edge

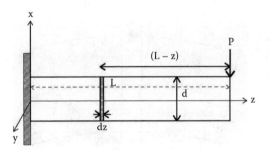

FIGURE 3.13 Cantilevered beam with a point load at the free end.

of the beam remains perpendicular to the wall, where the two abut no matter what deflection it might incur.

2. The length of the beam is much more than either the width or the thickness.
3. The deflection suffered by the beam because of the load P is much smaller than the length L.
4. The beam can be treated as a homogeneous elastic solid characterized by a modulus E and Poisson ratio v.
5. Deflections caused by the weight of the beam are entirely negligible.

Strictly speaking, we must now solve the field equations of continuum theory given in Table 3.1 to determine the stresses and deflections caused by the load P subject to the boundary conditions that the beam is cantilevered as shown in Figure 3.13. This is an exceedingly difficult problem in general; however, we expect that the simple geometry and loading condition of this problem will lead to considerable simplifications. In particular, we expect that the stress induced in the beam will be dominated by the axial component σ_{zz}. Therefore, we expect that $\sigma_{xx} = \sigma_{yy} = \sigma_{xy} = \sigma_{yx} = 0$. Furthermore, the dominant strain component will be the axial strain ε_{zz} associated with the axial stress σ_{zz}. Using the SOM approach, the next step is to guess the nature of the deformation induced by the bending load P. This is all quite legitimate because such guesses are usually guided by physical intuition and nearly always point in the correct direction.

The SOM method assumes that the deformation of the beam will be nearly circular, at least locally, if the end deflection is not too large. This is illustrated in Figure 3.14, which depicts the shape of the deformed beam as a segment of a circular arc. Assuming that the circular deformation of the beam is correct, then a number of facts become apparent from inspection of Figure 3.14. First, it is clear that elements of the beam at the outer circumference are stretched by some increment Δs, and for every such element there is a companion element near the inner circumference that is compressed by a similar amount. Second, it is clear that there is a line near the center of the beam* where there is no strain deformation at all, which will henceforth be referred to as the neutral line. Finally, it is clear that the upper half of the beam will be in tension and the corresponding lower half in compression.

---

* For the uniform rectangular beam under consideration, the neutral line lies exactly at the center. For more general cross sections, it lies on the centroid.

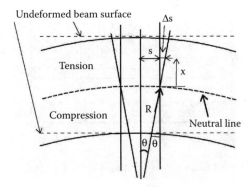

FIGURE 3.14 Section of a beam undergoing simple bending. Assuming small deflections, the local shape of the beam is assumed to be a circular arc.

For this very simple bending type of deformation, the precise amount of deformation can be estimated from simple geometry. Figure 3.14 shows three vertical lines corresponding to undeformed cross sections before bending and two radius vectors originating from the center of the circle, which defines the radius of curvature of the bend. If the distance between two of the vertical lines is taken as s, then at some height x above the neutral line this length is stretched to some greater length $s + \Delta s$. If we identify θ as the angle between the central vertical line and the radius vector to the neutral line, then simple geometry tell us that

$$s = R\theta \tag{3.53}$$

In addition, we know

$$\Delta s = x\theta \tag{3.54}$$

Taking the ratio of Equations (3.54) to (3.53) gives the following simple formula for the strain $\Delta s/s$:

$$\varepsilon(x) = \frac{\Delta s}{s} = \frac{x}{R} \tag{3.55}$$

where the distance x is measured up or down from the neutral line, giving a positive tensile strain above the axis and a negative compressive strain below.

Figure 3.15 illustrates an enlargement of the forces generated within the beam as a consequence of the bending. From the diagram, it is evident that for each tensile force above the neutral line there is an equal and opposite compressive force below the line. These forces come in pairs and set up a net torque about the neutral line. The concept of a force acting on some lever arm about some fixed center is called a moment or torque in the language of structural mechanics. Thus, referring to Figure 3.15, the increment of force ΔF acting at a distance x above the neutral line generates the element of moment ΔM as follows:

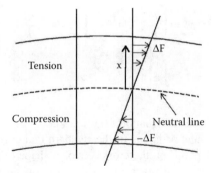

FIGURE 3.15 Tensile force induced in beam as a result of bending.

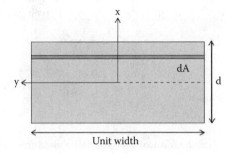

FIGURE 3.16 Cross section of beam as seen end on.

$$\Delta M = x\,\Delta F \tag{3.56}$$

Going to infinitesimal elements, the total moment about the neutral line for any particular cross section of the beam will be given by the integral of Equation (3.56) over the cross-sectional area, as shown in Figure 3.16:

$$M = \int_A x\,dF \tag{3.57}$$

By definition, the stress induced by the force increment dF is $\sigma = dF/dA$. Also, from the simplest form of Hooke's law, we have $\sigma = E\varepsilon$, where the strain ε is given by Equation (3.55). Thus, the increment of force in Equation (3.57) can be cast in terms of beam elastic properties and dimensions as follows:

$$dF = \sigma\,dA = E\varepsilon\,dA = E\frac{x}{R}\,dA \tag{3.58}$$

Equation (3.57) can now be written as

$$M = \int_A \frac{E}{R} x^2 \, dA = \frac{E}{R} \int_A x^2 \, dA = \frac{EI}{R}$$

$$I = \int_A x^2 \, dA$$

(3.59)

The quantity I defined in Equation (3.59) is known as the moment of inertia, and along with the moment M is one of the standard quantities of interest in beam mechanics. It clearly depends on the cross-sectional shape of the beam and is well defined for all beam shapes of practical interest. The quantity EI is known as the beam "stiffness" and is clearly a measure of how hard it is to bend the beam. Thus, beams of high modulus E and moment of inertia I will be hard to bend and vice versa.

Summarizing the results thus far from Equations (3.55) and (3.58), we have expressions for the stress and strain in the beam over a given cross section as follows:

$$\sigma(x) = E \frac{x}{R}$$

$$\varepsilon(x) = \frac{x}{R}$$

(3.60)

To determine the overall deflection of the beam, we need to specify precise loading conditions, as shown in Figure 3.17. For this simple case of a cantilevered beam with a concentrated load at one end, we see that the applied load generates an external moment at each cross-sectional element of the beam given by $M_{ext} = P(L - z)$. At equilibrium, this applied moment (or torque) is counterbalanced by the internal moment given by Equation (3.59), which is generated by the beam's internal elastic resistance to bending. The result is the following equilibrium condition:

$$\frac{EI}{R} = P(L - z)$$

(3.61)

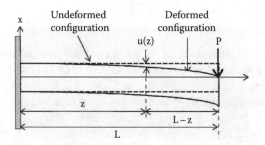

FIGURE 3.17 Deformation of cantilevered beam subjected to concentrated end load.

At this stage, everything is known except the radius of curvature R. This missing quantity can be cast in terms of the vertical deflection $u(z)$ (see Figure 3.17) by the following formula, usually derived from the calculus of plane curves:

$$\frac{1}{R} = \frac{\dfrac{d^2u}{dz^2}}{\left[1+\left(\dfrac{du}{dz}\right)^2\right]^{3/2}}$$

(3.62)

$$= \frac{d^2u}{dz^2} \quad \text{when} \quad \frac{du}{dz} \ll 1$$

Because one of the assumptions of beam theory is that the deflection of the beam is small, that is, $du/dz \ll 1$, one always uses the simplified second equation for $1/R$ given in Equation (3.62). Thus, Equation (3.61) may be written

$$\frac{d^2u}{dz^2} = \frac{P}{EI}(L-z)$$

(3.63)

We readily solve this equation by elementary methods using the cantilevered boundary conditions that $u(0) = 0$ and $du/dz = 0$ at $z = 0$, giving

$$u(z) = \frac{P}{EI}\left(L\frac{z^2}{2} - \frac{z^3}{6}\right)$$

(3.64)

We can now collect all our results for the cantilevered beam from Equations (3.60), (3.61), and (3.64):

$$\sigma_{zz} = \frac{P}{I}x(L-z)$$

$$\varepsilon_{zz} = \frac{P}{EI}x(L-z)$$

(3.65)

$$u(z) = \frac{P}{EI}\left(L\frac{z^2}{2} - \frac{z^3}{6}\right)$$

Equations (3.65) comprise the most elementary solution of the bending beam problem based on basic mechanical principles of equilibrium and an intuitive guess concerning the nature of the deformation induced by simple bending. These equations belong to SOM theory and as mentioned date to the earliest efforts in understanding the mechanics of materials before the modern theory of elasticity even existed. The main merit of Equations (3.65) is that they are simple, easy to derive and understand, and by and large accurate enough to "do the job" in terms of predicting the bending behavior of simple beams. The main drawback to Equations (3.65) is not so much in the final results as in the methodology used to derive them. As pointed out, SOM theory is a collection of results and methodologies collected over time and proven adequate to the task at hand. It is not clear in particular whether Equations (3.65) satisfy the more general field equations of continuum theory summarized in Table 3.1 (they do not) or how accurate they are in a given set of circumstances.

To overcome the above limitations of SOM theory, we revisit the bending beam problem using an approach based directly on the general field equations in Table 3.1. We start with the momentum balance equations given by Equation (3.22):

$$\mathbf{f_b} + \nabla \cdot \underline{\sigma} - \rho \frac{d^2 u}{dt^2} = 0$$

In the present case, body forces are assumed negligible, so $\mathbf{f_b} = 0$; furthermore, the problem is one of static equilibrium, so the time derivative is also zero, leaving

$$\nabla \cdot \underline{\sigma} = 0$$

The next step is to simplify the problem by noticing that the deformation and stresses of interest for the beam problem in Figure 3.13 are in the x-z plane, and the stress in the y direction is zero (plane stress problem). Almost no one attacks the full three-dimensional elasticity problem using analytical methods. This type of problem is almost exclusively the domain of powerful numerical methods such as the finite element technique and only then when accurate results are required for a complex geometry with complex boundary conditions. In two dimensions, the momentum balance equation simplifies to

$$\frac{\partial \sigma_{xx}}{\partial x} + \frac{\partial \sigma_{xz}}{\partial z} = 0$$

$$\frac{\partial \sigma_{zx}}{\partial x} + \frac{\partial \sigma_{zz}}{\partial z} = 0$$

(3.66)

Equations (3.66) give us two equations in three unknowns ($\sigma_{xz} = \sigma_{zx}$ because the stress tensor is symmetric), which is not sufficient to determine the stresses. In the parlance of structural mechanics, the problem is said to be statically indeterminate. This comes about because the body deforms under the given load conditions, so the elastic properties of the material come into play. This means that we have to consider the strains also. Referring to Table 3.1 and taking into account that only small strains are involved in the current problem, we see that the correct strain formula is given by Equation (3.29f):

$$\underline{\boldsymbol{\varepsilon}} = 1/2\left[\left(\nabla\mathbf{u}\right)+\left(\nabla\mathbf{u}\right)^{T}\right]$$

This can be written out in component form as

$$\underline{\boldsymbol{\varepsilon}} = \frac{1}{2}\begin{bmatrix} \dfrac{\partial u_x}{\partial x} & \dfrac{\partial u_x}{\partial y} & \dfrac{\partial u_x}{\partial z} \\[2mm] \dfrac{\partial u_y}{\partial x} & \dfrac{\partial u_y}{\partial y} & \dfrac{\partial u_y}{\partial z} \\[2mm] \dfrac{\partial u_z}{\partial x} & \dfrac{\partial u_z}{\partial y} & \dfrac{\partial u_z}{\partial z} \end{bmatrix} + \frac{1}{2}\begin{bmatrix} \dfrac{\partial u_x}{\partial x} & \dfrac{\partial u_x}{\partial y} & \dfrac{\partial u_x}{\partial z} \\[2mm] \dfrac{\partial u_y}{\partial x} & \dfrac{\partial u_y}{\partial y} & \dfrac{\partial u_y}{\partial z} \\[2mm] \dfrac{\partial u_z}{\partial x} & \dfrac{\partial u_z}{\partial y} & \dfrac{\partial u_z}{\partial z} \end{bmatrix}^{T} \qquad (3.67)$$

Note that our notation has been modified using (x, y, z) in place of $(1, 2, 3)$ to be consistent with the general literature and with Equation (3.66). Now, because of the plane stress conditions, all of the y components of Equation (3.67) are either zero or can be ignored.* The relevant equations for the elastic strains therefore collapse down to the following simple relations:

$$\text{a)}\quad \varepsilon_{xx} = \frac{\partial u_x}{\partial x}$$

$$\text{b)}\quad \varepsilon_{zz} = \frac{\partial u_z}{\partial z} \qquad (3.68)$$

$$\text{c)}\quad \varepsilon_{xz} = \frac{1}{2}\left[\frac{\partial u_x}{\partial z} + \frac{\partial u_z}{\partial x}\right]$$

We are now faced with what the mechanics people call the compatibility problem. Equations (3.68) are three equations relating three strains to only two displacements. If we were going about the problem by first finding the displacements and then determining the strains from Equations (3.68), then there would be no problem. A problem arises because our approach is first to compute the stresses and strains and then derive the displacements. Equations (3.68) therefore present a difficulty because they are overdetermined in the strains.

This problem is resolved by noting that Equations (3.68) are not all independent. In fact, a simple relation among the strains can be derived by the following clever stratagem. Differentiate both sides of Equation (3.68a) twice in respect to y, then differentiate both sides of Equation (3.68b) twice in respect to x, and finally differentiate both sides of Equation (3.68c) once each in respect to both x and y to obtain the following relations:

---

* Note that $\partial u_z/\partial y$ is not zero for the plane stress condition but is ignored in this problem because it does not contribute to the strain energy.

$$\text{a)} \quad \frac{\partial^2 \varepsilon_{xx}}{\partial z^2} = \frac{\partial^3 u_x}{\partial z^2 \partial x}$$

$$\text{b)} \quad \frac{\partial^2 \varepsilon_{zz}}{\partial x^2} = \frac{\partial^3 u_z}{\partial x^2 \partial z} \tag{3.69}$$

$$\text{c)} \quad \frac{\partial^2 \varepsilon_{xz}}{\partial x \partial z} = \frac{1}{2}\left[\frac{\partial^3 u_x}{\partial z^2 \partial x} + \frac{\partial^3 u_z}{\partial x^2 \partial z} \right]$$

It is now easy to see that Equations (3.69a) and (3.69b) add up to twice Equation (3.69c), thus giving rise to the following relation among the strain components:

$$\frac{\partial^2 \varepsilon_{xx}}{\partial z^2} + \frac{\partial^2 \varepsilon_{zz}}{\partial x^2} = 2\frac{\partial^2 \varepsilon_{xz}}{\partial x \partial z} \tag{3.70}$$

Equation (3.70) is referred to as the compatibility relation, and it ensures that the strains will be compatible with an admissible displacement field.* What remains to be done is to connect the strains in Equations (3.68) to the stresses in Equations (3.66). This is the job of the constitutive relations given in Table 3.1 as Equations (3.38):

$$\varepsilon_{xx} = \frac{1}{E}\left[\sigma_{xx} - v\left(\sigma_{zz}\right) \right]$$

$$\varepsilon_{zz} = \frac{1}{E}\left[\sigma_{zz} - v\left(\sigma_{xx}\right) \right] \tag{3.71}$$

$$\varepsilon_{xz} = \frac{\sigma_{xz}}{2G}; \quad \frac{1}{G} = \frac{2\left(1+v\right)}{E}$$

Equations (3.71) now complete the formulation where the (1, 2, 3) subscripts of Equation (3.38) have been replaced by (x, y, z), and the y components have been dropped out because of the plane stress condition. These equations can now be used to eliminate the strains in Equation (3.70) in favor of the stresses. After some algebraic manipulation and replacing the shear modulus G by the Young's modulus E, the following remarkably simple relation in terms of stresses alone emerges:

$$\left(\frac{\partial^2}{\partial y^2} + \frac{\partial^2}{\partial x^2} \right)\left(\sigma_{xx} + \sigma_{yy} \right) = 0 \tag{3.72}$$

---

* It should be noted that Equation (3.70) could have been derived by direct application of Equation (3.29g) from Table 3.1, which gives the most general definition of the compatibility condition for elastic strains. In this case, it is simpler to manipulate Equations (3.69) and avoid a lot of tedious algebra.

Adding Equation (3.72) to Equations (3.66) now gives the following complete set of field equations in terms of stresses alone:

Momentum Balance

$$\frac{\partial \sigma_{xx}}{\partial x} + \frac{\partial \sigma_{xz}}{\partial z} = 0$$

$$\frac{\partial \sigma_{zx}}{\partial x} + \frac{\partial \sigma_{zz}}{\partial z} = 0 \qquad (3.73)$$

Compatibility

$$\left(\frac{\partial^2}{\partial z^2} + \frac{\partial^2}{\partial x^2} \right)(\sigma_{xx} + \sigma_{zz}) = 0$$

Equations (3.73) now represent a closed set of three field equations for the three independent stress components σ_{xx}, σ_{zz}, and σ_{xz}. These equations plus the boundary conditions now specify a complete solvable boundary value problem for the cantilevered beam.

So, are we done yet? No, not quite. It still remains to come up with the solution for these equations, which is not straightforward. We need another trick to simplify things somewhat. At this stage, the continuum mechanic reaches into the toolbox and pulls out what is known as a stress function ϕ. It is not unlike what your local garage mechanic does in dealing with a tough nut. Do not force it; get a bigger hammer. In this case, the "bigger hammer" is the stress function. A somewhat similar stratagem is used in simple particle mechanics; instead of trying to solve the problem in terms of the force, which is a vector quantity, the force is defined as the gradient of a scalar potential ϕ, and the problem is solved in terms of ϕ first, from which the force components can be easily calculated by taking the gradient.

The case is similar but a little more involved for the stress problem. First, a scalar stress function is defined as follows:

$$\sigma_{xx} = \frac{\partial^2 \phi}{\partial z^2}$$

$$\sigma_{zz} = \frac{\partial^2 \phi}{\partial x^2} \qquad (3.74)$$

$$\sigma_{xz} = -\frac{\partial^2 \phi}{\partial x \partial z}$$

It can easily seen by direct substitution that the stress function defined by Equations (3.74) identically satisfies the momentum balance relations given in Equations (3.73). To satisfy the compatibility relation, it must also satisfy the following:

$$a) \left(\frac{\partial^2}{\partial x^2} + \frac{\partial^2}{\partial z^2} \right)\left(\frac{\partial^2 \phi}{\partial x^2} + \frac{\partial^2 \phi}{\partial z^2} \right) = 0 \qquad (3.75)$$

which can be written as

$$b) \ \nabla^2 \nabla^2 \phi = 0$$

or as

$$c) \ \nabla^4 \phi = 0$$

Thus, the stress function must satisfy what is known as the biharmonic equation, which shows up fairly often in problems involving the bending of beams, the torsion of bars, and even the flow of fluids.[6] For present purposes, its main virtue is that it is a scalar equation that can be readily solved. Knowing ϕ, Equations (3.74) then give all the stresses, and from the stresses, Equations (3.71) then gives all the strains, which finally yield the displacements from Equations (3.68). Now, all that remains to be done is to smoke out solutions of Equation (3.75).

We solve Equation (3.75) following the methodology employed by Timoshenko and Goodier,[7] who investigated the mechanics of beam bending in great detail using the stress function approach. It helps somewhat to write out Equation (3.75) in full component form:

$$\left(\frac{\partial^4 \phi}{\partial x^4} + 2\frac{\partial^4 \phi}{\partial x^2 \partial z^2} + \frac{\partial^4 \phi}{\partial z^4} \right) = 0 \qquad (3.76)$$

Looking at Equation (3.76), it is clear that any polynomial function with terms of the form $x^n z^m$ where $n + m$ is three or less will be an exact solution to this equation. In particular, the following is a solution to Equation (3.76):

$$\phi_2 = \frac{A_2}{2} x^2 + B_z xz + \frac{C_2}{2} z^2 \qquad (3.77)$$

Using Equations (3.74), we see that this stress function gives the following stresses:

$$\sigma_{xx} = C_2; \ \sigma_{zz} = A_2; \ \sigma_{zx} = B_2 \qquad (3.78)$$

Thus, Equation (3.76) gives constant stresses throughout the sample. This clearly is not going to do the job for us because from the elementary solution we know that σ_{zz} depends on both x and z. However, we do not have to stop at a second-order stress function. We can even go to fourth and higher-order stress functions if we allow for certain constraints. In the fourth-order case, we have:

$$\phi_4 = \frac{A_4}{12} x^4 + \frac{B_4}{6} x^3 z + \frac{C_4}{2} x^2 z^2 + \frac{D_4}{6} xz^3 + \frac{E_4}{12} z^4 \tag{3.79}*$$

Substituting Equation (3.79) into (3.76), we see that ϕ_4 will be a solution providing the following condition holds:

$$A_4 + E_4 + 2C_4 = 0 \tag{3.80}$$

Using Equations (3.74), the stresses come out to be

$$\sigma_{xx} = \frac{\partial^2 \phi_4}{\partial z^2} = C_4 x^2 + D_4 xz + E_4 z^2$$

$$\sigma_{zz} = \frac{\partial^2 \phi_4}{\partial x^2} = A_4 x^2 + B_4 xz + C_4 z^2 \tag{3.81}$$

$$\sigma_{xz} = -\frac{\partial^2 \phi_4}{\partial z \partial x} = -\frac{B_4}{2} x^2 - 2C_4 xz - \frac{D_4}{2} z^2$$

Now, the game is basically this: Using Equations (3.81), we have a range of choices as to what the stress field in the beam should look like by setting the coefficients subject to the condition Equation (3.80). For instance, we know that, for the cantilevered beam with end load, $\sigma_{xx} = 0$ because there is no load on the beam other than at the end. This right away dictates that $C_4 = D_4 = E_4 = 0$, leaving the following solution:

$$\sigma_{xx} = 0$$

$$\sigma_{zz} = \frac{\partial^2 \phi_4}{\partial x^2} = A_4 x^2 + B_4 xz \tag{3.82}$$

$$\sigma_{zx} = -\frac{\partial^2 \phi_4}{\partial z \partial x} = -\frac{B_4}{2} x^2$$

To satisfy the constraint Equation (3.80), we must also set $A_4 = 0$, which leaves

$$\sigma_{zz} = B_4 xz$$

$$\sigma_{zx} = -\frac{B_4}{2} x^2 \tag{3.83}$$

Now, Equations (3.83) are close to what we want but do not quite make it because we cannot satisfy our boundary conditions, which are

---

* Note that the numerical coefficients 1/12, 1/6, and so on are put in for convenience to make the stresses come out cleaner. Because the coefficients are arbitrary, you can clearly put in any numbers you want.

$$\sigma_{zz}(x,z) = 0 \quad \text{for} \ z = L; \ \text{all} \ x$$

$$\sigma_{zx}(x,z) = 0 \quad \text{for} \ x = \pm d/2; \ \text{all} \ z \tag{3.84}$$

However, we have one more maneuver we can call on to spring us out of the current impasse. Recall from Equation (3.78) that any constant stress distribution satisfies the field equations. Now, because the problem is linear, the sum of any two solutions is also a solution. Adding Equations (3.78) to (3.83) gives

$$\sigma_{zz} = A_2 + B_4 xz$$

$$\sigma_{zx} = -B_2 - \frac{B_4}{2} x^2 \tag{3.85}$$

The four undetermined constants in Equations (3.85) now give us sufficient latitude to satisfy the boundary conditions in Equations (3.84). Substituting Equations (3.85) into (3.84) and using a little algebra gives

$$A_2 = -B_4 xL$$

$$B_2 = -\frac{B_4}{2}\left(\frac{d}{2}\right)^2 \tag{3.86}$$

The beam stresses now come down to the following near-final form:

$$\sigma_{zz} = B_4 x (z - L)$$

$$\sigma_{zx} = \frac{B_4}{2}\left(\left(\frac{d}{2}\right)^2 - x^2\right) \tag{3.87}$$

What remains is to satisfy the equilibrium condition that the shear force at the end of the beam exactly balances the downward load P shown in Figure 3.17. This condition may be written as follows:

$$\text{Opposing shear force in beam} = \int_{-\frac{d}{2}}^{\frac{d}{2}} \sigma_{zx} dx = \tag{3.88}$$

$$\text{Downward end load} = -P$$

Substituting the shear stress σ_{zx} from Equations (3.87) into Equation (3.88) gives the following equilibrium condition:

$$\frac{B_4}{2} \int_{-\frac{d}{2}}^{\frac{d}{2}} \left(\left(\frac{d}{2}\right)^2 - x^2 \right) dx = -P \tag{3.89}$$

The integral in Equation (3.89) is elementary and is readily evaluated. Thus, the coefficient B_4 emerges after a short algebra exercise and noting that the moment of inertia of a beam of unit width and thickness d from Equation (3.59) is given by $I = d^3/12$:

$$B_4 = -\frac{P}{I} \tag{3.90}$$

Putting this result into Equation (3.87), our solution for the cantilevered beam may be summarized as follows:

$$\sigma_{zz} = \frac{P}{I} x (z - L)$$

$$\sigma_{zx} = \frac{P}{2I} \left(\left(\frac{d}{2}\right)^2 - x^2 \right) \tag{3.91}$$

The associated strains come directly from Equations (3.38), where we identify (1, 2, 3) as (x, y, z) and set $\Delta T = 0$:

$$\varepsilon_{xx} = \frac{1}{E} \left(\sigma_{xx} - \nu \left(\sigma_{yy} + \sigma_{zz} \right) \right)$$

$$\varepsilon_{yy} = \frac{1}{E} \left(\sigma_{yy} - \nu \left(\sigma_{xx} + \sigma_{22} \right) \right)$$

$$\varepsilon_{zz} = \frac{1}{E} \left(\sigma_{zz} - \nu \left(\sigma_{xx} + \sigma_{yy} \right) \right) \tag{3.92}$$

$$\varepsilon_{xy} = \frac{\sigma_{xy}}{2G}; \quad \varepsilon_{zx} = \frac{\sigma_{zx}}{2G}; \quad \varepsilon_{zy} = \frac{\sigma_{zy}}{2G};$$

Noting that $\sigma_{xx} = \sigma_{yy} = \sigma_{xy} = \sigma_{zy} = 0$, Equations (3.92) shrink to

$$\varepsilon_{xx} = -\frac{1}{E}\nu\sigma_{zz} = -\frac{\nu Px}{EI}(L-z)$$

$$\varepsilon_{zz} = \frac{1}{E}\sigma_{zz} = \frac{Px}{EI}(L-z) \qquad (3.93)$$

$$\varepsilon_{zx} = \frac{\sigma_{zx}}{2G} = \frac{-P}{4IG}\left(x^2 - \left(\frac{d}{2}\right)^2\right)$$

The beam deformations are obtained from the strains using the basic definition of the infinitesimal strain tensor given by Equation (3.29):

$$\underline{\underline{\varepsilon}} = 1/2\left[\nabla\mathbf{u} + \left(\nabla\mathbf{u}\right)^T\right] \qquad (3.94)$$

Adopting the common notation for the vector $\mathbf{u}$ as (u_x, u_y, u_z) where u_x, u_y, and u_z are generally functions of the coordinates x, y, z, Equation (3.94) can be written out in component form as

$$
\begin{bmatrix} \varepsilon_{xx} & \varepsilon_{xy} & \varepsilon_{xz} \\ \varepsilon_{yx} & \varepsilon_{yy} & \varepsilon_{yz} \\ \varepsilon_{zx} & \varepsilon_{zy} & \varepsilon_{zz} \end{bmatrix} = \frac{1}{2}\begin{bmatrix} \dfrac{\partial u_x}{\partial x} & \dfrac{\partial u_x}{\partial y} & \dfrac{\partial u_x}{\partial z} \\ \dfrac{\partial u_y}{\partial x} & \dfrac{\partial u_y}{\partial y} & \dfrac{\partial u_y}{\partial z} \\ \dfrac{\partial u_z}{\partial x} & \dfrac{\partial u_z}{\partial y} & \dfrac{\partial u_z}{\partial z} \end{bmatrix} + \frac{1}{2}\begin{bmatrix} \dfrac{\partial u_x}{\partial x} & \dfrac{\partial u_x}{\partial y} & \dfrac{\partial u_x}{\partial z} \\ \dfrac{\partial u_y}{\partial x} & \dfrac{\partial u_y}{\partial y} & \dfrac{\partial u_y}{\partial z} \\ \dfrac{\partial u_z}{\partial x} & \dfrac{\partial u_z}{\partial y} & \dfrac{\partial u_z}{\partial z} \end{bmatrix}^T \qquad (3.95)
$$

The strain matrix is clearly symmetric, and we can thus easily read out the six independent strain components as follows:

$$\varepsilon_{xx} = \frac{\partial u_x}{\partial x}; \; \varepsilon_{yy} = \frac{\partial u_y}{\partial y}; \; \varepsilon_{zz} = \frac{\partial u_z}{\partial z}$$

$$\varepsilon_{xy} = \frac{1}{2}\left(\frac{\partial u_x}{\partial y} + \frac{\partial u_y}{\partial x}\right)$$

$$\varepsilon_{yz} = \frac{1}{2}\left(\frac{\partial u_y}{\partial z} + \frac{\partial u_z}{\partial y}\right) \qquad (3.96)$$

$$\varepsilon_{zx} = \frac{1}{2}\left(\frac{\partial u_z}{\partial x} + \frac{\partial u_x}{\partial z}\right)$$

We can adapt Equations (3.96) to the current problem by redefining the components of the deformation vector $\mathbf{u}$ as $(u_x, u_y, u_z) \rightarrow (u, v, w)$ and further noting that it is only the x and z components that are of current interest:

$$\varepsilon_{xx} = \frac{\partial u}{\partial x}; \quad \varepsilon_{zz} = \frac{\partial w}{\partial z}$$

$$\varepsilon_{zx} = \frac{1}{2}\left(\frac{\partial w}{\partial x} + \frac{\partial u}{\partial z}\right)$$

(3.97)

The above relations may now be combined with Equations (3.93) to give a set of differential equations for the components of the displacement field:

a) $\dfrac{\partial u}{\partial x} = -\dfrac{\nu Px}{IE}(L - z)$

b) $\dfrac{\partial w}{\partial z} = \dfrac{Px}{IE}(L - z)$ (3.98)

c) $\dfrac{1}{2}\left(\dfrac{\partial w}{\partial x} + \dfrac{\partial u}{\partial z}\right) = -\dfrac{P}{2IG}\left(x^2 - (d/2)^2\right)$

Equations a) and b) above can be integrated directly to give

a) $u = -\dfrac{\nu Px^2}{2IE}(L - z) + f(z)$

b) $w = \dfrac{Px}{IE}\left(Lz - \dfrac{z^2}{2}\right) + g(x)$ (3.99)

where $f(z)$ and $g(x)$ are arbitrary functions that remain to be determined. The next step is to substitute the above expressions for u and w into Equation (3.98c) to obtain

$$\frac{\nu Px^2}{2EI} + f'(z) + \frac{P}{EI}\left(Lz - \frac{z^2}{2}\right) + g'(x) = -\frac{P}{2IG}\left(x^2 - (d/2)^2\right) \qquad (3.100)$$

The above relation may look rather hopeless, but it in fact holds the key to finding the displacement functions. First, note that any given term is either a function of x alone or z alone or constant. This suggests that we collect all like terms together and rearrange Equation (3.100) as follows:

a) $G(x) + F(z) = \dfrac{P}{2IG}(d/2)^2$ (3.101)

where

$$\text{b) } F(z) = f'(z) + \frac{P}{EI}\left(Lz - \frac{z^2}{2}\right)$$

$$\text{c) } G(x) = g'(x) + \frac{P}{2IG}x^2 + \frac{\nu Px^2}{2EI}$$

The key thing to note about Equation (3.101a) is that it asserts that a function of x alone plus a function of z alone is constant. A moments reflection gives the immediate conclusion that the only way this can happen is if each function is separately constant. Thus, if we set $G(x) = H$ and $F(z) = J$, Equations (3.101) convert to the following:

$$\text{a) } H + J = \frac{P}{2IG}(d/2)^2 \qquad (3.102)$$

where

$$\text{b) } J = f'(z) + \frac{P}{EI}\left(Lz - \frac{z^2}{2}\right)$$

$$\text{c) } H = g'(x) + \frac{P}{2IG}x^2 + \frac{\nu Px^2}{2EI}$$

Equations (3.102) are now a set of ordinary differential equations for the functions g and f. These are readily integrated to give the following solutions:

$$\text{a) } f(z) = -\frac{P}{EI}\left(L\frac{z^2}{2} - \frac{z^3}{6}\right) + Jz + N$$

$$ \qquad (3.103)$$

$$\text{b) } g(x) = -\frac{P}{2IG}\frac{x^3}{3} - \frac{\nu P}{2EI}\frac{x^3}{3} + Hx + M$$

These relations may now be directly substituted into Equation (3.99) to give the following expressions for the displacement functions:

$$\text{a) } u = -\frac{\nu Px^2}{2IE}(L - z) - \frac{P}{EI}\left(L\frac{z^2}{2} - \frac{z^3}{6}\right) + Jz + N$$

$$ \qquad (3.104)$$

$$\text{b) } w = \frac{Px}{IE}\left(Lz - \frac{z^2}{2}\right) - \frac{P}{2IG}\frac{x^3}{3} - \frac{\nu P}{2EI}\frac{x^3}{3} + Hx + M$$

Equations (3.104) now bring us close to home, giving two algebraic expressions for the displacement functions u and w with four undetermined constant coefficients J, N, H, and M. These constants are readily disposed of by invoking the boundary conditions on the displacements at the fixed end (also called the built-in end) of the cantilevered beam:

a) $w\left(x = 0, z = 0\right) = 0$ Fixed lateral displacement

b) $u\left(x = 0, z = 0\right) = 0$ Fixed vertical displacement   (3.105)

c) $\left.\dfrac{\partial u}{\partial z}\right|_{x=z=0} = 0$ Cantilevered condition

Conditions a) and b) above are self-evident. Condition c) further stipulates that because the beam is built into the retaining wall, it must approach at an angle of 90° to the surface. This can be taken as a definition of the cantilevered boundary condition. Applying the above conditions to Equations (3.104), it is clear that the constants M, N, and J must be zero. Further, by invoking Equation (3.102a), we see that H must be given by

$$H = \frac{P}{2IG}\left(d/2\right)^2 \tag{3.106}$$

The final solution for the displacements can now be written as

a) $u = -\dfrac{\nu P x^2}{2IE}\left(L - z\right) - \dfrac{P}{EI}\left(L\dfrac{z^2}{2} - \dfrac{z^3}{6}\right)$

$$\tag{3.107}$$

b) $w = \dfrac{Px}{IE}\left(Lz - \dfrac{z^2}{2}\right) - \dfrac{P}{2IG}\dfrac{x^3}{3} - \dfrac{\nu P}{2EI}\dfrac{x^3}{3} + \dfrac{P}{2IG}\left(d/2\right)^2 x$

Referring to Equations (3.65), we see that the above result is considerably more involved than the elementary SOM formula. In particular, several terms are missing in Equations (3.65) because the elementary approach implicitly assumes that the Poisson ratio is zero. Also, the fact that the beam deforms in the axial direction is neglected entirely in the elementary treatment. However, in most practical applications it is only the vertical deflection of the end of the beam that is of interest. Inserting $z = L$ into the formula for the vertical displacement u in Equations (3.107) gives $u = -PL^3/(3EI)$, which is the same as the elementary result except for a sign convention. Note that the elementary result will be a good approximation to the beam deformation for narrow long beams where $x < d \ll z \sim L$. The final solution for the cantilevered beam problem can be summarized as follows:

a) $\sigma_{zz} = \dfrac{P}{I} x (z - L)$

b) $\sigma_{zx} = \dfrac{P}{2I} \left(\left(\dfrac{d}{2} \right)^2 - x^2 \right)$

c) $\varepsilon_{xx} = -\dfrac{1}{E} \nu \sigma_{zz} = -\dfrac{\nu P x}{EI} (L - z)$

d) $\varepsilon_{zz} = \dfrac{1}{E} \sigma_{zz} = \dfrac{P x}{EI} (L - z)$ (3.108)

e) $\varepsilon_{zx} = \dfrac{\sigma_{zx}}{2G} = \dfrac{-P}{4IG} \left(x^2 - \left(\dfrac{d}{2} \right)^2 \right)$

f) $u = -\dfrac{\nu P x^2}{2IE} (L - z) - \dfrac{P}{EI} \left(L \dfrac{z^2}{2} - \dfrac{z^3}{6} \right)$

g) $w = \dfrac{P x}{IE} \left(L z - \dfrac{z^2}{2} \right) - \dfrac{P}{2IG} \dfrac{x^3}{3} - \dfrac{\nu P}{2EI} \dfrac{x^3}{3} + \dfrac{P}{2IG} (d/2)^2 x$

3.5 GENERAL METHODS FOR SOLVING FIELD EQUATIONS OF ELASTICITY

Despite the complexity and work involved in deriving the solution of the cantilevered beam problem represented by Equations (3.108), they still represent only an approximation to the exact solution. In particular, the y components of the stresses, strains, and displacements are notably absent. For most practical purposes, these quantities are of no significance, which justifies the use of Equations (3.108). However, be aware that these equations are in fact an approximation, and that one can obtain a more accurate description of the bending problem if necessary. Therefore, this section outlines a number of more general methods for the solution of the elasticity field equations so the reader can at least be aware of what is available in the general literature should the need arise. There are basically three approaches to solving the field equations summarized in Table 3.1 assuming constant temperature so that the temperature field can be ignored. Which approach is used depends on the type of boundary conditions involved. The three types are as follows:

1. *Displacement formulation*: The deformation of the object is specified everywhere on the boundary.
2. *Stress formulation*: The surface loads (formally the surface tractions or forces per unit area) are specified everywhere on the boundary of the object.
3. *Mixed formulation*: The deformations are specified over part of the boundary and the surface tractions over the remaining part.

3.5.1 DISPLACEMENT FORMULATION

Using the displacement formulation approach, one starts with the momentum balance relationship as given by Equation (3.22):

$$f_b + \nabla \cdot \underline{\sigma} - \rho \frac{d^2 \mathbf{u}}{dt^2} = 0$$

The stresses are eliminated by using the constitutive relation in the form

$$\underline{\sigma} = \frac{E}{1+v}\left(\underline{\varepsilon} + \frac{v\underline{I}}{1-2v} tr\underline{\varepsilon}\right) + E\alpha\Delta T\underline{I}$$

This gives the following intermediate relationship:

$$\frac{E}{1+v}\left(\nabla \cdot \underline{\varepsilon} + \frac{v\underline{I}}{1-2v} \cdot \nabla tr\underline{\varepsilon}\right) + \mathbf{f}_b = \rho \frac{d^2 \mathbf{u}}{dt^2}$$

The strain tensor is eliminated using the familiar small deformation definition:

$$\underline{\varepsilon} = 1/2\left[\nabla\mathbf{u} + \left(\nabla\mathbf{u}\right)^T\right]$$

After substitution and some formal manipulation, one finally arrives at the following expression:

$$G\nabla^2\mathbf{u} + \frac{E}{2(1+v)(1-2v)}\nabla(\nabla \cdot \mathbf{u}) + \mathbf{f}_b = \rho\frac{d^2\mathbf{u}}{dt^2} \qquad (3.109a)$$

One often sees this equation in the literature written using the classical Lamé constants instead of the Young's modulus and Poisson's ratio as follows:

$$\mu\nabla^2\mathbf{u} + (\mu + \lambda)\nabla\nabla \cdot \mathbf{u} + \mathbf{f}_b = \rho\frac{d^2\mathbf{u}}{dt^2} \qquad (3.109b)$$

As expressed above, Equations (3.109) are known as the Lamé equations. and the constants μ and λ are called the Lamé constants. The quantity μ is in fact identical to the shear modulus G, and λ is equivalent to $vE/[(1+v)(1-2v)]$. Given displacement boundary conditions, one solves Equations (3.109) for the displacement field $\mathbf{u}$ and then works back to compute all the strains from the strain displacement relation. The constitutive relation then gives all the stresses, and the problem is done. A big problem that arises is that one does not always have all the boundary displacements at one's disposal. For the cantilevered beam problem, for instance, the displacements were specified only at the built-in end of the beam, and only the load was specified at the other end. Equations (3.109) are of little help in such a situation. For these types of conditions, other formulations of the problem are more convenient. As a final note, Feynman[8] has given an interesting development of Equations (3.109) and applied them to the problem of wave propagation in solids. The solution method is quite simple, giving straightforward and elegant expressions for the velocity of transverse and compressional waves in a solid.

3.5.2 STRESS FORMULATION

The next option is to have all the boundary conditions expressed as stresses or surface tractions.* In this case, one starts with the equation of compatibility for the strains as given by Equation (3.29g) from Table 3.1:

$$\nabla \times \nabla \times \underline{\varepsilon} = 0$$

This can be rewritten in a more convenient form using the following identity,[9] recalling that for any tensor $\underline{\mathbf{A}}$ that $\operatorname{tr} \underline{\mathbf{A}} = A_{11} + A_{22} + A_{33} + \ldots A_{NN}$:

$$\nabla \times \nabla \times \underline{\varepsilon} = \nabla^2 \underline{\varepsilon} + \nabla\nabla\left(tr\underline{\varepsilon}\right) - \nabla\nabla\cdot\underline{\varepsilon} - \left(\nabla\nabla\cdot\underline{\varepsilon}\right)^T = 0$$

Combining this with the constitutive relation Equations (3.83) from Table 3.1 in the form

$$\underline{\varepsilon} = \frac{1+v}{E}\underline{\sigma} - \frac{v\,tr\underline{\sigma}}{E}\mathbf{I}$$

and further using the momentum balance relation from the same table for a body at equilibrium

$$\mathbf{f_b} + \nabla\cdot\underline{\sigma} = 0$$

gives the following field equation for the stresses in an elastic body at equilibrium:

$$\nabla^2\underline{\sigma} + \frac{1}{1+v}\nabla\nabla\left(tr\underline{\sigma}\right) + \nabla\mathbf{f_b} + \left(\nabla\mathbf{f_b}\right)^T + \frac{v}{1+v}\left(\nabla\cdot\mathbf{f_b}\right)\mathbf{I} = 0 \qquad (3.110a)$$

---

* Tractions and stresses are essentially the same entities. When considering points inside an elastic body, the term *stress* is generally used. When dealing with stresses at a surface, the term *traction* is preferred, especially when dealing with surface stresses arising from external forces.

Equation (3.110a) is commonly referred to as the Beltrami-Michell equation of compatibility in honor of the originators Eugenio Beltrami and John-Henry Michell. Stresses that satisfy this equation are guaranteed not only to satisfy the momentum balance condition, but also, when substituted into the constitutive equation, to give strains that satisfy the compatibility relations. The strains so derived will then be compatible with an allowable deformation field when substituted into the infinitesimal strain tensor relationship Equation (3.29f) from Table 3.1. When the body force vector $\mathbf{f}_b$ can be ignored, Equation (3.109a) simplifies greatly into

$$\nabla^2 \underline{\sigma} + \frac{1}{1+\nu} \nabla \nabla \left(tr \underline{\sigma} \right) = 0 \qquad (3.110b)$$

Equation (3.110b) is in fact the full three-dimensional generalization of Equations (3.73), which were used to solve the cantilevered beam problem in the two-dimensional plane stress approximation. It can be shown that the stresses derived from that approximation Equations (3.107a,b) do not completely satisfy Equation (3.110b) in that the shear stresses must be modified as follows[10]:

$$b') \quad \sigma_{zx} = \frac{1}{1+\nu} \left(\frac{P}{2I} \right) \left(\left(\frac{d}{2} \right)^2 - x^2 \right)$$

$$(3.107')$$

$$c') \quad \sigma_{yz} = \frac{-\nu}{1+\nu} \left(\frac{P}{I} \right) xy$$

Thus, the shear stress σ_{zx} is modified by a Poisson contraction effect, and the out-of-plane shear σ_{yz} now enters the picture. Strictly speaking, even including the above corrections, Equations (3.107) do not give the mathematically exact elasticity solution of the cantilevered beam problem because the details of the stress distribution near the built-in end and near the tip where the load is applied will be in error. In particular, there will be some complex stress distribution where the end load is applied, depending on the details of the loading conditions. For instance, one can apply the load using weights, hooks, pins, or any number of other devices, each of which will give a different local stress field. The situation is precisely identical to that shown in Figure 3.4, illustrating several different methods of loading a tensile sample. As in that case, it is St. Venant's principle that saves the day by ensuring that the errors incurred will be significant only close to the load point, and Equations (3.107) will accurately represent the actual stress field for the greater part of the beam.

3.5.3 MIXED FORMULATION

In the mixed elasticity problem, one has displacement conditions over one part of the body and force loads elsewhere. In this case, one has to deal with the field equations represented in Table 3.1 directly, and there is no single well-recognized formalism available for cranking out a solution. The situation is more like a free-for-all in which one takes a "hook-or-by-crook" approach by applying different methods to different sections of the body in question and then piecing it all together in the end. Perhaps the best known approach is what is known as the semi-inverse method, again developed by the patron saint of elasticity theory, Barre' de A. J. C. St. Venant, previously mentioned regarding his famous principle. This is essentially the approach used to tackle the cantilevered beam problem in Section 3.2.1. The basic approach is to start with some approximate solution to the problem that might be guessed at by physical intuition or borrowed perhaps from an approximate SOM calculation. One then makes appropriate adjustments using the full set of field equations to arrive at an improved solution.

3.6 NUMERICAL METHODS

3.6.1 INTRODUCTION

Given the workout involved in determining the stresses, strains, and displacements for the elementary cantilevered beam problem presented here, it should come as no surprise that more complex problems are simply beyond the reach of the analytical methods outlined in the previous sections. However, if accurate numerical solutions are what are required, then several powerful computational methods are available. The most common are the finite element approach followed by the boundary element method. Other methods are available, but these dominate the computing landscape. The finite element method in particular is the hands down most popular technique. Several commercial programs are available, going under names such as *ABACUS*, *ALGOR*, *ANSYS*, *MSC NASTRAN*, *PATRAN*, and so on. Commercial boundary element programs are available but are a distant second to the ever-popular finite element approach.

Using the finite element method, one essentially divides the material object under investigation into a large number of small regular elements of simple shapes, such as rectangles, triangles, cubes, and the like. The full stress-strain-displacement problem can be worked out exactly in an analytical form for these simple elements, and they thus form the elementary building blocks for constructing a more general model of the structure. Once the structure under study has been broken down into these simple elements, the program assembles all the pieces into one large matrix, which carefully preserves the connectivity of all the small elements and further incorporates all of the relevant loading and constraint conditions. In the case of linear problems, this matrix must be inverted to determine the displacements of all of the nodes that define the corners of the subelements. All of the strains and stresses then follow essentially from the formulae of Table 3.1. When the computer is done churning, one essentially has a huge database of displacements, stresses, and strains that constitute the "answer" sought.

In the early days of finite element work, one of the biggest problems was dealing with this monstrous collection of data. This impasse was solved in the 1980s with the development of very clever and powerful graphical methods that could give the user a stress distribution plot or a displacement diagram for any section of the model of interest. These days, it is common to come across full multicolor stress contour plots of three-dimensional objects that quickly reveal the stress distribution at a glance.

Although less popular than the finite element method, the boundary element approach does have a number of distinct advantages in certain circumstances. The first advantage is that one discretizes only the boundary of the object modeled, as opposed to the finite element approach, which divides up the entire structure into small elements. Thus, in the case of a two-dimensional model one simply divides the boundary of the object into a number of discrete line segments, and this constitutes the entire geometric model. Using finite elements, the entire area enclosed by the boundary must be discretized. Right away, we see that the boundary element approach reduces the dimensionality of the problem by 1, which greatly simplifies the problem of model generation. In addition, the resulting model is solved only for the boundary displacements, again greatly reducing the complexity of the problem. Stresses, strains, and displacements are then obtained for any desired interior region from a special formula central to the whole boundary element approach. This gives the distinct benefit of only having to compute solutions in regions of interest and greatly reduces the size of the numerical database that is generated.

So, the question is why the whole world is not using the boundary element approach as opposed to the finite element technique. The answer comes down to essentially two technical issues. One is the fact that the precise kernel function (Green's function in technical parlance) that is central to the boundary element method may not be known for the problem of interest. The second problem is the complication of nonlinear behavior. For nonlinear problems, the boundary element method is still valid, but now one must discretize the entire model to deal with the more complex material behavior, thus melting away one of the major advantages of the method. As a matter of practice, the boundary element method tends to find favor in dealing with problems of electromagnetism, for which the Green's function kernel is well known and nonlinear material behavior is not an issue. Thus, if one is designing electromagnetic lenses for an advanced electron microscope for which one is basically modeling a shell structure and the fields need to be precisely determined in only a small subregion, the boundary element approach would be the tool of choice.

3.7 DETAILED STRESS BEHAVIOR OF A FLEXIBLE COATING ON A RIGID DISK

In Section 3.3.2, we treated the elementary case of a coating on a rigid substrate subject to equibiaxial tension. The problem was quite simple because we essentially assumed the whole system was infinite, and thus the stresses and strains were constant throughout the structure. The question naturally arises regarding what happens with a finite sample. Thus, say we have a small disk 32 mm in diameter with a thin 1-mm thick coating on it, what do the stresses look like? It is clear now that we have to deal with what happens at the edge of the disk, and the assumption

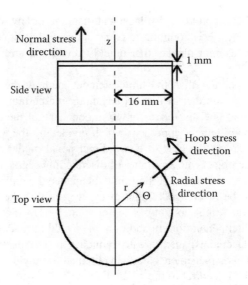

FIGURE 3.18 Uniform coating on a rigid substrate subjected to thermal strain loading. The stress distribution in the coating near the center of the disk is accurately predicted by the elementary membrane equation Equation (3.50).

of constant stresses and strains is out the window. This problem can be assaulted by analytical methods, but the mathematical complexity quickly gets out of hand. The interested reader might like to look at the paper by Aleck,[11] who applied analytical methods to the related problem of a clamped rectangular plate.

Figure 3.18 illustrates the geometry of the problem. The substrate is effectively a rigid slab, and the coating material has elastic properties close to those common for high-temperature insulator materials such as the polyimides. For this example, we take the modulus of the coating to be 3 GPa and the Poisson's ratio as 1/3. Further assuming that the stress in the coating is caused by thermal strains induced by cooling from the cure temperature of the material, we set the thermal expansion mismatch between the coating and substrate at 30 ppm/C. For the sake of this example, the cure temperature has been taken as 465 C, and the whole sample is now sitting at 20 C and has thus undergone a temperature swing of 445 C. Under these conditions, the simplified formula Equation (3.50) for the in-plane stress in the coating comes to close to 60 MPa.

The actual in-plane stresses are shown in Figure 3.19a and Figure 3.19c. The conclusion from these figures is fairly straightforward. The in-plane stress as represented by the radial and hoop stresses shown in these figures comes quite close to the prediction of the elementary formula given by Equation (3.50). Near to the center of the disk, the coating is in a state of equibiaxial tension at a level close to 60 MPa. Figure 3.19b and Figure 3.19d confirm that both the normal stress and the shear stress are zero near the center of the disk, which is also in accord with the equibiaxial condition assumed by Equation (3.50). As expected, however, all this breaks down as you approach the edge of the disk. The radial stress plummets to zero at the edge

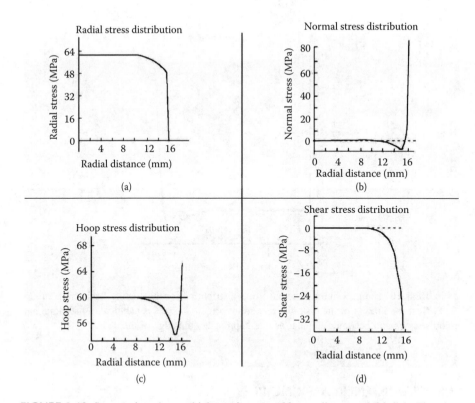

FIGURE 3.19 Stresses in a 1-mm thick coating on a 32-mm diameter rigid disk. The elementary membrane equation Equation (3.50) predicts an in-plane stress level of 60 MPa near the center of the disk away from the edge and zero normal and shear stresses in this region. These predictions are accurate once one is roughly five film thicknesses away from the edge region. Closer to the edge, the elementary formula breaks down; the normal, shear, and hoop stresses become singular; and the radial stress goes to zero. The singular behavior of the normal and hoop stresses explains why coatings tend to delaminate and crack preferentially near edges and other discontinuities.

of the disk, as required by the condition of no radial constraint on this boundary. The hoop stress, however, takes a slight dip and then increases dramatically. Elasticity theory in fact predicts that this stress should tend to infinity at the interface between the coating and the substrate as you approach the edge of the disk. The same holds true for both the normal and the shear stress. The singular stress condition in fact explains why coatings tend to delaminate at abrupt boundaries, edges, and other discontinuities.

Finally, Figure 3.20 exhibits the maximum principal stress in the coating, which also conforms to Equation (3.50) in the disk interior but becomes singular as you approach the edge. This stress component will act in both the normal and hoop directions and can cause either radial cracking or delamination, depending on the strength of the coating material and how well it is adhered to the substrate.

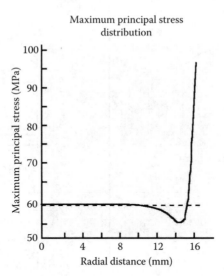

FIGURE 3.20 The maximum principal stress distribution for the coating depicted in Figure 3.18. Near the interior of the disk, the maximum principal stress is identical to the radial and hoop stresses and becomes singular as one approaches the edge of the disk.

3.8 STRAIN ENERGY PRINCIPLES

This chapter on continuum theory discusses the concept of elastic strain energy. This is a key topic because a great deal of the fracture mechanics of adhesion discussed more fully in the next chapter relies on the principle of stored elastic strain energy. A very important class of delamination problems involve what are known as self-loading systems, which through a variety of mechanisms build up a high level of elastic strain that sits like a time bomb ready to go off and drive both fracture and delamination processes. Unlike standard situations in which a structure fails because of readily apparent applied loads, these self-loading systems can superficially appear to be quite stable and then self-destruct at some inopportune moment.

One suspects that if a residual stress exists within an elastic body, then there must be an energy field associated with this stress. One intuitive way to derive what this energy field must be is to go back to elementary mechanics, by which the concept of energy is most easily derived. The reader should recall from introductory physics classes that one always starts with Newton's second law, which states that, for a particle of mass m residing in some force field $\mathbf{F} = \mathbf{F}(\mathbf{x})$, an acceleration $\mathbf{a}$ is imparted to the particle given by the classic formula $\mathbf{F} = m\mathbf{a}$. Here, of course, $\mathbf{x}$ is the position vector of the particle and the acceleration $\mathbf{a}$ is given by the second time derivative of $\mathbf{x}$ (also first-time derivative of velocity $\mathbf{v}$), which allows the second law to be written as a differential equation:

$$m\frac{d^2x}{dt^2} = m\frac{dv}{dt} = F(x) \qquad (3.111)$$

At this stage, one chooses to work with the velocity $\mathbf{v}$ because it involves only a first-degree equation as opposed to a second degree in $\mathbf{x}$. The next step is to take the dot product of both sides of Equation (3.11) with the velocity vector $\mathbf{v}$. It is not clear who was the first to come up with this nonobvious stratagem, which must remain one of the lost secrets of history, but doing it gives the following interesting relation:

$$m\frac{dv}{dt} \cdot \mathbf{v} = F(x) \cdot \mathbf{v} \qquad (3.112)$$

The next step is no more obvious but involves integrating both sides of Equation (3.112) with respect to time and using the following elementary identities from vector calculus:

$$\frac{dv}{dt} \cdot v = \frac{1}{2}\frac{dv^2}{dt}; \quad vdt = \frac{dx}{dt}dt = dx \qquad (3.113)$$

Performing the integration and employing the above identities while noting that the constants can be taken within the derivative operation gives

$$\int \frac{d}{dt}\left[\frac{m}{2}v^2\right]dt = \int F(x)dx + E \qquad (3.114)$$

The left-hand side of Equation (3.111) is a perfect differential and therefore is immediately integrated, giving the famous formula for the kinetic energy KE = $1/2mv^2$. The integral on the right-hand side is taken as the potential energy given by

$$V = -\int F(x)dx \qquad (3.115)$$

Finally, the constant of integration E is taken to be the total energy. With these definitions, Equation (3.114) becomes the classic expression for the conservation of energy:

$$KE + V = E \qquad (3.116)$$

Thus, the kinetic energy plus the potential energy is constant under the broad assumption of a non-time-varying force field $\mathbf{F}$. If, for instance, the force field is linear in the displacement x, as it would be if the particle were attached to some kind of fixed spring, $\mathbf{F}$ would have the form $-k\mathbf{x}$, and from Equation (3.115) the potential energy would be given by

$$V(x) = -\int -kx\ dx = \frac{k}{2}x^2 \qquad (3.117)$$

Equation (3.117) is the point particle mechanics analog of strain energy for a single particle trapped by simple harmonic potential. What we require, however, is the case of zillions and zillions of interacting particles that form the mass of some lump of matter. Each particle can be thought of as trapped in some kind of harmonic potential; therefore, the whole collection should behave collectively as an elastic continuum. However, to get at this problem in a general way, we can work by analogy with the above simple deviation of the single-particle potential energy, only instead of starting with the particle mechanics form of Newton's second law given by Equation (3.111), we must use the general momentum balance equation, Equation (3.22) from Table 3.1, given by

$$\mathbf{f_b} + \nabla \cdot \underline{\sigma} - \rho \frac{d^2\mathbf{u}}{dt^2} = 0$$

Following the same strategy as used for the point particle mechanics example, we take the scalar product of the velocity field vector $\mathbf{v}$ with the momentum balance equation and integrate the whole thing over the entire body, noting that $\mathbf{v} = d\mathbf{u}/dt$

$$\int \left[\mathbf{v} \cdot \mathbf{f_b} + \mathbf{v} \cdot \nabla \cdot \underline{\sigma} \right] dV = \int \rho \mathbf{v} \cdot \frac{d\mathbf{v}}{dt} dV = \int \frac{\rho}{2} \frac{d\mathbf{v}^2}{dt} dV \qquad (3.118)$$

We immediately recognize the right-hand side of Equation (3.118) as the rate of change of the total kinetic energy of the body, so we know we are on the right track. On the left-hand side, the term $\mathbf{v} \cdot \mathbf{f_b}$ represents the work done on the body by the external body forces and is not of direct concern at the moment. The internal strain energy we are after is hiding in the $\mathbf{v} \cdot \nabla \cdot \underline{\sigma}$, term so we need to look more closely at it. To obtain this term, however, we need another of our clever vector identities:

$$\nabla \cdot \left(\mathbf{v} \cdot \underline{\sigma} \right) = \nabla \mathbf{v} : \underline{\sigma} + \mathbf{v} \cdot \nabla \cdot \underline{\sigma} \qquad (3.119)$$

Equation (3.119) requires some additional explanation. Because it involves the divergence operator acting on the scalar product of a vector with a second rank tensor, it is somewhat more involved than the standard vector calculus type identity.* First, why go through the trouble of introducing such an entity in the first place?

---

* The interested reader should refer to Appendix A for more in-depth details on this type of analysis.

The main reason is that the expression on the left-hand side is a complete divergence expression that, when inserted into the volume integral in Equation (3.118), can be converted into a surface integral using the divergence theorem. This then has the effect of separating out the contribution of surface tractions to the overall energy equation. The first term on the right-hand side of Equation (3.119) is the double scalar product (also referred to as a double contraction) of two tensor quantities and is thus a scalar, as are the other terms in the equation. This is the term that contains the internal strain energy we seek. Finally, the second term on the right of Equation (3.119) is identical to the second term on the left of Equation (3.118), which allows us to eliminate it from that expression in favor of the remaining terms in Equation (3.119) as follows:

$$\frac{d}{dt}KE = \int \left[v \cdot \mathbf{f_b} + \nabla \cdot (v \cdot \underline{\sigma}) - \nabla v : \underline{\sigma} \right] dV \tag{3.120}$$

Equation (3.120) now has all the ingredients we want. We can now rearrange it to bring the internal energy term to the left-hand side with the kinetic energy and use the divergence theorem to convert the divergence term to a surface integral as follows:

$$\frac{d}{dt}KE + \frac{d}{dt}U = \int_V v \cdot \mathbf{f_b}\, dV + \int_S v \cdot t\, dS = \frac{d}{dt}W$$

$$\frac{d}{dt}U = \int_V \nabla v : \underline{\sigma}\, dV \tag{3.121}$$

$$t = \underline{\sigma} \cdot \mathbf{n} \quad \left(\text{Surface traction}\right)$$

Equation (3.121) is a generalization of the well-known conservation of energy theorem stating that the time rate of change of the kinetic plus internal energy of a solid is equal to the total time rate of work done by both body forces and surface loads.* What is of most interest to us is the internal energy term because it represents that portion of the total energy of the body that is locked up in the internal elastic strain of the body caused by a variety of residual stress mechanisms. Such residual stresses include, but are not limited to, the following:

1. Residual stress/strain in a coating resulting from thermal expansion mismatch with the substrate on cooling or heating.
2. Residual stress/strain caused by solvent evaporation and shrinkage of a coating.

---

* Note that the applied work expressions **v.f** dV and **v.t** dS are essentially force times velocity terms that amount to work per unit time or power. Thus, these terms indeed represent the time rate of change of energy or power input to the body in question.

3. Residual stress/strain caused by grain boundary inclusions/voids in poly-crystalline materials.
4. Residual stress/strain caused by frozen-in thermal gradients in glassy materials, such as amorphous thermoplastic polymers and common boro-silicate glasses.

These mechanisms are what lead to the self-loading/self-destructing behavior of coatings and other structures.* So, it is important to have a closer look at the internal energy term in Equation (3.121), which we single out as follows:

$$\frac{d}{dt}U = \int_V \nabla v{:}\underline{\sigma}\, dV \qquad (3.122)$$

Because we are interested more in conditions of static equilibrium than in time-dependent problems, we need to do something about the velocity gradient term in Equation (3.122). One problem that arises right away is the fact that the velocity gradient will have components that arise from rotational as well as stretching kinds of motion. We faced this problem with the displacement gradient also and were compelled to remove the rotational component of the deformation because it did not give rise to any stresses when we wanted to define a strain measure in terms of the gradient of the deformation. We need to do the same with the velocity gradient, which can be accomplished by using the following identity:

$$\nabla \mathbf{v} = \frac{1}{2}\left[\nabla \mathbf{v} + \left(\nabla \mathbf{v}\right)^T\right] + \frac{1}{2}\left[\nabla \mathbf{v} - \left(\nabla \mathbf{v}\right)^T\right] = \underline{\mathbf{Y}} + \underline{\mathbf{V}}$$

$$\underline{\mathbf{Y}} = \frac{1}{2}\left[\nabla \mathbf{v} + \left(\nabla \mathbf{v}\right)^T\right] \qquad (3.123)$$

$$\underline{\mathbf{V}} = \frac{1}{2}\left[\nabla \mathbf{v} - \left(\nabla \mathbf{v}\right)^T\right]$$

Equation (3.123) separates the velocity gradient into two parts, one symmetric given by the tensor $\underline{\mathbf{Y}}$ and one antisymmetric given by the tensor $\underline{\mathbf{V}}$. All of the rotational behavior is contained in the antisymmetric term $\underline{\mathbf{V}}$ and the stretching deformations are correspondingly contained in $\underline{\mathbf{Y}}$. Because $\underline{\mathbf{V}}$ is antisymmetric, it will not contribute to the internal energy because the double-dot product with the symmetric stress tensor is zero, that is,

---

* Note that none of this implies that the kinetic energy and other terms are unimportant because ballistic impact and excessive vibration are kinetic energy factors that can clearly destroy a structure. The point is that these are rather obvious mechanisms that do not have the insidious stealth property the internal stress processes.

$$\underline{V}:\underline{\sigma} = \sum_{ij} \left(\underline{\sigma}\right) \left(\underline{V}\right)_{ij} = 0$$

Since: $\left(\underline{\sigma}\right)_{ij} = \left(\underline{\sigma}\right)_{ji}$

and: $\left(\underline{V}\right)_{ij} = -\left(\underline{V}\right)_{ji}$ (3.124)

with: $\left(\underline{V}\right)_{ii} = 0$

Using the identity given above, Equation (3.118) can now be written in a more useful form:

$$\frac{d}{dt}U = \int_V \underline{Y}:\underline{\sigma}\,dV \qquad (3.125)$$

Using the definition of $\underline{Y}$ from Equation (3.123), we can eliminate the velocity gradient from Equation (3.125) using the following identity:

$$\underline{Y} = \frac{1}{2}\left[\nabla v + \left(\nabla v\right)^T\right] = \frac{1}{2}\left[\nabla\frac{du}{dt} + \left(\nabla\frac{du}{dt}\right)^T\right]$$

$$= \frac{d}{dt}\left[\frac{1}{2}\left[\nabla u + \left(\nabla u\right)^T\right]\right] = \frac{d\underline{\varepsilon}}{dt} \qquad (3.126)$$

Equation (3.126) can now be used to eliminate the velocity gradient $\underline{Y}$ from Equation (3.125), giving

$$\frac{d}{dt}U = \int_V \underline{\sigma}:\frac{d\underline{\varepsilon}}{dt}\,dV \qquad (3.127)$$

Equation (3.127) gets us almost what we want. The quantity under the volume integral is apparently the time rate of change of the local strain energy per unit volume or strain energy density, which we write as dw/dt. Furthermore, because we are interested mainly in quasistatic behavior, we can eliminate the time differential and write simply the local increment of strain energy density dw as follows:

$$dw = \underline{\sigma}:d\underline{\varepsilon} \qquad (3.128)$$

Equation (3.128) is the basic starting place for the calculation of the strain energy in an elastic solid. The next step is to assume that because of some fixed applied

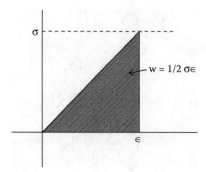

FIGURE 3.21 Buildup of strain energy in a linear elastic solid as the strain is increased from 0 to some finite level.

stress $\underline{\sigma}$, the strain in the material increases from 0 to some small but finite value $\underline{\varepsilon}$ as shown in Figure 3.21. The total strain energy density created at the point in question is given by the area under the curve, which is the integral of Equation (3.128), giving the following result:

$$w = \frac{1}{2}\underline{\sigma}:\underline{\varepsilon} \tag{3.129}$$

The above result is quite general for elastic solids and forms the starting point for further detailed analysis. The next step is to use the constitutive relation connecting the stress to the strain to eliminate one or the other from Equation (3.129). In the worst case, one can drag out the most general linear elastic relation given previously, which can be written as:

$$\underline{\sigma} = \underline{\underline{C}}:\underline{\varepsilon} \tag{3.130}$$

In Equation (3.130), $\underline{\underline{C}}$ is the fourth-order elasticity tensor we encountered first in Equation (3.30). Note that the elasticity tensor is doubly contracted with the strain tensor, leaving a second-order tensor as a result, which Equation (3.130) identifies as the stress tensor. If the fates were really cruel and we had to deal with the stress state of some exotic triclinic crystal with essentially no simplifying symmetry properties, then the tensor $\underline{\underline{C}}$ might have as many as 21 different parameters. Using Equation (3.130) to eliminate the stress tensor from Equation (3.129), the strain energy density in the solid becomes:

$$w = \frac{1}{2}\underline{\varepsilon}:\underline{\underline{C}}:\underline{\varepsilon} \tag{3.131}$$

If written out in detail for a low-symmetry crystal, Equation (3.131) can be a horrendously complex expression that would easily consume an entire page or more in dense hieroglyphics. If ever confronted with the problem of evaluating Equation (3.131) for a complex crystal, one can refer to dedicated works such as Kittel's text on solid state physics.[12] A somewhat lighter and eminently readable discussion of the problem has been given in the Feynman Lectures.[13] For current purposes, we settle for simple isotropic materials that show spherical symmetry and thus involve only two elastic constants. This turns out to be a surprisingly good approximation in many cases of practical interest because in many engineering applications one is dealing either with amorphous materials such as glasses or with polycrystalline materials with small, randomly oriented crystallites. In either case, the material behaves approximately like a homogeneous isotropic solid when examined on a macroscopic scale. In such cases, we can use the constitutive relation given by Equations (3.38) from Table 3.1 to eliminate the strain tensor from Equation (3.129) as follows:

$$w = \frac{1}{2}\underline{\sigma}{:}\underline{\varepsilon}$$

$$\underline{\varepsilon} = \frac{1+\nu}{E}\underline{\sigma} - \frac{\nu\ tr\underline{\sigma}}{E}\underline{I} \tag{3.132}$$

$$w = \frac{1}{2}\underline{\sigma}{:}\left[\frac{1+\nu}{E}\underline{\sigma} - \frac{\nu\ tr\underline{\sigma}}{E}\underline{I}\right] = \frac{1+\nu}{2E}\underline{\sigma}{:}\underline{\sigma} - \frac{\nu\left(tr\underline{\sigma}\right)^2}{2E}$$

The last expression in Equation (3.132) holds because $\underline{\sigma}{:}\underline{I} = tr(\underline{\sigma})$. Equation (3.132) gives us one of the most useful forms of the linear elastic strain energy of a material treated as a homogeneous isotropic solid in terms of the stress tensor and the basic elastic constants E and ν. In typical fracture mechanics calculations, one first computes the stress field $\underline{\sigma}$, and Equation (3.132) then gives the strain energy density. One can also go by the inverse route and calculate the strain field $\underline{\varepsilon}$, but now one has to use a different expression for the strain energy, which can be easily gotten from Equation (3.132) by eliminating the stress in terms of the strain using the inverse of the constitutive relation given in Equation (3.132). Because these are important and handy relations to have, we summarize them below, and because $\underline{\sigma}$ and $\underline{\varepsilon}$ are symmetric tensors, we can simplify the double-contraction operation using the following handy formula, valid for any symmetric second-rank tensor $\underline{A}$: $\underline{A}{:}\underline{A} = tr(\underline{A}^2)$*:

$$a)\ \ w = \frac{1+\nu}{2E}tr\left(\underline{\sigma}^2\right) - \frac{\nu\left(tr\underline{\sigma}\right)^2}{2E} \quad \left(\text{Stress formulation}\right)$$

$$\tag{3.133}$$

$$b)\ \ w = \frac{E}{1+\nu}\left[\frac{\nu tr\left(\underline{\varepsilon}^2\right)}{1-2\nu} + \frac{\left(tr\underline{\varepsilon}\right)^2}{2}\right] \quad \left(\text{Strain formulation}\right)$$

---

* Note that $\underline{A}^2$ is evaluated by standard matrix multiplication.

The above relations for the strain energy density w have been derived using rather formal and no doubt at times somewhat-mysterious methods. However, these relations can be reached using more basic force times distance arguments as developed in elementary physics courses. The main advantage of the formal approach is that we can see how they arise from the general momentum balance relation Equation (3.22) and are thus an integral part of the overall continuum mechanics formalism.

3.9   THE MARVELOUS MYSTERIOUS J INTEGRAL

We conclude our discussion of continuum theory with what definitely has to be one of the higher flights of the art of the mechanics of continuous media. What we discuss is a collection of conservation laws that bear directly on the behavior of solid bodies. The rationale for this work goes back to the elementary theory of the motion of point particles acted on by external force fields. The great success of Newton's laws of motion in describing the motion of celestial bodies is what gave this field its initial impetus. In particular, Johannes Kepler's original empirical laws of planetary motion were shown to be direct consequences of the laws of conservation of angular momentum and conservation of energy. The conservation laws also proved very useful in working out various problems of particle motion that were addressed at the time.

As an example, consider the conservation of energy principle as given by Equation (3.116). Here, we have the simple statement that, for a conservative system (i.e., one without friction or other dissipative influences), the sum of the kinetic plus potential energy of a particle is constant. This can be quite useful for computational purposes as follows. Say we have a compact mass of some sort that is supported either by a ramp or wire at some height h above a tabletop. We would like to know how fast the mass will be traveling if we let it slide down the ramp or wire by the time it reaches the table. The precise geometry of the ramp or wire can be any roller coaster shape provided that it is relatively frictionless and ultimately lets the mass reach the tabletop.

Use of Equation (3.116) allows us to answer this question without recourse to the detailed equations of motion or even knowing the particulars of the ramp or wire. Refer to Figure 3.22 and assume for the sake of concreteness we are dealing with a small rectangular block of mass m sliding down a frictionless ramp. Starting the mass off from an initial height h above the tabletop, we can immediately employ Equation (3.116) to arrive at the following relation:

$$\frac{1}{2}mv^2 + mgy = mgh \tag{3.134}$$

where

m = Mass of block (kg)
v = Velocity of block at any instant (m/sec)

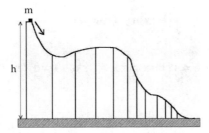

FIGURE 3.22 A compact mass m getting ready to slide down some arbitrary roller coaster-shaped frictionless ramp. The conservation of energy principle asserts that the final velocity of the mass when it reaches the end of the ramp will depend only on its initial height h and the local gravitational field strength g regardless of the detailed shape of the ramp.

> y = Height of block above tabletop at any instant (m)
> h = Initial height of block above tabletop at time 0 (m)
> g = Acceleration caused by gravity at Earth's surface = 9.8 m/sec^2

Because we are interested in the block's velocity when it reaches the tabletop, we set $y = 0$ in Equation (3.134), and further noticing that the block mass cancels out, we immediately determine the magnitude of the velocity to be given by the simple expression $(2gh)^{1/2}$. Note that had we attempted to solve the fundamental equation of motion of the block for its detailed motion on the slide depicted in Figure 3.22 and then evaluated the final solution as the block reached the tabletop. it would have involved a substantial exercise in solving a differential equation. Thus, the conservation on energy principle given by Equation (3.116) allows us to achieve rather simple solutions to what would otherwise be fairly complex problems in the realm of classical particle mechanics.

Because continuum theory is essentially an extension of Newtonian particle mechanics to continuous media, this immediately suggests the notion that similar kinds of conservation relations might exist in continuum theory that would allow us to gain useful solutions to the field equations documented in Table 3.1 without having to go through the exertion of grappling with the partial differential equations head on. Miraculously, there are in fact conservation relations that apply to the field equations and that can be used to easily derive useful solutions to problems that would otherwise involve considerable computational effort. These conservation relations appear in the form of what are called path-independent integrals.* With a path-independent integral, one essentially has some quantity defined in terms of the stresses and deformations occurring in the body that, when integrated over some closed path, always gives the same answer regardless of which path is taken, subject only to the constraint that the path be closed (i.e., one always returns to the same starting point). Three such integrals are known to exist:

---

* It should be noted that Equation (3.134) can be derived from a path-independent integral.

$$\text{a)} \quad J = \int_S \left(W(\mathbf{x})\mathbf{n} - \mathbf{T} \cdot \nabla \mathbf{u} \right) dS$$

$$\text{b)} \quad L = \int_S \left(W(\mathbf{x})\mathbf{n} \times \mathbf{x} + \mathbf{T} \times \mathbf{u} - \mathbf{T} \cdot \nabla \mathbf{u} \times \mathbf{x} \right) dS$$

$$\text{c)} \quad M = \int_S \left(W(\mathbf{x})\mathbf{x} \cdot n - \mathbf{T} \cdot \nabla \mathbf{u} \cdot \mathbf{x} - \frac{1}{2}\mathbf{T} \cdot \mathbf{u} \right) dS \qquad (3.135)$$

$$W(\mathbf{x}) = \int_0^\varepsilon \underline{\sigma} : d\underline{\varepsilon} \quad \left(\text{Strain energy density}\right)$$

$$\mathbf{T} = \underline{\sigma} \cdot \mathbf{n} \quad \left(\text{Applied surface tractions}\right)$$

The quantities in Equation (3.135) are defined in respect to Figure 3.23, which illustrates some arbitrary material body B that we can assume is in some well-defined state of stress and deformation. The integrations are over some interior surface S that surrounds an arbitrary subvolume V of the body. The variables are defined as follows:

$\mathbf{x}$ = Spatial coordinate within B referred to some appropriate coordinate system
$\mathbf{u}$ = Displacement field defined on all of B
$\underline{\varepsilon}$ = Strain tensor derived from the displacement field
$\underline{\sigma}$ = Stress tensor induced by the strain field
$\mathbf{n}$ = Unit normal vector on subsurface S
$\mathbf{T}$ = Traction vector acting on subsurface S

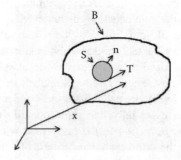

FIGURE 3.23 The general path-independent integrals of Equation (3.135) hold for some subregion of an elastic body B that is enclosed by a subsurface S. The vector **n** is the unit normal to S, and the vector **T** is the traction field that exists on S because of the general deformation and stress state of B. The vector **x** locates the points in B with regard to some convenient coordinate system.

Each of the integrals in Equation (3.135) is invariant in the sense that it is independent of the particular subsurface within B that is chosen to evaluate it. Given this information alone, one begins to see how these integrals might come in handy when the subsurface S can be judiciously chosen such that the field quantities are either zero or of a very simple form over most of its extent and complex only in a small region of interest, such as a local crack or other defect.

Eshelby[14,15] was apparently the first to use this stratagem to study the forces acting on cavitylike defects in elastic bodies. Subsequently, Knowles and Sternberg[16] demonstrated that Eshelby's work was a special case of a class of conservation laws for elastic continua given by Equations (3.135). For our purposes, the work of Rice[17,18] is most relevant. He recognized that Eshelby's calculations could be profitably brought to bear on problems of cracks in elastic bodies. The term J integral apparently originated with Rice's investigation in which he essentially worked with a two-dimensional version of Equation (3.135a) as follows:

$$J = \int_{\Gamma} \left(W(\mathbf{x})dy - \mathbf{T} \cdot \frac{\partial \mathbf{u}}{\partial x} ds \right)$$

$$W(\mathbf{x}) = \int_{0}^{\varepsilon} \underline{\sigma} : d\underline{\varepsilon} \quad \left(\text{Strain energy density} \right) \tag{3.136}$$

$$\mathbf{T} = \underline{\sigma} \cdot \mathbf{n} \quad \left(\text{Applied surface tractions} \right)$$

Equation (3.136) is what you get by taking the surface integral in Equation (3.135a) and shrinking it to a plane surface. As a consequence, all the quantities in Equation (3.136) are two dimensional, applying to conditions of plane stress or plane strain. The integration is carried out over a closed contour Γ as shown in Figure 3.24. All quantities are as defined in Equation (3.135), only taking into account that now everything is referred to a two-dimensional coordinate system, and the quantity ds is an infinitesimal increment taken along the contour Γ, which is comprised of four separate sections as shown in Figure 3.24. In the figure, the contour Γ surrounds a notch-type defect in the material body that is assumed to be in some state of deformation and stress. Because the integral J is path independent, it will evaluate to zero when evaluated over Γ or any other closed loop. The advantage of the particular contour shown in the figure is that it can be deformed to allow the evaluation of J over the notch tip represented by the contour segment Γ_t, for which the stress field can be quite complex or even singular. Thus, a typical strategy would involve expanding the contour segment Γ_1 to encompass the boundary of the body where the stresses and deformations are either 0 or well known; further, let the segment Γ_2 contract to coincide with the notch segment Γ_t. In this way, J is evaluated as the sum of the two parts J_1 and J_t, where the component J_1 is readily evaluated in terms of the known loading and boundary conditions imposed on the body, and J_t is a measure of the strain energy density in the region of the notch. Because J

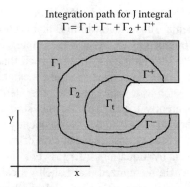

Integration path for J integral
$$\Gamma = \Gamma_1 + \Gamma^- + \Gamma_2 + \Gamma^+$$

FIGURE 3.24 The closed integration path for the two-dimensional J integral for some general notched body. Because the J integral is path independent, the total path Γ is broken down into four segments. This allows the path to be distorted in some convenient way so that it can be easily evaluated for all segments except for the line Γ_t, where the stress field is large and rapidly varying in some complex fashion. In this way, one can obtain valuable information about complex and singular defect fields without having to set up and solve an involved boundary value problem for the field equations in Table 3.1.

must evaluate to zero, J_1 and J_t must be equal and of opposite sign, which thus pins down the difficult integral in terms of the known one.

An elementary example will better serve to illustrate the uses of the J integral. Figure 3.25 illustrates a long bar of unit thickness with a long notch in the center extending from the left-hand end to the center. For the sake of simplicity, it is assumed that the bar is of such a length that the ends can be considered at plus or minus infinity and thus do not interact in any way with the notch tip. The top and bottom edges are considered clamped and given some small uniform displacement in the y direction that will induce some complex stress field in the vicinity of the notch tip. The conditions of the problem are now such that if we take the contour Γ to be the boundary of the specimen, then the integrand in Equation (3.136) is nonzero only at the notch tip and at the far right end of the bar. In particular, the terms $\partial u/\partial x$ and dy are both zero for the top and bottom surfaces, which thus give no contribution. The stresses and tractions on the edges to the left of the notch are also zero or very small, so that no contribution comes from those segments. This then leaves only the right-hand end and the notch tip as giving nonzero contributions. Further, they must give equal and opposite contributions because the total line integral must come to zero. The contribution at the right-hand end is easiest to evaluate because $\partial u/\partial x$ is zero there, and thus J has the value $W_\infty h$ because the energy density is constant due to the constant displacement condition. This must also be the value of J at the notch tip. Thus, assuming plane strain conditions and the elementary formula for W in an elastic medium with Young's modulus E and Poisson ratio ν under some constant displacement u, we get the following formula for J at the notch tip:

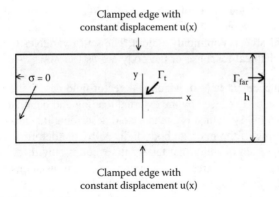

Clamped edge with
constant displacement u(x)

Clamped edge with
constant displacement u(x)

FIGURE 3.25 Elementary example of the use of the J integral to evaluate the strain energy release rate of a simply loaded specimen with a very long notch.

$$J = \int_{-\frac{h}{2}}^{\frac{h}{2}} \left(W(\mathbf{x}) \right) dy = W_\infty h = \frac{\left(1-v\right)Eu^2}{\left(1+v\right)\left(1-2v\right)h} \tag{3.137}$$

It should be noted that J has the dimensions of an energy per unit area because W has units of energy per unit volume. These are precisely the units of the strain energy release rate or driving force for propagating the notch as a crack in the sample. The concept of a strain energy release rate is discussed in more detail in Chapter 4. We simply note now that the concept of path-independent integrals, particularly the J integral, can be a useful tool in evaluating this quantity.

3.10 SUMMARY

This chapter provides a reasonably detailed overview of the continuum theory of solid bodies. The need for this arises from the fact that to carry out fully quantitative adhesion measurements one inevitably has to grapple with the stresses and deformations that all adhesion measurement tests induce in both the adherend and the adherate materials. For this purpose, the continuum theory is without doubt the most general and useful formalism capable of providing quantitative results testable by practical measurements. The continuum theory has its foundations in Newtonian particle mechanics and generalizes this formalism by introducing the concepts of the calculus of fields in three dimensions. Thus, solid bodies are treated as subsets of the three-dimensional Euclidean space, which the mathematicians refer to as R^3. This essentially generalizes the real line R^1 to cover the full spatial environment with which everyone is familiar. Material bodies are treated as subsets of R^3 and are assigned attributes such as mass density and elastic behavior, which allows them to serve as models for real solids once appropriate boundaries have been specified.

The picture is completed by supplying a set of field equations that describe how the bodies behave when subjected to a variety of loading conditions analogous to those that real material objects commonly endure. The set of variables that then describes the thermal-mechanical response of the body are as follows:

- The displacement field **u**, which is a vector field describing the deformation of the body caused by the external loading and boundary constraints.
- The strain field $\boldsymbol{\varepsilon}$, which is a tensor field describing the internal stretching and twisting of the material body induced by the deformation field. This is an important quantity that takes into account only those deformations that will give rise to stresses, as opposed to pure translations and rotations, which do not induce any stress within the body.
- The stress field $\boldsymbol{\sigma}$, which is a tensor field induced by the strain field because of the assumed elastic behavior ascribed to the body.
- The temperature T, which is a scalar field describing the thermodynamic state of the body. All real materials exist at some finite temperature, although in the current treatment the temperature is mostly assumed to be uniform throughout the body and thus does not give rise to any deformation, strain, or stress. A more general treatment of the continuum theory of solids would posit the existence of a free energy functional from which would follow equations of state and constitutive relations that relate all of the above-mentioned field quantities and thus give a complete thermodynamic description of the material body in question. This treatment omits covering this important aspect of the continuum theory mainly because it greatly increases the level of complexity involved and further finds little in the way of practical application to situations most commonly encountered in adhesion testing. The one exception is the thermodynamic treatment of the peel test given in Chapter 5, which simply introduces the free energy as an ad hoc assumption. The reader interested in more advanced treatment should consult the text by Billington and Tate,[19] who give a full development of continuum theory based on thermodynamic principles.

Having defined all the relevant field quantities required to describe whatever solid is of interest, one then requires the field equations, which are summarized in Table 3.1, to describe how things evolve once a set of relevant loading and boundary conditions is specified. Finally, a set of energy principles is discussed; these principles are useful in simplifying certain computations and further provide an important foundation for the all-important fracture mechanics theory, which is covered in Chapter 4.

Notes

1. "The Linear Theory of Elasticity," M.E. Gurtin, in *Mechanics of Solids*, Vol. 2, C. Truesdell, Ed. (Springer-Verlag, Heidelberg, Germany, 1984).
2. "Introduction to Linear Elasticity," P.L. Gould (Springer-Verlag, New York, 1983), Chapter 4.

3. *Strength of Materials*, J.P. Den Hartog (Dover, New York, 1961; first published by McGraw-Hill, 1949).

4. *History of Strength of Materials*, S.P. Timoshenko (Dover, New York, 1983; first published by McGraw-Hill, 1953). A highly recommended work for those interested in the history of materials and materials testing. We learn, for instance, that Archimedes was apparently the first to rigorously prove in the third century BC the conditions of equilibrium for a lever. This subject has a very long history indeed.

5. *Roark's Formulas for Stress and Strain*, 6th ed., W.C. Young (McGraw-Hill, New York, 1989). This is a veritable bible for the practicing engineer in need of a simple formula to estimate the deformation and stress in simple structures. The fact that it is in its sixth edition is silent testimony to the enduring relevance of this topic to modern technology. The first edition apparently came out in 1938. If the formula you need is not in this volume, chances are you really do not need it.

6. This is not an entirely coincidental accident because the stress field in a solid can be viewed as a sort of frozen flow field, quite similar to what is observed in fluids. In particular, contours of constant stress bear a striking resemblance to lines of constant flow in a fluid. A curious example is the grain pattern observed in wood. Trees constantly have to arrange their structural fibers to oppose the gravitational loads to which they are subjected and thus tend to align the grain to oppose the tensile stress acting on the branch in question. If you examine a knot in a board, you can see that the surrounding grain in a sense flows around this defect much as a fluid flows around a stationary obstacle. The mathematically inclined will notice that the stress equilibrium condition in solids $\nabla \cdot \mathbf{\sigma} = 0$ bears a close resemblance to the continuity of flow condition $\nabla \cdot \mathbf{J} = 0$ for the flow flux vector in a fluid.

7. *Theory of Elasticity*, 3rd ed., S.P. Timoshenko and J.N. Goodier (McGraw-Hill, New York, 1970), Chapter 3. This is a virtual Bible of early 20th century elasticity theory. The first edition was published in 1934. Many standard problems are worked out in this volume, which relies heavily on analytical techniques because it was published in the precomputer days of engineering science.

8. *The Feynman Lectures on Physics*, Vol. 2, R.P. Feynman, R.B. Leighton, and M. Sands (Addison-Wesley, Reading, MA, 1964), Chapter 40. Always a delight to read, these lectures focus mainly on problems of electromagnetism, but several interesting chapters deal with aspects of elasticity theory in Prof. Feynman's inimitable style and are highly recommended as supplementary material.

9. See Note 1, Section D, "Elastostatics," for full details. Gurtin develops a collection of powerful identities that lead to elegant and compact derivations.

10. For details, see Note 2, Chapter 6.3. Gould also derives the Beltrami-Michell equations using the versatile and powerful indicial notation. The reader should be warned, however, that tracking the proliferation of subscripts that arise by this approach can be a daunting task.

11. "Thermal Stresses in a Rectangular Plate Clamped Along an Edge," B.J. Aleck, *Journal of Applied Mechanics*, 16, 118 (1949).

12. *Introduction to Solid State Physics*, 2nd ed., C. Kittel (John Wiley and Sons, New York, 1956).

13. Note 8, Chapter 39. Detailed calculations of the elastic coefficients are carried out for a few simple crystals.

14. "A Continuum Theory of Lattice Defects," J.D. Eshelby, in *Solid State Physics: Advances in Research and Applications*, Vol. 3, F. Seitz and D. Turnbull, Eds. (Academic Press, New York, 1956), p. 79.

15. "The Energy Momentum Tensor in Continuum Mechanics," J.D. Eshelby, in *Inelastic Behavior of Solids*, M.F. Kanninen et al., Eds. (McGraw-Hill, New York, 1970), p. 77.
16. "On a Class of Conservation Laws in Linearized and Finite Elastostatics," J.K. Knowles and E. Sternberg, *Archive for Rational Mechanics and Analysis*, 44, 187 (1972).
17. "A Path Integral and the Approximate Analysis of Strain Concentrations by Notches and Cracks," J.R. Rice, *Transactions ASME, Journal of Applied Mechanics*, 35, 379 (1968).
18. "Mathematical Analysis in the Mechanics of Fracture," J.R. Rice, in *Fracture; An Advanced Treatise*, Vol. 2, H. Liebowitz, Ed. (Academic Press, New York, 1968), p. 19.
19. *The Physics of Deformation and Flow*, E.W. Billington and A. Tate (McGraw-Hill, New York, 1981).

4 Elementary Fracture Mechanics of Solids: Application to Problems of Adhesion

...yet it is true that the whole subject is littered with traps for the unwary and many things are not as simple as they might seem. Too often the engineers are only called in, professionally, to deal with the structural achievements of "practical" men at the same time as the lawyers and the undertakers.

J. E. Gordon, *Structures or Why Things Don't Fall Down*

4.1 INTRODUCTION

4.1.1 INTRODUCTORY CONCEPTS

Chapter 3 established the rudiments of the continuum theory of solids. Every solid body is idealized as a mathematical continuum field that can be characterized by a handful of elastic constants and other thermodynamic parameters. Elementary concepts of deformation and stress were established, leading to a collection of field equations that describe the thermomechanical state of a solid body under a wide variety of loading conditions. To deal with problems of adhesion, however, there has to be a way of dealing with the problem that solid bodies come apart by a variety of processes, including delamination of separate material layers or direct fracture within the bulk of the separate materials. From the point of view of fracture mechanics, however, the cracking of the bulk material and the delamination of two different material layers are considered to be different aspects of the same thing, namely, the sundering of the body because of an excessive loading condition.

Fracture mechanics, in essence, tries to accommodate fracture and delamination phenomena within the framework of established continuum theory. In doing this, the theory first has to reconcile the fact that cracks and delaminations introduce what are known as stress singularities into the material body under consideration. At the very tip of a crack or at the very leading edge of a delamination, the continuum theory predicts that certain components of the stress tensor become infinite. The main contribution of fracture mechanics is to reconcile this awkward prediction with the known fact that real material bodies do not harbor infinite stresses of any kind. The simple truth is that continuum theory harbors quite a number of stress singularities.

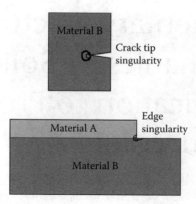

FIGURE 4.1 Common examples of stress singularities in solid structures.

So, what should be done about the problem of infinite stresses? Figure 4.1 shows two situations of direct relevance to the current work. The top figure shows a common crack configuration originating at the surface of the material. If a tensile load is applied to this body, then the stress field near the tip of the crack becomes infinite according to standard continuum theory. The bottom figure is even more interesting. A laminate of two dissimilar materials with a sharp edge can harbor a stress singularity at the right angle corner even in the absence of any external loads. The most common example is a coating applied to a substrate at an elevated temperature and subsequently cooled to room temperature. If the thermal expansion coefficient of the coating is higher than that of the substrate, then the coating will develop a biaxial tensile stress as given by Equation (3.50). However, the stress at the right-angle corner is infinite according to continuum theory. These two observations help to explain why in a solid failure tends to originate from surface flaws, and coatings tend to delaminate from the edges.

The problem of how to handle the stress singularity remains. In real materials, the stress field near crack tips and sharp corners is very high but finite because no crack tip or corner is perfectly sharp. There is always some rounding, and this then gives rise to a finite stress level. The straightforward solution would then be always to assume some finite radius of curvature for crack tips and corners. However, this approach would be very awkward from a computational viewpoint, and experimentally it would be very difficult, if not impossible, to determine accurately just what the correct radius of curvature should be in any given situation. The fracture mechanics approach to solving this problem relies on the observation that, even though the stress at a sharp crack tip or corner may be infinite, the strain energy contributed by this singularity is finite. This can be seen by looking at the analytical solution of the simple biaxially loaded crack problem shown in Figure 4.2. An elementary treatment of this problem has been given by Broek[1] using the Airy stress function approach. The main result of this analysis is that the components of the stress field near the crack tip can be written in the following simple form:

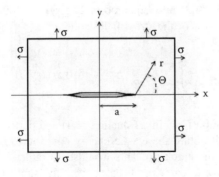

FIGURE 4.2 Crack flaw in a biaxially loaded sheet sample.

$$\sigma_{ij} = \frac{K_I}{\sqrt{2\pi}} \frac{f_{ij}(\theta)}{\sqrt{r}} \tag{4.1}$$

where

    σ_{ij} = Stress component i, j
    r = Radial coordinate measured from crack tip
    θ = Angular coordinate about crack tip
    $K_I = \sigma\,(\pi a)^{1/2}$ = Mode I stress intensity factor
    $f_{ij}(\theta)$ = Elementary function of trigonometric expressions

The main feature to note about Equation (4.1) is the reciprocal square root dependence of the stress components on the radial distance from the crack tip. This is the source of the singular behavior. However, what is important is that this is what is known as a weak singularity in that if the stress function in Equation (4.1) is integrated over some small volume containing the crack tip, the result is finite. For an elastic solid, the energy associated with a given stress field can be written from Equation (3.129) as

$$U = \frac{1}{2} \int_V \sum_{ij} \sigma_{ij} \varepsilon_{ij} dV \tag{4.2}$$

The volume V covers a small region containing the crack, and the strain field near the crack tip ε_{ij} is given by the standard constitutive relation for an elastic solid:

$$\varepsilon_{ij} = \frac{1+v}{E}\sigma_{ij} - \frac{v}{E}\delta_{ij}tr(\boldsymbol{\sigma}) \tag{4.3}$$

where $tr(\boldsymbol{\sigma}) = \sigma_{11} + \sigma_{22} + \sigma_{33}$.

Inserting Equation (4.3) into Equation (4.2) gives the following expression for the strain energy in terms of the stress field alone:

$$U = \int_V \sum_{ij} \left[\frac{1+v}{E} \sigma_{ij}^2 - \frac{v}{E} \delta_{ij} \sigma_{ij} tr(\underline{\boldsymbol{\sigma}}) \right] dV \tag{4.4}$$

If we insert Equation (4.1) into Equation (4.4) and perform the indicated sums and integrals, we will obtain the elastic energy contributed by the singular stress field associated with the crack tip. This would be a rather onerous chore that we fortunately do not have to carry out because we only want to demonstrate that the energy contributed by the crack singularity is finite. This is easy to see from Equation (4.4) because it is clear that after substituting Equation (4.1), every term in the expanded sum will be of the following form*:

$$U = C \int_0^{r_0} \sigma_{ij}^2 r dr = C \int_0^{r_0} \frac{1}{r} r dr = C r_0 \tag{4.5}$$

The integration has been performed in polar coordinates, and the factor C is a constant containing the elastic constants and the result of the angular integration. Thus, the energy contributed by the stress singularity is finite because every term in Equation (4.4) is finite.

The fact that the strain energy contributed by the singular stress field near a crack tip or sharp corner is finite basically settles the singularity problem. We take the attitude that it is okay that the stress goes to infinity at the crack tip, but we do not worry because the singularity is weak and makes only a finite contribution to the overall strain energy, which is the important quantity. That the stress is infinite is an error introduced by the continuum assumption that cracks and corners are perfectly sharp, which they are not. We only hope that our expression for the stress field given in Equation (4.1) gives an adequate approximation to the stress field near the crack tip and is sufficiently accurate to estimate the strain energy in that region.

Another complication that enters the picture is the tendency for real materials to develop a state of plastic flow once the local stress state exceeds the yield stress. Thus, close enough to the crack tip, Equation (4.1) will no longer hold within some closed region encompassing the crack tip of radius R. The magnitude or R can be estimated by the following simple formula:

$$R = \frac{1}{2\pi} \left(\frac{K_I}{\sigma_y} \right)$$

---

* The volume element is in polar coordinates of the form $r\, dr\, d\theta\, dz$. The θ and z integrations involve no singularities, and they can be lumped into the constant C.

The tacit assumption we make is that R is negligibly small compared to the region size over which Equation (4.1) holds, and thus the discrepancy caused by the local yielding can be ignored. This is what is commonly referred to as the small-scale yielding assumption and needs to be checked when attempting careful calculations. The whole dilemma of how to deal with the singular stress problem near crack tips has generated a fair size body of literature on the subject. Dugdale[2] addressed the problem in detail for metals, and Barenblatt[3] engaged the problem for brittle materials. The reader wishing to explore these issues in more detail should consult the advanced text by Kanninen and Popelar.[4] For the purposes of this treatment, we assume that the assumption of small-scale yielding holds.

4.1.1.1 Strain Energy Approach

There are now two basic approaches that can be taken in dealing with fracture and delamination problems that involve singular stress fields. One approach deals exclusively with the strain energy because it is always finite. Fracture and delamination processes are driven by elastic energy built up within the solid. By fracturing or delaminating, the solid dissipates this energy and thus achieves a more stable thermodynamic state. The basic assumption in this approach is that a given solid can sustain the presence of a crack flaw if the elastic energy present is insufficient to cause the crack to propagate. In other words, it requires a certain amount of energy to propagate a crack in a solid because new surface area will be created, and there is an energy cost for this. If the elastic energy stored in the solid near the crack tip is not high enough, then the crack is stabilized and will not propagate further. If, however, the local strain energy is increased beyond a certain critical level, then sufficient energy is available to create the new surface area, and the crack can propagate. The concept of a strain energy release rate is introduced to quantify the amount of elastic strain energy that is released when an existing crack is allowed to propagate by some increment.

Thus, consider the crack illustrated in Figure 4.2. Assume the plate has some thickness w, and the crack propagates to the right by some increment da. Then, an increment of area dA = wda is created. Assume also that on propagating the strain, energy in the plate drops by an amount $-dU$. The strain energy release rate G is now defined as

$$G = -\frac{dU}{dA} = -\frac{1}{w}\frac{dU}{da} \tag{4.6}$$

The quantity G is thus a measure of the driving force available within the solid caused by built up elastic energy for extending an existing crack. Another way of looking at it is that G is a measure of how efficiently the body can convert stored up elastic energy into the surface energy required for crack propagation. It is also assumed that, under an assumed set of circumstances, G must reach a certain critical level G_c for the crack to propagate. G_c is assumed to be a material property of the solid that can be determined by experiment. Thus, the following simplified criteria for crack propagation in a solid are derived:

$G < G_c$ Crack stabilized

$G = G_c$ Marginal stability　　　　　　　　　　　　　　　　　(4.7)

$G > G_c$ Unstable crack propagation

In a practical engineering design calculation, Equations (4.6) and (4.7) may be used to predict whether a certain coating will adhere sufficiently to some well-defined substrate. The first step is to use the field equations of continuum theory to calculate the stress field in the coating-substrate composite system. In most practical applications, numerical methods such as finite element analysis must be used to accomplish this task. When this step is completed, all the components of the stress tensor will be known everywhere in the system under investigation. For a coating on a substrate, close attention will be paid to the edges of the coating, where a singular stress condition can exist. Once all the stress components are known, Equation (4.4) can then be used to compute the total elastic energy locked up in the system.

Assuming that a particular edge region of the coating has been identified as a likely spot for delamination to occur, a second complete calculation of the stored elastic energy is performed by which the boundary condition of the problem is modified to allow the coating to delaminate some small increment at the edge location under investigation. If the coating is under a tensile stress, then it will be found that allowing the coating to delaminate at the edge will cause the calculated stored elastic energy to drop by some amount ΔU. Because the geometry of the situation is fully known, the amount of surface area created ΔA can also be readily calculated; thus, from Equation (4.6) the strain energy release rate can be estimated as

$$G = \frac{\Delta U}{\Delta A} \quad (\lim \Delta A \to 0) \Rightarrow \frac{dU}{dA} \qquad (4.8)$$

Two important pieces of information can be derived from Equation (4.8). The first is the magnitude of the driving force for a delamination event to occur in the system under investigation. This has to be compared to an experimental determination of the adhesion strength of the coating made by using one or more of the experimental methods discussed in Chapter 2. For example, G might be obtained from blister test data, which will give an estimate of G_c that is the minimum value of G required for delamination to occur. Equations (4.7) can then be used to predict whether delamination is possible.

The second piece of information that can be obtained from Equation (4.8) concerns the stability properties of the crack or delamination studied. If it turns out that increasing the length of the crack causes G to increase, then an unstable situation exists regardless of the magnitude of G. For example, a particularly dangerous situation can exist in glass and ceramic samples when stress corrosion mechanisms are at work. Water vapor and glass are classic examples of this type of situation. It is well known that for a glass sample under stress the presence of moisture can cause the fracture resistance of the glass to be dramatically lower than it would be under dry conditions.

If we imagine a glass example under stress with a preexisting surface crack, then the following type of situation can occur. Assume at time zero the crack is arrested because the driving force for propagation is less than the fracture strength G_c of the glass. Next, assume that the sample is introduced into a moist environment where water can condense on the surface of the glass and subsequently diffuse to the tip of the surface crack. Under these conditions, the local fracture resistance of the glass near the crack tip is significantly reduced to the point at which the crack can propagate by some small increment. Because the new glass surface is dry, it has a higher fracture strength, and the crack again is arrested. However, the crack is now somewhat longer, and the driving force G is now larger than it was before by our initial assumption. Furthermore, more moisture can diffuse in from the surface again, weakening the glass near the crack tip and thus allowing further lengthening of the crack. It is easy to see at this point that this process is autocatalytic in that the crack will grow at an increasing rate until G exceeds the critical strength of the dry glass, at which point catastrophic failure will occur.

4.1.1.2 Stress Intensity Factor Approach

A second school of thought insists on working with the singular stress concept as opposed to dealing with the strain energy. The stress approach has the advantage of revealing more of the details of what is happening at the crack front as opposed to the energy approach, which tends to cover over details. Going back to Equation (4.1), the stress analyst will point out that what is important is not the inverse square root singularity, which is common to all cracks, but the prefactor K_I, which is defined in terms of the applied stress and critical geometric factors, such as the crack length. It is these factors that determine whether a crack is dangerous because it is well known that the higher the stress level and the longer the crack, the more likely the occurrence of catastrophic failure is. The factor K_I, commonly called the mode I stress intensity factor, is taken as a measure of the strength of the crack singularity and is used as a parameter to distinguish between different types of cracks. It is assumed that the larger K_I becomes, the more likely it is that a given crack will propagate. Furthermore, each material is assumed to have a certain critical fracture toughness K_{I_c} that marks the boundary between cracks that are arrested and cracks that can cause catastrophic failure. Thus, similar to Equation (4.7), we have the following criteria for crack stability in terms of the stress intensity parameter:

$$K_I < K_{I_c} \quad \text{Crack stabilized}$$

$$K_I = K_{I_c} \quad \text{Marginal stability} \tag{4.9}$$

$$K_I > K_{I_c} \quad \text{Unstable crack propagation}$$

Adhesion Measurement Methods: Theory and Practice

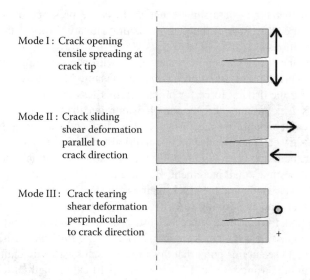

Mode I : Crack opening
          tensile spreading at
          crack tip

Mode II : Crack sliding
           shear deformation
           parallel to
           crack direction

Mode III : Crack tearing
            shear deformation
            perpindicular
            to crack direction

FIGURE 4.3 Three basic modes of crack propagation: Mode I represents tensile pulling apart of the crack surfaces. Mode II is a shear sliding in the direction of crack propagation. Mode III is a shear sliding perpendicular to the direction of crack propagation. Modes I and II are often referred to as the plane problem and Mode III as the antiplane problem.

The stress intensity approach to engineering design differs significantly from the strain energy method. By this method, one focuses on the stress field in a small region near the crack tip instead of computing the stress everywhere, as in the strain energy approach. As usual for most problems of practical interest, numerical methods must be used to compute the stress field. When using the finite element method, a special class of singular elements is used at the crack tip that takes into account the inverse square root dependence of the stress field. Special features of the code also estimate the best value of the stress intensity factor as part of the final result. Equations (4.9) are then used to compare the resulting value of K_I to the critical fracture toughness of the material K_{I_c}, which can be measured by a number of methods, such as the double cantilevered beam technique.

When using the stress analysis approach, a number of details are apparent that are buried by the energy method. The most significant detail is the fact that crack propagation can proceed by three separate mechanisms, commonly referred to as modes I, II, and III, with corresponding stress intensity factors K_I, K_{II}, and K_{III}. Figure 4.3 illustrates these modes for the case of fracture in the bulk of a solid material.

It is remarkable, but after reflection not too surprising, that there is a definite connection between the energy and stress approaches to crack propagation. In fact, the stress intensity factors can be related to corresponding strain energy release rates by the following simple formulae*:

---

* See page 119 of Broek[1] for an elementary derivation.

$$G_I = \left(1 - v^2\right)\frac{K_I^2}{E}$$

$$G_{II} = \left(1 - v^2\right)\frac{K_{II}^2}{E} \qquad (4.10)$$

$$G_{III} = \left(1 + v\right)\frac{K_{III}^2}{E}$$

The total strain energy release rate G is the simple sum of the three components given in Equation (4.10).

4.2 FRACTURE MECHANICS AS APPLIED TO PROBLEMS OF ADHESION

4.2.1 ELEMENTARY COMPUTATIONAL METHODS

Chapter 3 devotes many pages to developing the lore of continuum mechanics as applied to solid elastic bodies. Everything developed there forms the foundation for fracture mechanics theory, which in turn is the basis for all quantitative analysis of adhesion strength. This section explores in some detail the delamination of a thin coating from a rigid substrate due to the internal stress in the coating. This can apply to paints on steel, dielectrics on silicon, or any of a large variety of situations involving a thin elastic coating on an effectively rigid substrate. The problem is easy to state and has a simple answer, but the full development from basic principles is surprisingly involved. However, by going through this one example in detail, the reader should acquire a full appreciation and understanding of what is involved in carrying out a nontrivial fracture mechanics calculation.

4.2.1.1 Basic Model of Thin Coating on Rigid Disk

Figure 4.4 illustrates the conceptual model for treating a thin coating on a rigid substrate. We follow the development given by Thouless[5] with the exception that Equations (4) from that reference will be derived in some detail here as opposed to conjuring them "deus ex machina."* The following assumptions are made throughout:

1. A thin film of a homogeneous, isotropic, elastic material is deposited on a massive and rigid substrate that is effectively immovable.
2. A state of biaxial uniform stress σ_0 exists because of processing or other intrinsic condition. An example would be residual thermal expansion mismatch stress caused by some high-temperature curing process as given by Equation (3.50).
3. From Items 1 and 2, we conclude that the coating is in a state of plane stress as defined by Equation (3.16).

---

* Note that Equation (4) from Dugdale[2] is clearly wrong because of a misprint. The equation for σ_r should have a plus sign.

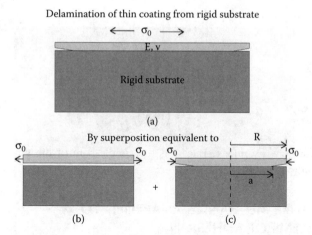

Delamination of thin coating from rigid substrate

FIGURE 4.4 Coating on a rigid substrate treated by the superposition method. Configuration (a) is equivalent to the superposition of configurations (b) and (c).

In Figure 4.4, we have a thin coating on a rigid disk that is delaminating from the edges because of the internal stress σ_0. The problem is to determine the driving force for delamination G as defined by Equation (4.6) so that the stability of the coating can be evaluated according to Equations (4.7). The first step has to be the determination of the elastic energy in the coating as a function of the delamination crack length. This requires that an appropriate boundary value problem be set up to determine all the stresses and strains so that the elastic energy can be evaluated using the standard formula

$$U^T = \frac{1}{2} \int_V \sigma_{ij} \varepsilon_{ij} dV \qquad (4.11)$$

In Chapter 3, the film stresses for the undelaminated coating are derived, leading to the simple formula Equation (3.50). This case was very simple because there were no edges or other discontinuities, and the continuum equations given in Table 3.1 collapsed to essentially three simple constitutive relations. Things are very different now in that edges and boundaries cannot be avoided, and we have to take into account the full set of field equations given in Table 3.1.

At this point, realize that in solving these kinds of problems there is rarely, if ever, a straightforward formulaic approach that can be blindly followed to determine the final solution. Rather, one inevitably has to rely on some physical insight or other trick to hammer the problem into a manageable form. The accomplished continuum mechanic carries around a large toolbox full of such insights and tricks that have filled the mechanics literature for the past 100 years.

In the current problem, the insight is shown in Figure 4.4b, which relies on the fact that, because the problem is linear, it can be reduced to the sum of two simpler ones. Step 1 is to consider the coating lifted off the substrate and put into biaxial

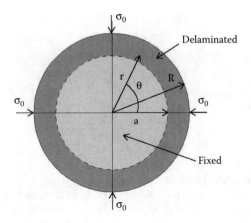

FIGURE 4.5 Top view of Figure 4.4.

tension by applying a tensile stress σ_0 to the boundary. This gets us to the simple biaxial tension case mentioned above, which is readily solved but is not of interest by itself. In the next step, the coating is reattached to the substrate but only out to the delamination radius **a** as shown in the figure. In addition, a compressive stress $-\sigma_0$ is applied to the boundary to counter the tensile stress applied in Step 1 and leave the boundary in the required state of zero radial stress. Figure 4.5 depicts the situation in more detail. Because eventually we will take the derivative of the strain energy to compute the driving force G as specified by Equation (4.6), we can thus ignore the constant stress configuration shown in Figure 4.5 and focus on Figure 4.4c alone. Figure 4.4c has two regions, an inner core with zero stress because of the applied boundary condition of fixed radial displacement at $r = a$ and an outer detached annulus region. The problem thus reduces to determining the stress distribution in the annulus ring shown in Figure 4.5 subject to the boundary conditions that the inner diameter at radius a is fixed and therefore has zero displacement, and the outer diameter is subject to a radial compressive stress $-\sigma_0$.

The problem is now reduced to the classic situation of a ring with a fixed inner diameter and subjected to a uniform external pressure σ_0. The boundary conditions can thus be written explicitly as

$$u_r(r)\Big|_{r=a} = 0$$

$$\sigma_r(r)\Big|_{r=R} = \sigma_0$$

(4.12)

Now, the only thing remaining is to determine a general solution for the stresses and displacements in a ring. For this, the starting point is the basic field equations in Table 3.1. We again use the stress function approach to solve this problem as done in Chapter 3, Section 3.4, for the cantilevered beam. We recall from Equations (3.74) and (3.75) that the stresses may be defined in terms of a scalar stress function ϕ with the following properties:

$$\sigma_{xx} = \frac{\partial^2 \phi}{\partial y^2}$$

$$\sigma_{yy} = \frac{\partial^2 \phi}{\partial x^2} \qquad (4.13)$$

$$\sigma_{xy} = \frac{-\partial^2 \phi}{\partial x \partial y}$$

where $\nabla^4 \phi = 0$.

Because our problem involves the stresses in a disk, it is a very good idea to convert from Cartesian (x, y, z) to cylindrical (r, θ, z) coordinates, which makes the critical task of satisfying the boundary conditions much easier. The form of this operator in cylindrical coordinates can be looked up in any number of mathematical handbooks,[6] and it is certainly one of the basic tools that the well-prepared continuum mechanic should have in the toolbox.* The Laplacian operator in three dimensions has the following form:

Cartesian Coordinates

$$\text{a)} \quad \nabla^2 \phi = \left(\frac{\partial^2 \phi}{\partial^2 x} + \frac{\partial^2 \phi}{\partial^2 y} + \frac{\partial^2 \phi}{\partial^2 z} \right)$$

$$(4.14)$$

Cylindrical Coordinates

$$\text{b)} \quad \nabla^2 \phi = \frac{1}{r} \frac{\partial}{\partial r} \left(r \frac{\partial \phi}{\partial r} \right) + \frac{1}{r^2} \frac{\partial^2 \phi}{\partial \theta^2} + \frac{\partial^2 \phi}{\partial z^2}$$

Because our problem is symmetric about the z axis, the stress function will be a function of the radial coordinate r alone, and the expression for $\nabla^4 \phi$ reduces to

$$\frac{1}{r} \left[\frac{\partial}{\partial r} \left(r \frac{\partial}{\partial r} \left(\frac{1}{r} \frac{\partial}{\partial r} \left(r \frac{\partial \phi}{\partial r} \right) \right) \right) \right] = 0 \qquad (4.15)$$

Equation (4.15) is convenient because, as luck would have it, the nested derivatives can be directly integrated in the following sequence:

Multiply both sides of Equation (4.15) by r and integrate the resulting exact differential to obtain

$$r \frac{\partial}{\partial r} \left[\frac{1}{r} \frac{\partial}{\partial r} \left(r \frac{\partial \phi}{\partial r} \right) \right] = C_1 \qquad (4.16)$$

---

* See also Appendix A, Formula 29, in Table A.2.

Next, divide both sides of Equation (4.16) by r and integrate both sides once more:

$$\int \frac{\partial}{\partial r}\left(\frac{1}{r}\frac{\partial}{\partial r}\left(r\frac{\partial \phi}{\partial r}\right)\right)dr = \int \frac{C_1}{r}dr$$

$$\left[\frac{1}{r}\frac{\partial}{\partial r}\left(r\frac{\partial \phi}{\partial r}\right)\right] = C_1 \ln r + C_2$$

(4.17)

Again, follow the same procedure: Multiply both sides of Equation (4.17) by r and integrate again:

$$\int \left[\frac{\partial}{\partial r}\left(r\frac{\partial \phi}{\partial r}\right)\right]dr = \int \left(C_1 r \ln r + C_2 r\right)dr$$

$$\left(r\frac{\partial \phi}{\partial r}\right) = C_1\left(r^2 \ln r - \frac{r^2}{2}\right) + C_2 \frac{r^2}{2} + C_3$$

(4.18)

One last procedure finishes the job. Divide both sides of Equation (4.18) by r and integrate to get

$$\int \frac{\partial \phi}{\partial r}dr = \int \left(C_1\left(r \ln r - \frac{r}{2}\right) + C_2 \frac{r}{2} + \frac{C_3}{r}\right)dr$$

$$\phi = C_1\left(\frac{r^2}{4}\ln r - \frac{r^2}{4}\right) + C_2 \frac{r^2}{4} + C_3 \ln r + C_4$$

(4.19)

Now, because all of the constants of integration are arbitrary and will ultimately be determined by the boundary conditions, we can collect all terms of like functionality in r together and combine and absorb all the constants into a new set of redefined constants. This leads to the following general solution for the stress function that satisfies Equation (4.13), as can also be seen by direct substitution:

$$\phi = C_1^* r^2 \ln r + C_2^* r^2 + C_3 \ln r + C_4$$

(4.20)

Equation (4.20) is quite likely the easiest nontrivial solution one will ever obtain for a fourth-order differential equation. However, just when you think it is safe to quit and go have a beer, it turns out that there is one more difficulty to be overcome. It turns out that the term involving the constant C_1^* gives rise to nonphysical multivalued displacements; therefore, this constant must be set to zero. The pioneers in the field had to find this out the hard way[7] by computing the stresses and strains and then integrating to determine the displacements. However, we are sparing you that trauma by eliminating this pesky term right away. Thus, we can write the final general form of the stress function suitable for the ring problem as follows and further eliminate the superscript * to redefine the arbitrary constants:

$$\phi = C_1 r^2 + C_2 \ln r + C_3 \tag{4.21}$$

The next step is to compute the stresses and displacements using Equations (4.13). Because we are working in cylindrical coordinates, these relations have to be converted to polar form. This is a standard procedure and can be looked up in standard reference works.[7] The result for the two-dimensional cylindrical problem is

$$\sigma_{rr} = \frac{1}{r}\frac{\partial \phi}{\partial r} + \frac{1}{r^2}\frac{\partial^2 \phi}{\partial \theta^2}$$

$$\sigma_{\theta\theta} = \frac{\partial^2 \phi}{\partial r^2} \tag{4.22}$$

$$\sigma_{r\theta} = \frac{1}{r^2}\frac{\partial \phi}{\partial \theta} - \frac{1}{r}\frac{\partial^2 \phi}{\partial r \partial \theta}$$

These equations can now be used to compute the ring stresses from Equation (4.21):

$$\sigma_r = C_1 + \frac{C_2}{r^2}$$

$$\sigma_\theta = C_1 - \frac{C_2}{r^2} \tag{4.23}$$

In the above equations, we dropped the double index notation (rr) and ($\theta\theta$) in favor of the simpler r and θ indices. Note also that $\sigma_{r\theta}$ is zero because the stress function is independent of θ. We can now determine the elastic strains from the constitutive relations Equations (3.31) written in polar form[7]:

$$\varepsilon_r = \frac{1}{E}\left[\sigma_r - \nu\sigma_\theta\right]$$

$$\varepsilon_\theta = \frac{1}{E}\left[\sigma_\theta - \nu\sigma_r\right] \tag{4.24}$$

The strains are now determined by eliminating the stresses using Equations (4.23):

$$\varepsilon_r = \frac{1}{E}\left[C_1(1-v)+(1+v)\frac{C_2}{r^2} \right]$$

$$\varepsilon_\theta = \frac{1}{E}\left[C_1(1-v)-(1+v)\frac{C_2}{r^2} \right]$$

(4.25)

The final quantity we need is the radial displacement required to satisfy the boundary conditions Equations (4.12). This is easily computed using the radial strain from above and the following relatively intuitive relation for u_r[7]:

$$\varepsilon_r = \frac{du_r}{dr}$$

(4.26)

The radial displacement is now readily computed by integrating Equation (4.26) using ε_r from Equation (4.25).

$$u_r = \frac{1}{E}\left[C_1(1-v)r-(1+v)\frac{C_2}{r} \right]$$

(4.27)

Notice that the constant of integration introduced in Equation (4.27) can be a function of θ only and is therefore set to zero by axisymmetry. At this stage, it is helpful to collect the results derived to this point in one place:

a) $\sigma_r = C_1 + \dfrac{C_2}{r^2}$

b) $\sigma_\theta = C_1 - \dfrac{C_2}{r^2}$

c) $\varepsilon_r = \dfrac{1}{E}\left[C_1(1-v)+(1+v)\dfrac{C_2}{r^2} \right]$ (4.28)

d) $\varepsilon_\theta = \dfrac{1}{E}\left[C_1(1-v)-(1+v)\dfrac{C_2}{r^2} \right]$

e) $u_r = \dfrac{1}{E}\left[C_1(1-v)r-(1+v)\dfrac{C_2}{r} \right]$

Equations (4.28) form a fairly complete set of results for the elastic properties of the axisymmetric ring problem. Equations (4.28a) and (4.28e) are now substituted into the boundary conditions Equations (4.12) to determine the as-yet-undetermined constants C_1 and C_2 as follows:

$$\left[C_1 + \frac{C_2}{R^2} \right] = \sigma_0$$

$$C_1 \frac{(1-v)a}{E} - \frac{(1+v)}{E} \frac{C_2}{a} = 0$$

(4.29)

Equations (4.29) are readily solved for C_1 and C_2:

$$C_1 = \frac{\sigma_0}{D}; \, C_2 = \beta a^2 \frac{\sigma_0}{D}$$

(4.30)

where

$$\beta = \frac{(1-v)}{(1+v)}; \, D = 1 + \beta \left(\frac{a}{R} \right)^2$$

Equations (4.30) combined with Equations (4.28) complete the solution for the stresses, strains, and displacements for the ring with fixed inner radius subjected to an external pressure σ_0. The next step is to calculate the elastic strain energy in the ring caused by the applied stress. First, start with the strain energy density for an elastic body under strain:

$$U = \frac{1}{2} \sum_{ij} \sigma_{ij} \varepsilon_{ij}$$

(4.31)

For the current problem, only the main diagonal components of the stress tensor in the r and θ directions are nonzero, so Equation (4.31) simplifies to

$$U = \frac{1}{2} \left[\sigma_r \varepsilon_r + \sigma_\theta \varepsilon_\theta \right]$$

(4.32)

Using the constitutive equations, Equations (4.24), the strains can be eliminated in favor of the stresses, giving:

$$U = \frac{1}{2E} \left[\sigma_r^2 + \sigma_\theta^2 - 2v\sigma_r \sigma_\theta \right]$$

(4.33)

The relations for the stresses given in Equations (4.28) can now be substituted into Equation (4.33) using the values of C_1 and C_2 given by Equation (4.30):

$$U = \frac{(1-v)\sigma_0^2}{ED^2} \left(1 + \beta \left(\frac{a}{r} \right)^4 \right)$$

(4.34)

where

$$\beta = \frac{(1-v)}{(1+v)}; \quad D = 1 + \beta \left(\frac{a}{R}\right)^2$$

Remembering that U is only the elastic energy density stored in the ring (energy per unit volume), we still need to integrate Equation (4.34) over the entire ring volume to get the total elastic energy U^T stored in the ring:

$$U^T = \frac{2\pi h (1-v)\sigma_0^2}{ED^2} \int_a^R \left[1 + \beta \left(\frac{a}{r}\right)^4\right] r\, dr \qquad (4.35)$$

where h = ring thickness.

The integration in Equation (4.35) is carried out in cylindrical coordinates where the θ and z integrations are immediate giving the factor $2\pi h$, leaving only the radial component to be finished off. This integration is elementary, if somewhat messy, and gives the following final result for the total strain energy:

$$U^T = \frac{\pi h R^2 (1-v)\sigma_0^2}{ED} \left[1 - \left(\frac{a}{R}\right)^2\right] \qquad (4.36)$$

Coming into the home stretch, we can now derive the driving force for delamination G of the coating in Figure 4.5 using the definition Equation (4.8). Referring to the figure, we see that the delaminated area is

$$A = \pi \left(R^2 - a^2\right)$$

$$\mapsto dA = -2\pi a\, da$$

$$(4.37)$$

Combining Equations (4.8), (4.36), and (4.37), we get the following formula for the driving force G

$$G = \frac{-1}{2\pi a} \frac{dU^T}{da} = \frac{-1}{2\pi a} \frac{\pi h R^2 (1-v)\sigma_0^2}{E} \frac{d}{da} \left[\frac{\left(1 - \left(\frac{a}{R}\right)^2\right)}{\left(1 + \beta \left(\frac{a}{R}\right)^2\right)}\right] \qquad (4.38)$$

The differentiation implied in Equation (4.38) is another little workout in elementary calculus that on completion gives the much-awaited final result:

$$G = \frac{h\sigma_0^2}{E}\left[\frac{2\beta}{\left(\left(1+\beta\left[\frac{a}{R}\right]^2\right)^2\right)}\right] \tag{4.39}$$

$$\beta = \frac{1-\nu}{1+\nu}$$

One might think that after such an involved development the final result would be much more complicated than Equation (4.39). The key factor complicating this calculation is that we started ab initio from the fundamental equations of continuum theory, which lead to the derivation of Equations (4.28). From the point of view of the practicing fracture mechanic, these equations would be considered as "well known" because they are basically a homework problem for a first-year graduate student. The main point to be made here is, for at least one problem, to present a fairly complete derivation starting from first principles. In the remainder of this chapter, we take the same attitude as the accomplished fracture mechanic and use existing results as a starting point for this type of calculation, fully mindful that the trail leading to the starting point is likely to have been long and tortuous.

Surprisingly, we can derive a bonus result from this calculation by contemplating the related problem of delamination at the edge of a circular cut in the central region of a thin uniform film on a rigid substrate as illustrated in Figure 4.6. Proceeding as in the edge delamination case, we imagine the circular hole cut as the superposition of two simpler situations as illustrated in Figure 4.6b and Figure 4.6c. In Figure 4.6b, the unattached coating with the circular hole is stretched in uniform biaxial tension to simulate the presence of a uniform film stress. In Figure 4.6c the film is reattached

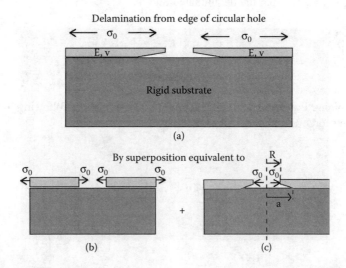

FIGURE 4.6 Delamination proceeding from a central hole in a constrained coating. This problem is essentially the flip side of Figure 4.4.

leaving a delaminated region out to radius "a" and a compressive stress σ_0 is applied to the hole circumference to cancel the uniform stress applied in Figure 4.6b and thus satisfy the stress free boundary condition at the edge of the hole. Now a moment of consideration will convince the reader that not only is this problem similar to the one shown in Figure 4.4 but, taking account of the revised definitions of the parameters R and a, the boundary conditions are also identical to Equations (4.12). Thus again we have a ring under compression at radius "R" and with fixed radial displacement at radius "a". The field equations will again have the same solution. The only thing that has changed is the interpretation of the quantities "R" and "a," which does not affect the formal mathematics used to derive our final result Equation (4.39). Thus we have the convenient and somewhat amusing result that Equation (4.39) can be used to describe either the delamination of a coating from an outer boundary or from the edge of an interior cut depending on how one interprets the parameters "R" and "a". Thus even though we worked hard to derive Equation (4.39) we wound up solving two interesting problems in a single sweep, which does not happen all that often in life. This is a fairly common property of formal mathematics in that equations per se have no particular meaning. The meaning is always supplied by the user who interprets the symbols according to some concrete model. Another example from elementary physics is the case of the damped harmonic oscillator. It can physically represent a number of different situations such as a mass attached to a spring or to a simple LRC circuit. Even though these are dramatically different physical situations the formal mathematics used to describe them is identical in every way, and by investigating one system we can draw valid conclusions concerning the other.[8]

Figure 4.7 and Figure 4.8 plot Equation (4.39) in normalized form to represent an edge delamination in the former and a delamination from a circular cut in the latter. The dimensionless driving force [G/G_0, where $G_0 = (h\sigma_0/E)$] is plotted versus the dimensionless crack length [$(1 - a/R)$ for the edge delamination and $(a/R - 1)$ for the circular cut].[9] Poisson's ratio v has been set to 1/3, which is a common value for most organic coatings such as paints and polymer insulators. The main feature to note from these plots is that the edge delamination is one of the dangerous failure modes in that the driving force increases with crack length. In such cases, subtle stress corrosion effects can cause a seemingly arrested crack to propagate stealthily to

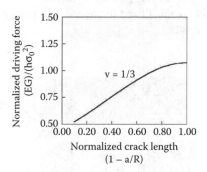

FIGURE 4.7 Plot of normalized driving force for an edge delamination according to Equation (4.39).

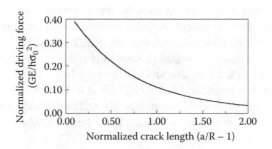

FIGURE 4.8 Plot of normalized driving force for delamination from a circular cut again using Equation (4.39) but reinterpreting the parameters R and a.

catastrophic failure. The delamination from an interior circular hole represents the opposite case of decreasing driving force with increasing crack length. Thus, one expects all delaminations of this kind eventually to become arrested.

4.2.2 DECOHESION NUMBER APPROACH OF SUO AND HUTCHINSON

The preceding section dealt with what might be called an elementary problem in fracture mechanics theory despite the lengthy and involved treatment. In what follows, a more advanced "industrial strength" model of fracture and delamination is considered based on the work of Suo and Hutchinson.[10] Figure 4.9 gives a schematic of the basic model. What is shown is a small section of a long crack in a coated substrate. Because the general problem of a short transient crack initiating from the very edge of a coating is quite complicated, Suo and Hutchinson choose to deal with the much simpler and also more practical case of a long crack that has reached steady-state conditions. The problem with short cracks initiating at the very edge of the coating is that they may start out as a simple delamination but then can easily dive into the substrate or the coating as cohesive cracks, depending on a variety of conditions. What is observed experimentally and confirmed theoretically is that such cracks tend to reach a fixed depth and then propagate along at that depth parallel to the sample surface.

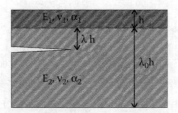

FIGURE 4.9 Schematic of the decohesion model of Suo and Hutchinson.

The model depicted in Figure 4.9 assumes that the crack has reached the equilibrium depth and will propagate parallel to the coating-substrate interface. The coating thickness represents the characteristic length of the system, and all other lengths are expressed as multiples of this length using the scaling parameters λ and λ_0. The parameter λ represents the depth of the crack, and λ_0 is the substrate thickness. Setting λ to zero gives a pure delamination problem. The final result of most significance to the delamination of coatings from substrates is the concept of a decohesion number for various coating-substrate systems that relates the mode I fracture toughness of the substrate material to the coating thickness and coating residual stress. An outline of the development of this relation is given here. Refer to Suo and Hutchinson's[10] original article for more complete details.

Using analysis techniques similar to those used in Section 4.2.2.1, Suo and Hutchinson developed equations for the strain energy release rate and stress intensity factors for the crack depicted in Figure 4.9. The analysis follows the work of Dundurs,[11] who introduced the following parameters for a bilayer laminate structure:

$$\alpha = \frac{\Gamma\left(\kappa_2 + 1\right) - \left(\kappa_1 + 1\right)}{\Gamma\left(\kappa_2 + 1\right) + \left(\kappa_1 + 1\right)}$$

$$\beta = \frac{\Gamma\left(\kappa_2 - 1\right) - \left(\kappa_1 - 1\right)}{\Gamma\left(\kappa_2 + 1\right) + \left(\kappa_1 + 1\right)}$$

$$(4.40)$$

These parameters are essentially measures of the elastic dissimilarity between the coating and the substrate materials, and both vanish when the elastic properties of the coating and substrate are the same. It turns out that, for current purposes, the main results of interest are independent of or only very weakly dependent on the parameter β, so it is ignored in the remaining discussion. The parameters κ and Γ are defined in terms of the shear moduli and Poisson ratios of the material layers:

$$\kappa_i = 3 - 4\nu_i \ \left(\text{plane strain}\right)$$

$$\kappa_i = \frac{3 - \nu_i}{1 - \nu_i} \ \left(\text{plane stress}\right)$$

$$\nu_i = \text{Poisson ratio of } i\text{th material } i = 1, 2 \qquad (4.41)$$

$$\Gamma = \frac{\mu_1}{\mu_2}$$

$$\mu_i = \text{Shear modulus of } i\text{th material } i = 1, 2$$

A final material parameter is the stiffness ratio, which is related to Dundurs' parameter α as follows:

$$\Sigma = \frac{c_1}{c_2} = \frac{1+\alpha}{1-\alpha}$$

$$c_i = (\kappa_i + 1)/\mu_i \quad (i = 1, 2)$$

(4.42)

The main results of Suo and Hutchinson's analysis are the following formulas for the strain energy release rate G and the Mode I and Mode II stress intensity factors K_I and K_{II}, respectively:

(Strain energy release rate)

$$G = \frac{c_2}{16}\left[\frac{P^2}{Uh} + \frac{M^2}{Vh^3} + 2\frac{PM\sin(\gamma)}{h^2\sqrt{UV}}\right]$$

(Mode I and Mode II Stress Intensity Factors)

(4.43)

$$K_I = \frac{P}{\sqrt{2Uh}}\cos(\omega) + \frac{M}{\sqrt{2Vh^3}}\sin(\omega + \gamma)$$

$$K_{II} = \frac{P}{\sqrt{2Uh}}\sin(\omega) - \frac{M}{\sqrt{2Vh^3}}\cos(\omega + \gamma)$$

where

$$G = \frac{c_2}{8}\left(K_I^2 + K_{II}^2\right)$$

Equations (4.43) appear to be fairly simple and compact, but they in fact hide a lot of complexity. In the first place, a number of subsidiary relations are required to define precisely the as-yet-unspecified quantities P, M, U, V, ω, and γ in terms of the basic elastic constants and scale factors λ and λ_0. The following subsidiary functions are required, which depend on the substrate crack depth parameter λ and the elastic constants:

Substrate neutral axis position:

$$\Delta(\lambda) = \frac{\lambda^2 + 2\Sigma\lambda + \Sigma}{2(\lambda + \Sigma)}$$

(4.44)

Dimensionless beam moment of inertia:

$$I(\lambda) = \frac{1}{3}\left[\Sigma\left[3(\Delta - \lambda)^2 - 3(\Delta - \lambda) + 1\right] + 3\Delta\lambda(\Delta - \lambda) + \lambda^3\right]$$

(4.45)

Dimensionless effective cross section:

$$A(\lambda) = \lambda + \Sigma \tag{4.46}$$

The following quantities related to the substrate thickness factor λ_0 are defined in terms of Equations (4.44) to (4.46) as follows:

$$\Delta_0 = \Delta(\lambda_0)$$

$$I_0 = I(\lambda_0) \tag{4.47}$$

$$A_0 = A(\lambda_0)$$

Further required functions of λ and λ_0 are defined as follows:

$$\frac{1}{U} = \frac{1}{A} + \frac{1}{\lambda_0 - \lambda} + \frac{12\left[\Delta + (\lambda_0 - \lambda)/2\right]^2}{(\lambda_0 - \lambda)^3}$$

$$\frac{1}{V} = \frac{1}{I} + \frac{12}{(\lambda_0 - \lambda)^3} \tag{4.48}$$

$$\frac{\sin(\gamma)}{\sqrt{UV}} = \frac{12\left[\Delta + (\lambda_0 - \lambda)/2\right]^2}{(\lambda_0 - \lambda)^3}$$

The model set up by Suo and Hutchinson is quite general and can be configured to include a varied set of loading conditions involving both bending (M) and tensile (P) loads. For present purposes, we specialize to the case for which the load is caused by the uniform residual stress σ existing in the coating layer of thickness h. For this condition to be realized, the loads P and M must be set as follows:

$$P = \sigma h \left[1 - C_1 - C_2 \left(1/2 + \lambda_0 - \Delta_0 \right) \right]$$

$$M = \sigma h^2 \left[\left(1/2 + \lambda - \Delta \right) - C_3 \left(1/2 + \lambda_0 - \Delta_0 \right) \right] \tag{4.49}$$

$$C_1 = \frac{A}{A_0}; \, C_2 = \frac{A}{I_0}\left[\left(1/2 + \lambda_0 - \Delta_0 \right) - \left(\lambda - \Delta \right) \right]; \, C_3 = \frac{I}{I_0}$$

Equations (4.43) to (4.49) conclude the formal definition of the strain energy release rate and the stress intensity factors K_I and K_{II}; however, we are not out of the woods yet, not by a long shot. Careful inspection of these equations reveals that there are still two quantities that remain to be defined in terms of either the elastic constants or the model geometry, and they are the constants ω and γ. Curiously, the constant ω is the most difficult to determine* but causes us the fewest problems because fortunately most results of interest are insensitive to its exact value. Suo and Hutchinson heroically studied the dependence of the crack depth and other quantities on ω and compiled useful tables for computational purposes. The final upshot is that for most purposes ω can be set to 52° and then forgotten, which is the path we follow from this point.

The parameter γ is not nearly as difficult to determine but does require a significant amount of work. As pointed out above, the crack investigated started out at the edge of the sample and through a process of seeking out conditions of maximum energy release rate settled down at some equilibrium depth below the coating/substrate interface. From general fracture mechanics, we know that the steady state of crack propagation is pure mode I crack opening. Thus, the condition for this to occur is that the mode II stress intensity should be zero, which can be written as

$$K_{II} = 0 \qquad (4.50)$$

Referring to the third of Equations (4.43), we see that this condition can be satisfied if

$$K_{II} = \frac{P}{\sqrt{2Uh}}\sin(\omega) - \frac{M}{\sqrt{2Vh^3}}\cos(\omega + \gamma) = 0 \qquad (4.51)$$

or

$$\frac{\cos(\omega + \gamma)}{\sin(\omega)} = \frac{Ph}{M}\sqrt{\frac{V}{U}}$$

Equation (4.51) can now be used to determine γ, which determines the equilibrium crack depth at which the crack propagates. Using Equation (4.51) to determine the equilibrium crack depth and fixing ω at the suggested value of 52°, one can now proceed to examine delamination and crack propagation in coating-substrate systems of interest. Figure 4.10 plots equilibrium crack depth versus substrate thickness for two systems of practical interest: nickel on glass and polyimide on glass-ceramic (GC). Recall that what is plotted are the scaling parameters λ and λ_0, which when multiplied by the actual coating thickness h give the actual crack depth and substrate thickness.

Table 4.1 summarizes the elastic properties of the materials of interest. Again referring to Figure 4.10, we see that the nickel/glass system gives a positive Dundurs'

---

* Determining ω precisely requires an exact solution of the crack problem involving the tackling of a rather involved and technical integral equation. It is all doable but very heavy sledding indeed. The ambitious and fearless will have to study Suo and Hutchinson[10] directly for details.

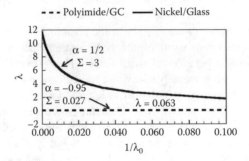

FIGURE 4.10 Plot of the crack depth scale factor versus the inverse of the substrate thickness scale factor for two common coating/substrate systems. Note that the crack depth for the polyimide/GC system is wholly insensitive to the substrate thickness, whereas the nickel/glass system is sensitive for relatively thick substrates.

TABLE 4.1
Nominal Thermoelastic Properties of Materials Considered in Figure 4.10

Material	Young's Modulus E (GPa)	Shear Modulus μ^a (GPa)	Poisson Ratio ν	Thermal Expansion Coefficient α (ppm/K)
Nickel	210	80.8	0.3	13.0
Glass	70	28	0.25	0.5
Standard[b] polyimide (PI)	3	1.15	0.3	30.0
Low TCE polyimide	14	5.38	0.3	6
Glass-ceramic (GC)	110	44	0.25	3.0

[a] Estimated from the classic formula $\mu = E/2(1 + \nu)$.
[b] These are values typical of a wide range of what could be referred to as "run of the mill" polyimides.

parameter of 0.5, indicating the high stiffness of the nickel relative to the glass. It is remarkable how sensitive the steady-state crack depth is for this system, especially for thick substrates. The case of polyimide on GC illustrates the opposite case of a flexible material on a rigid substrate, giving a negative Dundurs' parameter of –0.95. This system shows almost no dependence on substrate thickness and gives a near-constant value of λ near 0.063. For thin films, the decohesion of the GC substrate can easily be mistaken for clean delamination on casual inspection.

4.2.2.1 The Decohesion Number

From the point of view of the applications engineer interested in solving design problems for which delamination processes can pose a serious obstacle, the most useful result of the analysis by Suo and Hutchinson is the concept of a decohesion number. The basic idea is to come up with a method for determining whether a particular coating/substrate system will be stable regarding delamination/decohesion behavior. The first step is to focus on the steady-state behavior of large cracks as opposed to the transient behavior of small flaws for two reasons: (1) The transient behavior of small flaws is extremely tricky, and we really do not have a fully adequate theory at present; (2) the large steady-state crack is what will in the end destroy any particular fabricated item, and if you can demonstrate that large cracks cannot propagate, then only small arrested flaws can exist, which presumably will only amount to cosmetic defects and not affect the use and functionality of the intended product.

With the focus now on steady-state cracking, we can refer to Equations (4.43), which give the stress intensity factors for crack propagation. At steady state, K_{II} is zero, and K_I is given by

$$K_I = \frac{P}{\sqrt{2Uh}}\cos(\omega) + \frac{M}{\sqrt{2Vh^3}}\sin(\omega + \gamma) \tag{4.52}$$

Further, from Equations (4.49) we see that $P \sim \sigma h$ and $M \sim \sigma h^2$. Inserting these proportionalities into Equation (4.52), we see that K_I must have the following functional form:

$$K_I = \sigma\sqrt{h}\,\Omega(\alpha, \lambda_0) \tag{4.53}$$

Thus, the mode I stress intensity factor for steady-state crack propagation is equal to the coating stress times the square root of the coating thickness times some dimensionless function Ω that depends on the elastic constants through the Dundurs' parameter α and the substrate thickness scale factor λ_0. The function Ω is called the decohesion number and must be calculated using the procedure outlined above. This is not terribly difficult but is definitely a pain and time consuming to boot. Fortunately, help is close at hand. Suo and Hutchinson determined that, for a fixed value of the Dundurs' parameter, Ω is quite insensitive to λ_0 for values between 10 and infinity. The decohesion number is, however, sensitive to the stiffness ratio Σ defined in Equations (4.42). To alleviate this problem, they gave the following handy numerical fitting formulae to determine Ω over a wide range of values of Σ:

$$\Omega = (3.50\Sigma - 0.63)^{1/2} \quad 1 \le \Sigma \le 10$$
$$\Omega = 0.944\Sigma^{-0.4} - 0.63 \quad 0.1 \le \Sigma \le 1 \tag{4.54}$$

4.2.3 BACK-OF-THE-ENVELOPE CALCULATIONS

With Equations (4.53) and (4.54) in hand, we are now in a position to do some real-world engineering design calculations with the intent of preventing delamination problems before they occur. If you talk with nearly anyone in product development and manufacturing, you learn that it is much cheaper by orders of magnitude to prevent manufacturing defects from occurring in the development stage rather than let them pop up when full-scale manufacturing is under way. And lord help you if they occur after the product has been shipped to real customers. So, with this in mind let us look at a "hypothetical" problem that actually arose during my long experience with development problems in the microelectronics industry. It came to pass that in those days it was decided that it would be a good idea to use a low dielectric constant polyimide material as a high-performance insulator on top of a GC substrate for a particular chip-packaging application. There was significant concern about doing this because it was known from previous hard experience that polyimde materials could develop high residual stresses because of their curing process and that the GC material, although an excellent dielectric, was not particularly strong mechanically. Management wanted to know whether this product could be manufactured safely; furthermore, they wanted an answer quickly because time was always an issue.

4.2.3.1 Polyimide on Glass-Ceramic

The following problem was put to the thin film design engineers:

> It is proposed to coat a glass-ceramic substrate approximately 1.4-mm thick with a dielectric layer of polyimide approximately 175-μm (0.175-mm) thick. What kind of problems can be expected in terms of potential delamination or cracking? What steps can be taken to avoid these problems?

We can approach these questions in a step-by-step fashion. From Equation (4.53), we know that the stress intensity factor K_I for mode I cracking in the GC substrate depends on the thickness of the polyimide coating h and the residual stress in the coating σ. We can use this formula to calculate whether a residual stress in the polyimide coating will cause a crack to propagate in the GC substrate. The thickness has been specified as part of the proposed design specification, so we need to determine the residual stress. This can be estimated on the basis of past experience with these materials or by consulting with a polyimide guru. The most general class of polyimide materials develops internal stress when coated on a stiff substrate because of the thermal expansion mismatch strain induced during the curing process. These materials are generally coated as a viscous liquid [known to chemists as a poly(amic acid)] by spin coating on a substrate. At this stage, the stress is zero but so is the usefulness of the coating. To develop full mechanical and electrical properties, the poly(amic acid) is allowed to dry and then, through a series of curing steps, is heated to near 400°C for some prescribed time period to allow for the formation of the desired polyimide material.

The stress is still essentially zero at this stage, but the trouble starts during the cool-down process. The polyimide is no longer a liquid but a fully formed solid material with a fairly high thermal expansion coefficient. If coated on a rigid substrate of low thermal expansion material, it will want to shrink faster than the substrate will allow. Given the constraint of good adhesion between the polyimide and the substrate, the net result is a buildup of stress in the polyimide as the coating/substrate laminate is cooled to ambient conditions. A quick glance at Table 4.1 shows that the polyimide/GC system meets all the requirements for stress buildup in the poly-imide coating. To estimate what the residual stress might be for the polyimide/GC system, we reach into our mechanics toolbox for an appropriate formula. In fact, an appropriate formula was derived in Chapter 3 (Equation 3.50):

$$\sigma = \frac{E \Delta \alpha \Delta T}{(1 - \nu)} \tag{4.55}$$

The elastic constants in Equation (4.55) are those of the polyimide material and can be taken directly from Table 4.1. $\Delta \alpha$ is the difference in expansion coefficient between the polyimide and the GC, and from Table 4.1 comes to $30 - 3 = 27$ ppm/K. The temperature difference ΔT can be taken as the highest cure temperature minus ambient or $400 - 20 = 380$ K. Putting all the numbers into Equation (4.55) gives $\sigma = 3 \times (2710^{-6}) \times 380.7 = 0.044$ GPa, which is a fairly substantial stress level considering that the ultimate tensile strength of these materials is in the range of 0.07 to 0.1 GPa.

We now have nearly enough data to use Equation (4.53) to determine what the mode I stress intensity factor of a crack propagating in the GC would be. All we need is an estimate for the decohesion number Ω. Using the elastic constants in Table 4.1, we calculate the Dundurs' parameter from Equation (4.40) and the film stiffness ratio from Equation (4.42). The decohesion number can now be estimated from the second of Equations (4.54), giving a value $\Omega = 3.63$. Equation (4.53) now gives the mode I stress intensity factor K_I as $(44 \text{ MPa})(0.000175 \text{ m})^{1/2}$ $(3.63) = 2.11 \text{ MPa m}^{1/2}$. This is an alarmingly high value, especially in light of the measured fracture toughness of the GC material as roughly $1.4 \text{ MPa m}^{1/2}$. The prediction is completely unambiguous. If you try to coat 175 µm of standard polyimide material on the proposed GC substrate, the residual stress in the polyimide will drive a crack in the GC, lifting off the polyimide coating. Furthermore, from Figure 4.10 we can estimate the crack depth factor as roughly 0.063, which for a 175-µm coating gives a crack depth of 11 µm, which at first glance will look like a delamination event.*

We have thus answered the first of the questions listed above. Putting relatively thick layers of standard polyimide material onto a GC substrate is definitely not a good idea. However, we do not have to let the matter rest there. Nobody likes a negative answer, and it tends to put a damper on morale. We can ask the further question regarding which steps can be taken to make the process work. Equation (4.53) gives us two levers with which to play. One is the coating thickness, and the other is the film stress. If we can reduce either of these parameters, we may reduce the crack-driving force down to a safe level. Reducing the thickness is generally ruled out for reasons of electrical performance. The electrical design people need a

---

* It should be noted that these results were verified experimentally, albeit unintentionally, on a real manufacturing line.

certain thickness to get their signals to propagate correctly. This leaves the coating stress. What can be done there? Fortunately, it turns out that the polyimide materials form a fairly broad class with a significant range of thermal-mechanical behaviors. In particular, there is a class of low thermal expansion polyimide materials with thermal expansion coefficients near 5 ppm/K as opposed to the more standard material treated in the example above. With a little extra work, we can see whether using one of these materials will result in a stable laminate when coated on GC.

Table 4.1 lists the elastic properties of a typical low thermal expansion polyimide material, so we can redo the above analysis using these properties in place of those for the standard polyimide. In this case, the decohesion number works out to be 1.8, and the mode I stress intensity factor is calculated as $K_I = (7.6 \text{ MPa})(0.000175 \text{ m})^{1/2}$ $(1.8) = 0.18 \text{ MPa m}^{1/2}$. This is now an acceptable number less than 10 times the known fracture toughness of the GC. Thus, a rather simple solution is found to the coating problem using elementary formulae and knowledge of the material and fracture strength properties of the coating and substrate materials.

One final calculation should be done before calling it a day. It is clear that standard polyimide materials are ruled out for this application because they develop too high a stress for the substrate to endure. The low thermal expansion materials are clearly a good option because they develop a low internal stress caused by curing and will not fracture the substrate. The question remains, however, whether the low-TCE (thermal coefficient of expansion) polyimide will adhere properly to the GC. That is, can we rule out clean delamination as opposed to decohesion? If the adhesion of the coating is too weak, it may just cleanly delaminate from the substrate.

To put this question to rest, we go back to the first of Equations (4.43) and use the formula for the strain energy release rate G evaluated at zero crack depth (a clean delamination with $\lambda = 0$). The strain energy release rate is the best quantity to look at because it can be measured or closely estimated from fairly standard measurements, such as the peel test. All of the parameters are the same as used for estimating K_I in the previous calculation except that the crack depth is set at exactly zero. Note that in this case K_{II} is not zero as it was above. Delamination is a mixed-mode process, which is another reason for preferring to use the strain energy release rate, which is a scalar property. Doing the arithmetic is not nearly as convenient as using Equation (4.53), but it is straightforward nonetheless. The result comes to a driving force of roughly 2.64 J/m², which would require a very modest level of adhesion to resist. However, polyimide films coated without an adhesion promoter can exhibit levels of adhesion in this range or lower, so we would definitely recommend that an adhesion promoter such as one of the silane materials be used. In this case, it is easy to achieve adhesion levels of 20 J/m² or higher, so clean delamination can also be ruled out as a possible failure mode.

4.2.3.2 Nickel on Glass

As an example of a system in which the stiffness of the coating is much greater than the stiffness of the substrate, we consider the case of nickel on glass. Nickel is a useful material and is used many times in the microelectronics industry as a diffusion barrier. It also has uses in the optics industry for mirrors and lenses. Regardless of the application, however, one has to be very careful when coating nickel onto a brittle substrate because it can develop very high levels of internal stress. It also turns out that the internal stress level of the nickel coating can be very sensitive to the deposition conditions. For instance, if small amounts of nitrogen leak into the vacuum chamber during deposition, the nitrogen will segregate out in the grain boundaries of the coating and thereby induce a significant mismatch strain between grains, which can lead to a very high residual stress level. The engineering design person needs to know just how much residual stress can be tolerated in a particular coating for a given desired coating thickness.

The analysis of Suo and Hutchinson gives us a handy way to estimate this. Let us assume, for example, that we have a deposition process for nickel that consistently gives a residual stress level of 1000 MPa as estimated by x-ray measurements.* The question is how thick a coating of this material can be deposited on a substrate of silica glass? Pulling out the elastic constants for nickel and glass from Table 4.1, we quickly obtain the stiffness ratio for the nickel/glass system as $\Sigma = 3.0$. Using this value in the first of Equations (4.54), we easily calculate the decohesion number Ω as 3.14. Now, it is a simple matter to pull up Equation (4.53), which relates the stress in the nickel to the strength of the glass and the nickel coating thickness. Rewriting to obtain the coating thickness as a function of the other parameters, we get

$$h = \left(\frac{K_I}{\sigma \, \Omega} \right)^2 \qquad\qquad (4.56)$$

The mode I fracture toughness of the glass is typically in the range of 1.2 to 1.4 MPa m$^{1/2}$. Assuming the high value for the glass, we estimate the maximum nickel thickness as $[1.4/(1000 \times 3.14)]^2 = 0.2 \times 10^{-6}$ m or 0.2 μm. We would like to be a little cautious because we realize that, in a humid environment, the fracture toughness of glass can drop by 20% or so, which would put the fracture toughness at the lower value of 1.2 MPa m$^{1/2}$. In this case, our formula gives 0.15 μm as the maximum nickel thickness. Thus, a prudent recommendation would be to keep the nickel coating thickness under 0.15 μm for this application. If it is necessary to go higher in thickness for other engineering purposes, then one either has to use a stronger glass substrate or somehow reduce the stress in the nickel. In either case, this simple analysis gives us a rough-and-ready method of knowing where things stand regarding the possibility of the nickel coating fracturing the glass substrate.

The ability to perform calculations similar to the two examples given above is quite likely the most useful application of the quantitative methods presented thus far. The interested reader should refer to Appendix D, which gives a table of simple formulae for estimating the strain energy release rate for a variety of common failure modes in several types of laminate structures. These formulae, when coupled with the material property data given in Appendix C, provide a handy set of tools for performing "quick-and-dirty" calculations, which can be very useful in making

---

* Those used to dealing with polymer materials will find this a rather whopping number, but for a material like nickel it is rather middling and unimpressive.

decisions regarding whether certain laminated structures will be stable regarding delamination and fracture phenomena.

Another useful application of these formulae presents itself in diagnosing parts and structures that have already failed. If the observed failure mode can be found in the table given in Appendix D, then one can use the formulae to estimate the driving force responsible for the failure. These data, coupled with the elastic constants of the materials involved, can then provide some useful information on the stress level causing the failure by giving a lower bound estimate of this critical quantity.

4.3 SUMMARY

This chapter introduced enough of the concepts of fracture mechanics to enable a fully quantitative evaluation of delamination processes through the use of the concepts of the stress intensity factor and the strain energy release rate. As pointed out, the need for this approach is because all attempts to obtain a truly quantitative measure of the adhesion strength between two solid bodies must inevitably contend with the thermal-mechanical behavior of the materials in question. Fracture mechanics is firmly grounded in the continuum theory of solids and further extends this theory to deal with the problem of singular stress conditions induced by sharp crack tips in fractured or delaminated bodies or by sharp corners in multilevel laminates.

A detailed calculation of the strain energy release rate for a coating delaminating from the edge of a rigid substrate is given to demonstrate, for one case at least, how everything fits together. The stresses and displacements are computed from the general principles of continuum theory, and fracture mechanics concepts are then applied to compute the driving force for delamination from these results. Finally, a number of simple calculations are carried out using formulae taken from the literature to demonstrate how one can perform useful and important estimates of the structural stability of structures one might be intent on fabricating.

Notes

1. *Elementary Engineering Fracture Mechanics*, 3rd rev. ed., D. Broek (Martinus Nijhoff, Dordrecht, The Netherlands, 1984), p. 69.
2. "Yielding of Steel Sheets Containing Slits," D.S. Dugdale, *Journal of the Mechanics and Physics of Solids*, 8, 100 (1960).
3. "The Mathematical Theory of Equilibrium Cracks in Brittle Fracture," G.I. Barenblatt, *Advances in Applied Mechanics*, 7, 55 (1962).
4. "Advanced Fracture Mechanics," M.F. Kanninen and C.H. Popelar (Oxford University Press, New York, 1985).
5. "Decohesion of Films With Axisymmetric Geometries," M.D. Thouless, *Acta Metallica*, 36, 3131 (1988).
6. A very useful handbook from a physics point of view is *Physics Vade Mecum*, H.L. Anderson, Ed. (American Institute of Physics, 335 East 45th Street, New York, NY, 1981). See Section 1.05 for a comprehensive collection of formulas in Cartesian, cylindrical, and spherical coordinates.

7. The text to consult here for an elementary treatment of the multivalued displacement field problem is *Theory of Elasticity*, 3rd ed., S.P.T. Timoshenko and J.N. Goodier (McGraw-Hill, New York, 1970). See in particular p. 77.

8. This phenomenon is discussed in a most engaging and enlightening way by R. Feynman in *The Feynman Lectures on Physics*, Vol. 2, R. Feynman, R. Leighton, and M. Sands (Addison-Wesley, Reading, MA, 1964). See in particular Section 12-1, "The Same Equations Have the Same Solutions." This chapter applies formal mathematical results from electrostatics to problems of heat flow, membrane deformations, neutron diffusion, fluid flow, and uniform lighting of a plane. Even though this is old material, the development is a delight to read and demonstrates how a little imaginative thinking can be used to stretch the rewards of one's labors in one area to surprisingly different situations.

9. The normalized crack length is fairly straightforward to derive. For the edge delamination case, it is clearly given by $\Delta = R - a$. To make this dimensionless, we divide both sides by the sample radius R, giving the normalized version $\Delta/R = 1 - a/R$. Similar reasoning applies to the circular cut.

10. "Steady-State Cracking in Brittle Substrates Beneath Adherent Films," Z. Suo and J.W. Hutchinson, *Int. J. Solids Structures*, 23, 1337 (1989).

11. J. Dundurs, "Elastic Interaction of Dislocation With Inhomogeneties," in *Mathematical Theory of Dislocations*, (ASME, New York, 1969), pp. 70–115.

5 Applied Adhesion Testing

The man in the street, or the man in the workshop, thinks he needs virtually no theoretical knowledge. The engineering don is apt to pretend that to get anywhere worth while without the higher mathematics is not only impossible but that it would be vaguely immoral if you could. It seems to me that ordinary mortals like you and me can get along surprisingly well with some intermediate — and I hope more interesting — state of knowledge.

J. E. Gordon, *Structures or Why Things Don't Fall Down*

In the spirit of Prof. Gordon's remarks, we proceed in this chapter to investigate several applications of adhesion testing that are both informative and potentially useful. Among the plethora of adhesion tests that can be performed, a small subset tends to dominate in terms of actual implementation in practice. The most popular tests share the common properties that they are relatively easy to implement and give semiquantitative data that is relevant to a particular practical application. For flexible coatings, the peel test has to head the popularity poll in terms of number of publications in the literature and the number of different variations and applications reported. For hard brittle coatings, the prize goes to the scratch test for roughly the same reasons. We allow the pull test onto the chart because it can be applied to both flexible and brittle coatings, although in terms of overall popularity it ranks at number 3. In what follows, detailed examples of each of these tests are covered to give a more in-depth view of real-life adhesion testing, what the goals are, how the tests are implemented, and what results are achieved.

5.1 THE PEEL TEST

As a young engineer in the microelectronics industry, I was given the mission of devising an adhesion test for polyimide films, which at the time were under consideration as dielectric layers in advanced microcircuit devices and chip-packaging applications. The polyimide materials had a number of perceived advantages over the then-common sputtered glass materials, including a low dielectric constant, flexibility, and relative ease of application. Unfortunately, these materials were also highly prone to delamination when applied to silicon wafers and ceramic substrates, and a method of measuring their adhesion was urgently needed so that steps could be taken to overcome this problem. A simple and effective technique was needed, and there really was no time to purchase capital equipment, which required a minimum of 3 months just to get into the capital plan and obtain financial approval.

The test would have to be devised quickly using parts and materials more or less readily available in a standard chemistry laboratory. The peel test immediately suggested itself because the polyimide coatings were quite flexible and readily peeled off the substrates to which they were supposed to adhere. Two problems then remained: how to prepare the samples and which test equipment to use.

5.1.1 SAMPLE PREPARATION

The substrates of interest were silicon wafers that had a top layer of SiO_2 or silicon nitride. A number of possible adhesion promoters were considered to alleviate the delamination problem, and the adhesion test had to be devised to test the effectiveness of these promoters. A wafer sample was devised with a blanket coating of the polyimide cut into thin strips on which half of the wafer would be treated with adhesion promoter and the other half left untreated. Figure 5.1 shows a representative sample. This type of sample has several advantages, including

1. Use of one part to create several peel samples
    a. Avoids sample-to-sample variations associated with different parts
    b. Multiple peel samples available for before and after treatment experiments
    c. Multiple peel strips available for good sample statistics
2. Untreated area serves as a natural reference frame for evaluating the relative effectiveness of the adhesion promoter

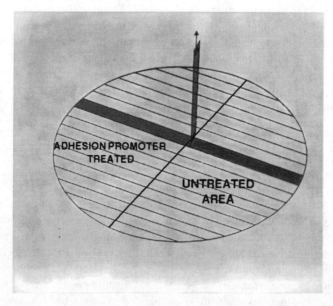

FIGURE 5.1 Peel test sample for evaluating the relative effectiveness of adhesion promoters. Half of a wafer is treated with adhesion promoter, and the other half is left untreated. A blanket coating is then applied and cut into strips to the depth of the coating thickness, leaving the wafer substrate intact. Peeling a strip from the untreated half into the treated region clearly and unambiguously reveals the effectiveness of the adhesion promoter treatment.

Creating the peel strips from the blanket coating can be done in a number of ways, including photolithography or micromilling techniques. Use of a dicing saw with the blade set to a depth just sufficient to cut the coating and leave the wafer uncut is also very effective. For those just starting and without any of the aforementioned capabilities, there is always the tried-and-true Exacto® knife and ruler method, which although crude is quite effective nonetheless.

5.1.2 TEST EQUIPMENT

Figure 5.2 illustrates a homemade version of a very effective piece of peel test equipment that can be assembled from parts and equipment available in nearly any chemistry lab. The test frame is set up using what is known as a Unislide® assembly, which is essentially a small platform arm mounted on a screw drive, which can be turned either by hand or by an electric motor. These are commonly used as x-y positioners for adjusting the position of samples under a microscope or for raising and lowering heavy apparatus in a precisely controlled way. As shown in Figure 5.2, the Unislide takes on the role of a tensile test machine. By attaching to the end of the peel strip with a common alligator clip and through the use of a pulley to maintain the peel angle at 90°, a weighted slide box can be pulled along a lubricated surface (a piece of Teflon sheet is handy), effectively emulating a standard 90° peel test. The role of the load cell is provided by a common lab balance, which records the apparent decrease in weight of the slide box sample system as the Unislide starts to pull up on the peel strip.

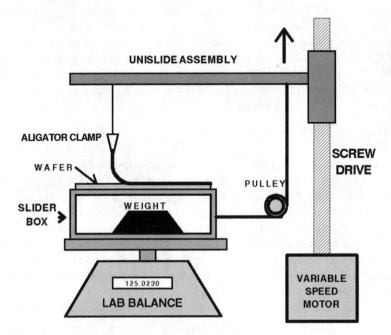

FIGURE 5.2 Homemade peel test apparatus assembled from parts available in nearly any chemistry lab.

If you are really desperate for time, you can simply read the peel load off the balance display at constant time intervals and thus obtain load-versus-stroke data knowing the peel rate from the speed of the Unislide. A more sophisticated way is to direct the output of the balance to a common strip chart recorder and thus have a permanent record of the peel experiment. To achieve a final touch of sophistication, the whole apparatus can be put inside a polyethylene glove bag under a constant gas pressure of known humidity. Thus, the sensitivity of the peel strength to humidity can be evaluated. This entire apparatus can be set up within a week from commonly available parts at a capital cost of essentially zero.

For the more traditional commercial route, one needs a test frame of the type commonly used for tensile testing. These normally go for about $50,000 and up because they have extensive capabilities for complex load cycles and can go to very high loads. Most of the extra capabilities, however, are not required for the simple peel test. Instead of the slide box shown in Figure 5.2, one would use a custom platform mounted on a rail using ball bearings with a small pulley to maintain the 90° peel angle. Such an item can run for several thousand dollars if it has to be fabricated from scratch. The lab balance and strip chart recorder are replaced by a load cell and a personal computer, which can accumulate the load-versus-stroke data of the peel test and store it on a magnetic disk for later analysis and plotting. Environmental control can be implemented by enclosing the apparatus in a Plexiglas enclosure. The entire apparatus can easily run to about $100,000 depending on capabilities and options. However, the quality of the data logged by such a system will be no better than the jury-rigged apparatus shown in Figure 5.2.

A commercial system will have distinct advantages, however, if a large number of routine tests must be performed and the data saved and stored for subsequent quality control purposes. The commercial system will in effect save a considerable amount of time and labor, which is important under conditions of mass production. However, for purposes of development in the laboratory, where one may not even be certain whether the peel test will work out in the long term, the apparatus shown in Figure 5.2 has definite advantages.

5.1.3 PEEL TESTING IN ACTION

The peel test setup shown in Figure 5.2 was used to test the efficacy of silane adhesion promoters using the sample configuration shown in Figure 5.1. The Unislide assembly depicted in Figure 5.2 was also used to dip coat half of a silicon wafer in a solution of silane in water, which further demonstrates the versatility and flexibility of the jury-rigged setup. After the silane treatment, the wafer went through a variety of drying and baking steps in accordance with the prescribed recipe for depositing silane materials. Next, the wafer was spin coated with a blanket coating of the polyamic acid precursor solution of the polyimide material* under testing. More drying and baking according to a prescribed recipe finally yielded a polyimide film a few micrometers thick on the wafer. Finally, a sharp Exacto knife was used to

---

* Standard polyimide materials start out as polyamic acid polymers, which are soluble enough to form viscous solutions that can be spin coated.

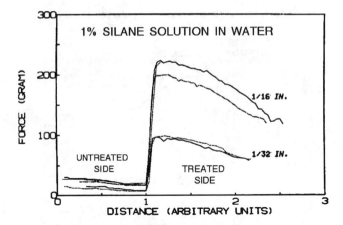

FIGURE 5.3 Peel test data for a polyimide coating on a silicon wafer prepared according to the configuration in Figure 5.1 and tested using the apparatus shown in Figure 5.2. Two strip widths were tested, 1/16″ and 1/32″. The peel force scales as expected with varying strip width.

slice the polyimide coating into strips of variable width. After converting the Unislide assembly back to peel test mode, several strips were peeled, giving the data shown in Figure 5.3. This figure clearly shows the effectiveness of the adhesion promoter in increasing the apparent adhesion of the polyimide to the underlying substrate. In this case, the silicon wafer had a thin blanket coating of silicon nitride that was the actual surface tested. The peel force goes up by about a factor of 5 as the peel strip passes into the treated half of the wafer.

Figure 5.4 and Figure 5.5 illustrate the flexibility and versatility of the multiple-strip sample design shown in Figure 5.1. When used as insulator layers in micro-electronic devices, the polyimide coatings have to endure several rounds of thermal

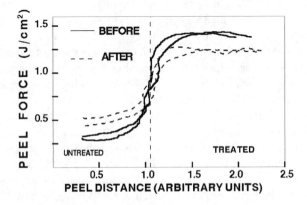

FIGURE 5.4 Use of the peel test to evaluate the effect of thermal cycling on the adhesion of polyimide to a silicon nitride-coated wafer. The sample was heated to 400°C and back 10 times.

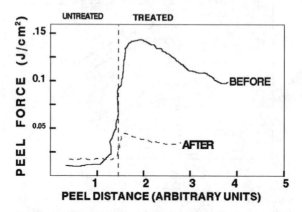

FIGURE 5.5 Same as Figure 5.4 showing the effect of 40 thermal cycles.

cycling between room temperature and 400°C; these are required for the deposition of metals and various soldering and brazing operations. The question naturally arises regarding how the adhesion of the polyimide will stand up under these conditions. Using a multistrip sample, it is quite straightforward to perform before and after experiments as shown in Figure 5.4, which illustrates the effect of 10 thermal cycles to 400°C and back. Because there is no question of sample-to-sample variation, the figure clearly shows that there is an apparent increase of adhesion of the polyimide film on the untreated half of the wafer and a concomitant decrease on the treated side. The data raise a concern that the silane adhesion promoter may be breaking down and causing the apparent adhesion loss on thermal cycling.

Even more disturbing are the data shown in Figure 5.5, which illustrates the effect of 40 thermal cycles on the polyimide adhesion. It turned out that at the time this work was done, 40 thermal cycles to 400°C was a manufacturing specification that anticipated not only the usual product fabrication steps, but also additional cycles for rework purposes. The 40 thermal cycles brought about an apparent adhesion loss of nearly a factor of 10. This result definitely raised some eyebrows, and further investigation was carried out to determine the underlying cause of the adhesion loss. As it turned out, there was a leak in the thermal cycling furnace that admitted parts per million of oxygen. Sealing the furnace leak and redoing the adhesion measurements then revealed an adhesion loss of at most 10% after 40 thermal cycles. This was quite likely the first example of use of the peel test as a leak detector.

One final example reveals the usefulness of the peel test in improving the manufacturing process. This example again deals with the silane adhesion promoters. In the early days, there was a recipe that these materials were to be put down from a 1% solution, which was subsequently rinsed because using 1% apparently created too thick a layer, which reduced the adhesion of the coating because it acted more as a weak boundary than as an adhesion promoter. However, after the rinse step the thickness was just right, and the silane material gave excellent results. A question

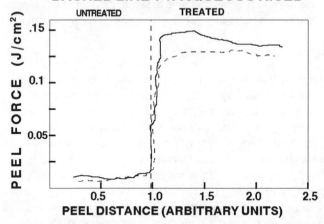

FIGURE 5.6 Peel test data comparing two different methods of applying silane adhesion promoters. The dashed line shows the effect of coating a 1% solution followed by a rinse step. The solid line shows the results obtained using a 0.1% solution with no rinse step. The data clearly show that the rinse step can be eliminated by using a 0.1% solution.

naturally arose, however, as to why the 1% was used in the first place. Why not just use a 0.1% solution and skip the rinse step. It was a good question, but anyone familiar the psychology of the manufacturing line knows that once a process is working those in charge are extremely loathe to change anything. Any suggestion to change any of the steps is treated like a dead raccoon. However, there is another side to manufacturing psychology that is attracted to the idea of eliminating steps in the overall process because this reduces both complexity and cost provided it can be unambiguously demonstrated that the required change will not cause problems.

Figure 5.6 directly compares the two separate adhesion promotion schemes discussed above. The dashed line data show the case of a 1% rinsed application, and the solid line shows the case of a 0.1% application with no rinse step. This data sealed the case for eliminating the rinse step, and the peel test again demonstrated its worth.

5.1.4 Advanced Peel Testing

The above discussion demonstrates that practical peel testing can be carried out with fairly elementary laboratory equipment and requires only a modicum of time and effort to set up. The data generated can be as valid and useful as that logged on far more expensive commercial apparatus. However, the data are always semiquantitative. Peel results can be used to unambiguously rank different adhesion promoters, but as implemented above, this type of test cannot give a fully quantitative measure of the adhesion strength between a coating and its substrate. The discussion that follows looks deeper into the mechanics and thermodynamics of the peel test to understand more fully what is happening in terms of energy flow and deformation processes as the coating is peeled. One major question that arises is whether the peel test can be made to yield fully quantitative data that can be used to support modeling efforts to understand the adhesion stability of a given coating substrate system.

5.1.4.1 Thermodynamics of the Peel Test

It is intuitively clear that energy is required to remove a coating by peeling and that this energy must be supplied by the test apparatus. It would be convenient to be able to say that the energy required to separate the coating from the substrate is just the energy of peeling, which can be simply calculated as the peel load times the length of peel strip removed. Dividing this number by the surface area peeled would then give a direct measure of the surface fracture energy. However, early experimenters quickly became aware that this is not the case.[1] Those familiar with the mechanics of bending knew that very large strains were induced at the bend in the peel strip, and that these strains easily exceeded the yield strain of most coating materials. The inescapable conclusion was that a considerable amount of viscoplastic deformation had to be occurring during the peel process and that a nontrivial fraction of the work done by the test machine was going into irreversible work of deformation. Clearly, neither the peel test apparatus described above nor any of the far more expensive setups is capable of addressing this problem.

5.1.4.2 Deformation Calorimetry

Clearly, a more innovative approach to the problem was required, and this was supplied in a seminal piece of work carried out by Farris and Goldfarb.[2] These authors used what is known as a deformation calorimeter, which is essentially a device that can measure heat loss or gain by a sample undergoing mechanical deformation. A schematic illustration of the device used in this work is shown in Figure 5.7.

The key to the operation of the deformation calorimeter lies in its ability to detect very small differences in pressure between two separate pressure cells using a differential pressure transducer. The peel sample rests in the central cell surrounded by a gas at some uniform temperature and pressure. A second reference cell sits without any sample in the identical thermodynamic state. The two cells are connected via tubing to a special transducer that can very sensitively detect any pressure difference between the two cells. The entire unit is immersed in a constant-temperature bath that is held at a uniform temperature.

A wire passes through a special mercury seal into the sample cell, where it is clamped to the peel strip. The other end of the wire goes to a suitably modified tensile test rig. As the peel test proceeds, any heat generated in the sample is conducted into the surrounding gas, causing a temperature rise. Because the sample chamber is held under adiabatic conditions, the temperature rise in the gas forces the pressure up, as dictated by the ideal gas law. The differential pressure transducer immediately detects the difference in pressure between the reference cell and the

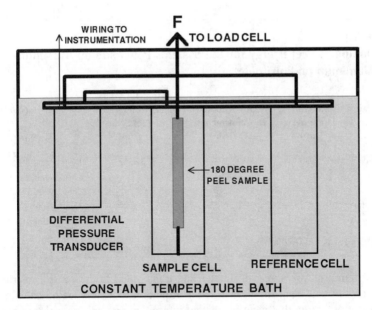

FIGURE 5.7 Schematic of a deformation calorimeter capable of measuring heat dissipation occurring during the peel test.

sample cell and transmits these data to an external data-logging device. The pressure change can then be translated into a heat flow via an appropriate calibration experiment using a wire of known resistance or other well-known heat source.

Needless to say, no one is going to conjure up a deformation calorimeter from common parts laying about the laboratory. This is a complex, one-of-a-kind setup requiring a high level of expertise in both mechanical and calorimetric design, definitely an instrument for the dedicated research laboratory or government-supported research institution. However, the deformation calorimeter has the capability of revealing aspects of the peel process beyond the capabilities of standard peel test setups. The energetic processes of the peel test as revealed by this instrument can be understood directly in terms of the first law of thermodynamics, which is essentially a restatement of the principle of conservation of energy:

$$\Delta W = \Delta U - \Delta Q \qquad (5.1)$$

where

ΔW = Total energy expended by tensile test apparatus

ΔU = Fraction of total energy transformed into latent internal energy of peel strip

ΔQ = Fraction of total energy dissipated as heat

The total work done by the tensile machine is simply the integral of the peel load over the entire displacement history and is readily logged by computer-supported instrumentation. The heat dissipated during the peel process ΔQ is measured

TABLE 5.1

Partitioning of Peel Energy for the Systems Polyimide on Aluminum and Aluminum on Polyimide[a]

Material Being Peeled	Total Work Done ΔW (J/m²)	Heat Dissipated to Calorimeter ΔQ (J/m²)	Latent Internal Energy Imparted to Peel Strip ΔU (J/m²)	Comments
Polyimide	625 ± 9	−302 ± 9	323 ± 13	Roughly 50/50 partition between heat and stored energy
Aluminum	847 ± 35	−852 ± 42	−5 ± 55	Nearly 100% heat dissipation

[a] Data from Farris and Goldfarb.[2]

by the differential pressure calorimeter and transmitted to the control instrumentation. With the work and the heat in hand, Equation (5.1) allows the calculation of the amount of energy that went into latent internal energy of the peel strip material. This is an interesting quantity that characterizes the nature of the deformation processes that different materials undergo during the peel process. As noted, the strains at the peel bend can get as high as 50%, which is well above the yield point of most materials. With metals, most of the deformation above the yield point goes into plastic flow. The metal behaves essentially like a highly viscous liquid, and nearly all of the work of deformation goes into heat. On the other hand, when dealing with polymeric materials the deformation processes are primarily viscoelastic/viscoplastic, which means that there are both elements of viscous flow and plasticity. Roughly half of the total work of deformation goes into heat and the other half into latent internal energy. This comes about because of the huge number of degrees of freedom of a polymer molecule compared to a relatively simple metal atom. Polymer chains can become entangled with one another, setting up bond distortions that can lock up large amounts of elastic energy when the material is below its glass transition temperature. Thus, one would expect to see large differences between the thermodynamic behavior of peeling metals as opposed to peeling polymers. This is dramatically illustrated in the data of Farris and Goldfarb on the adhesion of polyimide to aluminum and aluminum to polyimide, as summarized in Table 5.1.[3]

The data in Table 5.1 were taken on polyimide samples spun coated onto thick aluminum foil and fully cured to a standard recipe. The resulting laminate was then cut into strips roughly 1/2 cm wide by 5 cm long, which could easily be mounted in the deformation calorimeter. A rigid, thin stiffening rod was attached with epoxy to the side that was to serve as the substrate, preventing any bending. The side serving as the peel strip was attached to the grip in the calorimeter and peeled at 180°. Thus, separate strips cut from the same sheet could serve as either aluminum peel samples or polyimide peel samples, ensuring that the intrinsic aluminum-to-polyimide adhesion would be identical in both cases regardless of which material was peeled. The following conclusions follow immediately from the data in Table 5.1:

1. The work of peeling the aluminum off the polyimide goes almost entirely into heat, indicating that it is absorbed by plastic flow of the aluminum because of the high strain level in the peel bend.
2. The polyimide material is sharply different from the aluminum, with only about 50% of the total peel energy going into heat and the remaining presumably into latent internal energy.
3. Certainly, for the aluminum case the energy required just to separate the interface (surface fracture energy) is a very small fraction of the total peel energy and is completely disguised in the measurement errors.

The data in Table 5.1 effectively throw a wet blanket on the notion of trying to determine intrinsic surface fracture energies from peel test data. The basic problem is of course the monstrous strains induced in the peel bend, which take just about any material well beyond the yield point and into the realm of viscoplastic flow. A final example illustrated in Table 5.2 further dramatized the difficulty by comparing the work required to peel an epoxy material off an aluminum substrate as compared to the measured surface fracture energy estimated from contact angle measurements.

Contact angle measurements have long been recognized as a method for estimating the surface free energy of solid materials.[4] Reading from the Table 5.1, we conclude that the total surface energy associated with 1 m² of the epoxy material plus 1 m² of oxidized aluminum should be 0.7 J. However, according to the last entry in the table, the energy associated with producing the same amount of surface area by peeling apart the same materials should be about 400 J. This gives a clear indication that during the peel test only a small fraction of the measured work of adhesion goes into actually separating the materials at their common interface, and the lion's share goes into viscoplastic deformation processes. In the next section, we explore a continuum mechanics approach to the peel test and see what can be determined from a purely theoretical point of view.

TABLE 5.2
Comparison of Peel Energy Versus Surface Fracture Energy of Epoxy on Aluminum Substrate[a]

Measurement	Measured Energy (J/m²)	Comments
Surface free energy of epoxy	0.05	Contact angle measurement
Surface free energy of Al_2O_3	0.65	Contact angle measurement; oxidized aluminum surface assumed
Total estimated surface fracture energy of epoxy and aluminum	0.7	Sum of surface free energies of alumina and epoxy
90° peel energy	400	Total work of adhesion to peel epoxy of aluminum substrate

[a] Data from R. Farris, private communication.

5.2 FULLY QUANTITATIVE PEEL TESTING

5.2.1 Earliest Work, Elastic Analysis

A review of the commonly available literature reveals that apparently the earliest attempt to quantify the peel test was documented in the article by Spies.[5] In this article, Spies referred to a technical report put out by Aero Research Limited of Cambridge, England, dated 1947. Thus, we can date the earliest documented appearance of the peel test to 1947 in the aircraft industry.* Although the article fell substantially short of providing an accurate theoretical picture of what happens during a peel test, this work nonetheless broke some important ground and made a number of points that are as relevant today as they were 50 years ago given the limitations in available material property data and computing capabilities. The article attempted to compute the adhesion strength of two long, thin aluminum strips bonded together by Redux adhesive.** The basic peel test setup used was a variant of the rolling drum technique in which one of the bonded strips is firmly attached to a drum that freely rotates on a spindle, and the mated strip is grasped by a grip attached to a load cell within the test frame. The underlying assumption of the analysis is that the ultimate bonding strength is determined by the strength of the Redux adhesive. Spies readily acknowledged that this analysis was flawed because of the following problems:

1. The thermomechanical properties of the Redux adhesive were poorly known and depended sensitively on the curing process.
2. The adhesive layer was not uniform.
3. Failure may have occurred by delamination of the adhesive from the aluminum as well as cohesively within the adhesive itself.
4. The effect of internal stresses within the sample were not taken into account.
5. The aluminum peel strip deformed plastically as well as elastically, and plastic deformation could not be adequately modeled at the time.

These caveats notwithstanding, the author derived formulae for the adhesion strength and the peel force. These formulae are of little practical significance except for possible historical interest. However, the article does draw some important conclusions that are still relevant to the art of adhesion testing. The most important findings are as follows:

1. The mode of loading in the adhesion test should be as close as possible to that affecting the actual structure of interest. This is a basic axiom of

---

* Any reader who is aware of an earlier date should contact the author, who will promptly revise this informal history of the peel test for future editions.
** Little information is given on Redux in the article except that it comes in both liquid and powder forms and must be cured at elevated temperature and pressure. Quite likely, it was one of the early epoxy adhesives. Again, the author would appreciate any information that any reader cognizant of the history of aircraft glues might have.

all adhesion testing and is quite likely stated in Spies's article for the first time.

2. The peel test is very sensitive to the strength of the adhesive. It was pointed out that a 1% increase in the true adhesion strength gives rise to a 6% increase in the apparent peel force. This observation is equally true for thin film adhesion as it is for adhesives.

3. The peel force is dependent on the mechanical properties of the peel strip. This conclusion is clearly borne out by the deformation calorimetry work of Farris and Goldfarb discussed in Section 5.1.4.2. Recall that the apparent peel force for a polymer bonded to aluminum depended strongly on whether the polymer was peeled off the aluminum or whether the aluminum was peeled off the polymer, notwithstanding the fact that the actual adhesion strength was identical in both cases.

Following the work of Spies, a short article appeared by Bikerman[6] that gave an elementary analysis of peeling a strip adhesively bonded to a perfectly rigid surface. Figure 5.8 illustrates the geometry of the test configuration. The following parameters and variables define the problem:

Model parameters
- E = Modulus of peel strip
- E_1 = Modulus of adhesive
- σ_u = Ultimate tensile strength of adhesive
- w = Width of strip
- l = Length of strip
- δ = Thickness of strip
- I = Moment of inertia of strip cross section
- y_0 = Adhesive thickness

Model variables
- W = Peel load
- y = Strip deflection (function of x coordinate)

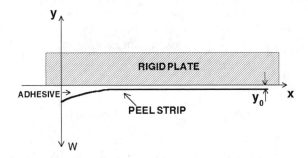

FIGURE 5.8 Peel strip glued to a rigid plate by an adhesive layer.

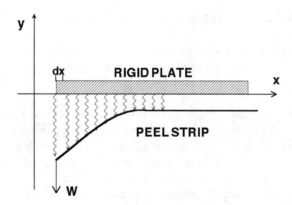

FIGURE 5.9 Simplified mechanical model of the peel strip sample in Figure 5.8. Note that the adhesive layer is modeled as a collection of linear springs.

Bikerman took the simplest approach possible. Rather than become involved with esoteric formulae based on continuum theory, why not treat the adhesive as a collection of simple springs that hold the peel strip onto the supporting plate? Figure 5.9 illustrates the essential idea. Each increment dx of the adhesive is treated as a Hookean spring. Because we know the elastic modulus of the adhesive, we can deduce the effective spring constant of each increment of the adhesive starting from Hooke's formula:

a) $\sigma = E_1\varepsilon$ (Hooke's Law)

b) $\sigma = \dfrac{F}{A}$ (F = applied force, A = adhesive cross section)   (5.2)

c) $\varepsilon = \dfrac{y}{y_0}$ (Definition of strain)

Because the adhesive cross section is given by wdx, we can write the standard force displacement relation of the effective elastic spring by substituting Equations (5.2b) and (5.2c) into Equation (5.2a):

$$F = \frac{E_1 w\, dx}{y_0} y \text{ (Force on element of length } dx)$$ (5.3)

or

$$f = \frac{F}{dx} = \frac{E_1 w}{y_0} y \text{ (force per unit length)}$$

Equation (5.3) gives the force that the adhesive exerts on the peel strip if it is displaced some amount y because of the action of peeling. The next step is to treat the peel strip as a simple beam and use standard beam theory. The mechanics of simple bending of uniform beams is outlined in Appendix B. At this point, we need formula (B.10d), which relates the deflection of a simple beam in terms of the applied load per unit length.*

$$\frac{d^4 y}{dx^4} = \frac{f}{EI} \tag{5.4}$$

Substituting f from the second relation in Equation (5.3) into Equation (5.4), we obtain the following differential equation for the peel strip deflection y**:

$$\frac{d^4 y}{dx^4} = -\frac{E_1 w}{EI\, y_0}\, y \tag{5.5}$$

The moment of inertia I of the peel strip of width w and thickness δ is $I = w\delta^3/12$. Substituting this into Equation (5.5) gives

$$\frac{d^4 y}{dx^4} = -\frac{12 E_1}{E\delta^3 y_0}\, y$$

$$\text{Defining } n^4 = \frac{3 E_1}{E\delta^3 y_0} \tag{5.6}$$

$$\frac{d^4 y}{dx^4} = -4 n^4 y$$

Equation (5.6) is an elementary fourth-order differential equation for the peel strip deflection y. The solution for the current problem is quite simple and is given as***

$$y = A e^{-nx} \cos nx \tag{5.7}$$

It remains to determine the arbitrary coefficient A. This is done by noting that at equilibrium the total force exerted by the adhesive holding the peel strip to the rigid plate must balance the total applied load W. From Equation (5.3), the force exerted by a small element of the adhesive of length dx and width w is given by

$$F = \frac{E_1 w\, dx}{y_0}\, y \tag{5.8}$$

---

* Note in Equation B.10d that the deflection is designated as u instead of y and the load as w instead of f.
** Note the negative sign because the load exerted by the adhesive is acting in the opposite direction of the deflection y, as shown in Figure 5.9.
*** The careful reader may wonder about the reason for the curious definition of the prefactor n in Equation (5.6). The simple reason is so that the solution Equation (5.7) will have the simple form that it does.

The total force exerted by the adhesive is just the integral of Equation (5.8) over the entire length of the peel strip, and this must equal the total applied load W. Thus, equating W to the integral of Equation (5.8) over the length of the peel strip and substituting Equation (5.7) for the deflection y, we get

$$W = \frac{E_1 wA}{y_0} \int_0^l A e^{-nx} \cos nx \, dx$$

$$= \frac{E_1 wA}{2ny_0} \left[e^{-nl} \sin(nl) - e^{-nl} \cos(nl) + 1 \right]$$

(5.9)

Because the length of the peel strip l is very long compared to any other dimension in the problem, the terms in e^{-nl} in Equation (5.9) are very small compared to 1; thus, this equation reduces to the following simple relation between the coefficient A and the total load W:

$$W = \frac{E_1 wA}{2ny_0}$$

(5.10)

The solution for the peel strip deflection can now be written entirely in terms of known quantities as follows:

$$y = \frac{2ny_0 W}{E_1 w} e^{-nx} \cos nx$$

(5.11)

What we are interested in is the load W_{max} at which the adhesive gives way and peeling starts. This will occur at $x = 0$, which is the point of maximum deflection. Therefore, setting $x = 0$ in Equation (5.11) and noting that y/y_0 is the strain ε in the adhesive, we get the following relation between the applied load and the strain in the adhesive at the edge of the peel strip where delamination is expected to start:

$$W = \frac{w}{2n} E_1 \varepsilon = \frac{w}{2n} \sigma$$

(5.12)

Equation (5.12) assumes Hookean behavior of the adhesive; thus, $E_1\varepsilon$ is equal to the stress σ in the adhesive. Peeling starts when the stress in the adhesive reaches the ultimate tensile strength σ_u. This occurs at the critical load W_{max}. If we insert σ_u and W_{max} into Equation (5.12) and replace the factor n by its definition given in Equation (5.6), we get the following final relation for the critical load for peeling:

$$W_{max} = \frac{w\sigma_u}{2} \left(\frac{E}{3E_1} \right)^{1/4} \delta^{3/4} y_0^{1/4}$$

(5.13)

TABLE 5.3
Parameters Appearing in Equation (5.13) for Two Adhesive Materials

Material	w (meter)	y_0 (meter)	δ (meter)	E_1 (Pa)	E (Pa)	σ (Pa)
Polyethylene (epolene)	0.01	9×10^{-5}	7.6×10^{-5}	3.2×10^8	7×10^{10}	5×10^6
Generic epoxy	0.01	9×10^{-5}	7.6×10^{-5}	2.4×10^9	7×10^{10}	5.5×10^7

Equation (5.13) gives us a rough-and-ready formula for estimating the initial peel strength of an adhesively bonded strip. The main limitation of this formula is the assumption of Hookean behavior of the adhesive. The dominating material parameter is the ultimate tensile strength of the adhesive σ_u. Given the underlying assumptions, one would expect that Equation (5.13) would tend to hold for hard brittle adhesives and give poor results for more rubbery viscoelastic materials. Bikerman in fact tested Equation (5.13) for a series of polyethylene-based adhesives.[7] The formula parameters are given in Table 5.3 for this adhesive and for a generic epoxy.

Inserting the parameters for the polyethylene-based adhesive into Equation (5.13) gives a value for W_{max} of 5.79 N. Bikerman reported that the measured value is in fact 1 N. Thus, we see a rather large error in estimating the peel force for this material, which is to be expected because the polyethylene materials violate the Hookean behavior assumption in the extreme. The polyethylene materials exhibit highly nonlinear deformation behavior because of their polycrystalline morphology and highly flexible chain architecture. The main reason for the low value of the peel strength is undoubtedly because, when stretched, the polyethylene material will go into viscoplastic flow well below the assumed Hookean stress level. Substituting the parameters for the epoxy material gives a peel strength of 34.9 N. The much larger value derives from the fact that the ultimate tensile strength of the epoxy is more than 10 times larger than the polyethylene material. Unfortunately, Bikerman did not try to test an epoxy material, but if he had the results would have turned out much better than for the polyethylene because this material will behave much more like a brittle solid and follow the Hookean assumption much more closely than the polyethylene. The conclusion is that Equation (5.13) will greatly overestimate the peel strength of strips glued with soft rubbery adhesives but should give much closer results for hard brittle adhesives. Thus, from the analysis we conclude that Equation (5.13) is a handy relation for estimating the force required just to initiate peel failure in a thin, flexible strip glued to a rigid substrate with a thin layer of adhesive. If the adhesive is of the rubbery viscoelastic type, then Equation (5.13) will substantially overestimate the peel force, so this equation should be used only with hard brittle adhesives such as the epoxies.

After the work of Bikerman, the peel problem was taken up again by Kaelble,[8] who extended the analysis to take into account more accurately the bending mechanics of the peel strip, which will turn out to be a very important part of the peel problem. In Bikerman's analysis, we essentially treated the peel strip as a beam glued to a rigid substrate and applied simple beam mechanics to work out what the

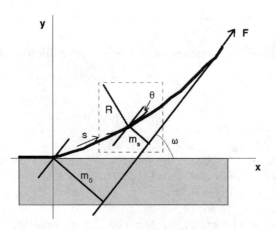

FIGURE 5.10 Peel strip modeled as a flexible ribbon.

peel load ought to be. Kaelble, on the other hand, now dealt with a long ribbon peeled off a rigid substrate as shown in Figure 5.10. This figure shows that the peel force F not only exerts a simple tension along the central axis of the peel strip, but also exerts a bending moment about the point of debonding, acting over a moment arm of length m_0. The main goal of Kaelble's analysis was to relate the peel force at steady-state stripping to the observed peel angle, the mechanical properties of the peel strip, and the adhesive forces holding the peel strip onto the rigid substrate.

We now develop Kaelble's analysis with the help of Figure 5.10 and Figure 5.11. For this analysis, we need to go beyond the simple analysis of Bikerman and consider a steady-state peeling process as shown in Figure 5.10, which shows a long section of delaminated peel strip that is steadily peeling off the substrate at an angle ω, which is the angle between a line parallel to the peel force vector and the substrate surface. The figure shows that the peel force vector is offset from the point of

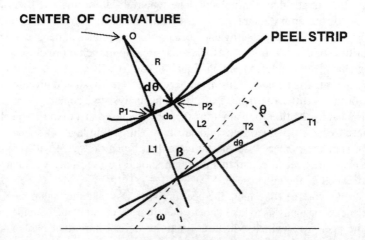

FIGURE 5.11 Blowup of a small region of length ds of the peel strip in Figure 5.10.

debonding by a perpendicular distance m_0, which indicates that the peel force exerts a moment ($F\, m_0$) about this point. The parameter s measures the distance along the peel strip out to some arbitrary point that is offset from the peel force vector by a distance m_s and thus experiences a moment ($F\, m_s$).

The task at hand is to determine a relationship between the peel force and the peel angle that will involve the distance m_0 (hereafter call the moment arm) and various properties of the peel strip. We start by first developing a relationship between the local bending of the peel strip at some point a distance s along the strip and the applied peel force F. The basic relationship we require is Equation (3.59), which was derived in Chapter 3 for the analysis of the cantilevered beam:

$$M(s) = \frac{EI}{R(s)} \tag{5.14}$$

This relationship keeps reappearing when analyzing the bending of simple beams and is often referred to as the *beam equation*. The quantities appearing in Equation (5.14) have the usual definitions but are redefined here for the sake of convenience:

s = Axial coordinate of arbitrary point on beam measured from some
      convenient reference point (the debonding point in this case)
$M(s)$ = Applied bending moment at s
$R(s)$ = Radius of curvature at s
E = Young's modulus of peel strip
I = Cross-sectional moment of inertia of peel strip

Figure 5.11 shows an enlargement of the bending geometry occurring about an arbitrary point P1 a distance s from the debonding point. At this point, the peel strip has a radius of curvature R centered at point O as shown in the figure. Loosely speaking, this means that a circle of radius R centered at O will be precisely tangent to the peel strip at point P1. For the sake of clarity, Figure 5.11 shows the tangent line T1 displaced from the peel strip curve by a conveniently chosen line L1, which is collinear with the radius R. The diagram also shows that the tangent line T1 makes an angle θ with the applied force vector, which makes an angle ω with regard to the horizontal. Progressing along the peel strip by a small increment ds, the radius R sweeps out a small increment of angle $d\theta$, arriving at the adjacent point P2. The tangent line T2 at the point P2 swings a small increment of angle $d\theta$ with regard to the initial tangent T1. From elementary geometry, we know that $ds = Rd\theta$ or for our purposes $1/R = d\theta/ds$. We can therefore rewrite Equation (5.14) as:

$$M(s) = EI\,\frac{d\theta}{ds} \tag{5.15}$$

Note from Figure 5.11 that θ is the angle between the tangent line at P1 and the tangent line at the tip of the peel strip where the peel force vector acts. It thus varies from 0 at the tip of the peel strip where the peel force acts to the peel angle ω at the debonding point. From Figure 5.10, it can be seen directly that $M(s) = F\,m_s$, which can be substituted into Equation (5.15) to give

$$EI\frac{d\theta}{ds} = F\,m_s \qquad (5.16)$$

At this stage, we would like to be able to integrate Equation (5.16) to get the peel force in terms of the peel angle, but we are stuck because we do not know how m_s depends on s because we do not know the shape of the peel strip. What we can determine, however, is a local differential relation between m_s and θ. For this, we refer to Figure 5.12, which shows how m_s varies as we move from point P1 to P2 along the peel strip. As we can see from the diagram, m_s decreases by an amount dm_s in going from P1 to P2. Figure 5.13 shows a blowup of the small increment in

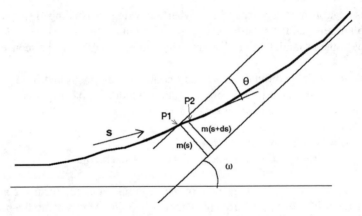

FIGURE 5.12 Behavior o. local bending moment m(s) along the peel strip as one proceeds from point P1 to P2. Note m(s) is decreasing and will become 0 at the tip of the peel strip where the peel force is acting.

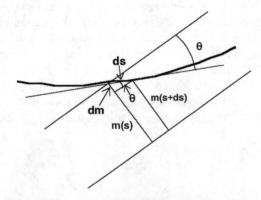

FIGURE 5.13 Expanded view of the peel strip near point P1 to help clarify the relationship between the increment in bending moment dm, the increment in length along the peel strip ds, and the angle θ.

peel strip of length ds between P1 and P2. From this diagram, it is clear that $dm_s = ds \sin \theta$, where θ is the angle between the tangent line at P1 and the peel force vector F as shown in Figure 5.12. The next step is to take the differential of both sides of Equation (5.16) and replace dm_s with $ds \sin \theta$:

$$EI \, d\left[\frac{d\theta}{ds}\right] = F \sin \theta \, ds \tag{5.17}$$

Equation (5.17) gets us closer to what we want, but the right-hand side of this equation still presents a problem because it mixes θ and ds. We need to eliminate ds in terms of $d\theta$. This is done by multiplying both sides of Equation (5.17) by $d\theta/ds$ and eliminating ds on the right-hand side using the calculus expression $(d\theta/ds)ds = d\theta$ to obtain

$$EI \, \frac{d\theta}{ds} \, d\left[\frac{d\theta}{ds}\right] = F \sin \theta \, d\theta \tag{5.18}$$

Equation (5.18) can now be integrated directly from $\theta = 0$ to $\theta = \omega$. Thus, referring to Figure 5.12, the integration proceeds from the tip of the peel strip where $\theta = 0$ to the debonding point where $\theta = \omega$, remembering that θ is the angle between the tangent to the peel strip and the peel force vector. The integration is straightforward and leads to the following result:

$$\frac{EI}{2}\left[\frac{d\omega}{ds}\right]^2 = F(1 - \cos \omega) \tag{5.19}$$

The next step is to eliminate the term $d\omega/ds$ from the above equation in terms of the peel force F and the moment arm at the debonding point m. This is achieved using Equation (5.16) evaluated at $\theta = \omega$ as follows:

$$EI \, \frac{d\omega}{ds} = F \, m \tag{5.20}$$

After performing the substitution and carrying out some algebraic manipulation, we arrive at the following expression relating the moment arm m to the applied peel force F:

$$m = \left[\frac{2EI(1 - \cos \omega)}{F}\right]^{1/2} \tag{5.21}$$

Equation (5.21) gives us a nice relationship between quantities that can in principle be measured conveniently except for the moment arm m, which presents some awkward measurement problems. However, m can be eliminated if we note that Fm, which is the moment applied by the peel force at the debonding point, must be balanced by a countervailing moment M_a exerted by the adhesive, giving the simple relationship $Fm = M_a$. Using this equation to eliminate m from Equation (5.21), we arrive at an expression relating the peel force to the peel angle plus quantities related to the mechanical properties of the peel strip and the binding forces exerted by the adhesive, all of which are constant during a peel test.

$$F = \frac{\left(M_a^2/2EI\right)}{\left(1 - \cos\omega\right)} = \frac{K}{\left(1 - \cos\omega\right)} \tag{5.22}$$

Equation (5.22) at last gives a nice, simple relationship between the peel force and the peel angle that can be easily tested experimentally. Figure 5.14 shows a plot of F versus $(1 - \cos\omega)$ for a spring metal peel strip glued to a rigid substrate with a rubber-based adhesive. The data are replotted from Kaelble's[8] Figure 9. Data for two peel rates are plotted, and the fits look quite reasonable, at least for the smaller peel angles between 40° and 90° [larger values of $1/(1 - \cos\omega)$]. According to Equation (5.22), the slopes of these lines should reflect the strength of the adhesive and the elastic properties of the spring steel. Because the peel strip properties are identical for both sets of data, the plots clearly indicate that the apparent strength of the adhesive is higher at the higher peel rate. This is entirely consistent with the rheological properties of rubbery materials.

Finally, because Equation (5.22) takes into account only the elastic properties of the peel strip, it cannot be expected to hold for viscoelastic or viscoplastic materials. Thus, we can attribute the relatively good fits shown in Figure 5.14 to the

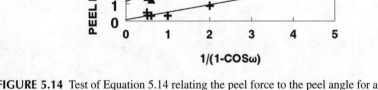

PEEL FORCE VS PEEL ANGLE

Replotted data of Kaelble ref. 8

+   0.02 in/min   ▲   200 in/min

FIGURE 5.14 Test of Equation 5.14 relating the peel force to the peel angle for a metal strip glued to a rigid substrate.

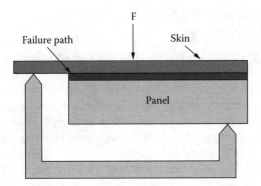

FIGURE 5.15 Novel approach to peel testing first studied by Yurenka.[9] Note that this approach is best adapted to peeling a relatively rigid skin material off a rigid substrate.

fact that the peel strip material is a highly elastic spring steel. In fact, the scatter in the data at high peel angles near 180° [low values of $1/(1 - \cos \omega)$] can at least in part be attributed to plasticity effects in the peel strip caused by the high bending strains.

The peel test was further analyzed by Yurenka,[9] who essentially expanded on and generalized the work of Bikerman.[6] Yurenka in addition analyzed a variant of the peel test that is advantageous for measuring the adhesion of thick-skinned bonded panels. A schematic of this test is shown in Figure 5.15. Yurenka gave an elementary analysis of this test based essentially on an expanded version of Bikerman's work. In particular, he took the critical bending moment for crack initiation M_c as a definition of the adhesion strength of the bond and arrived at the following equation:

$$M_c = \frac{F_c}{\left(F_c y_c / 2EI\right)^{1/3}} \tag{5.23}$$

where

M_c = Critical bending moment at joint failure
F_c = Critical applied load
y_c = Maximum deformation of adhesive at failure
E = Modulus of panel skin
I = Cross-sectional moment of inertia of panel skin

Use of Equation (5.23) is not recommended except possibly as a rough-and-ready estimate of the adhesion strength. A modern analysis of the problem would use the fracture mechanics analysis developed by Suo and Hutchinson discussed in Section 4.2.2. This test has much to recommend it in terms of ease of implementation and the relevance of the data to common use conditions endured by the actual structure. The main limitation, as can be inferred from Figure 5.15, is that this test is limited to relatively thick adherends with sufficient stiffness to support the required loading conditions. Those interested in pursuing this type of analysis further should also consult the work of Gardon,[10] who also extended and applied the previous work of Bikerman and Kaelble.

5.2.2 ELASTIC-PLASTIC ANALYSIS

To this point, the assumption of elastic behavior of both the peel strip and the adhesive has been made to simplify greatly the theoretical analysis by allowing the use of elementary beam theory. It was obvious, however, even to the early workers in adhesion measurement that the assumption of elastic behavior simply did not hold for many situations of practical importance.

Thus, Gent and Hamed[11] came forth with one of the first analyses of the peel test that specifically took into account the nonlinear behavior of the peel strip. They approached the problem both theoretically and experimentally. From the theoretical side, they used essentially standard bending beam theory modified to account for plasticity behavior. Experimentally, they employed a number of special peel test setups designed to control the radius of curvature of the peel strip and thus also the level of strain. Figure 5.16 illustrates the apparatus for peeling at a controlled radius of curvature. This setup cleverly controls the radius of curvature of the peel strip by symmetrically opposing two strips inside a rigid test frame.

A second method of controlling the peel strip curvature was through the use of a roller (Figure 5.17). The peel strip material chosen was Mylar [DuPont trade name for poly(ethylene terephthalate)], a flexible polycrystalline polymer that exhibits tensile stress strain behavior close to that of an ideal elastic-plastic solid. Figure 5.18 shows a schematic diagram of the stress distribution expected in a beam of an ideal elastic-plastic solid deformed in bending just beyond the critical strain required for plastic flow to occur in the most highly strained portion of the beam. As is well

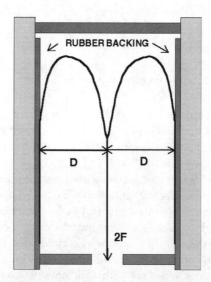

FIGURE 5.16 Modified peel test apparatus for peeling at a controlled bending curvature. Adapted from work of Gent and Hamed.[11]

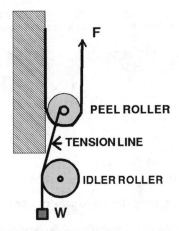

FIGURE 5.17 Peel testing using a roller to fix the radius of curvature of the peel bend. Adapted from work of Gent and Hamed.[11]

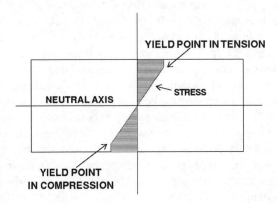

FIGURE 5.18 Stress distribution in a peel strip in which the stress has just exceed the yield point at the outer surfaces of the strip.

known, a beam of rectangular cross section in simple bending will have a neutral axis located at the center where the bending strain is zero. Above the neutral axis, the beam is in longitudinal tension, and below the neutral axis it is in compression, assuming the beam is bent downward.

Figure 5.18 shows the expected stress profile of the Mylar peel strip treated as an elastic perfectly plastic material. Moving up from the neutral axis, the stress in the Mylar ramps up in a linear fashion until it reaches the yield point of the material; then, the Mylar begins to flow at a constant level of stress. Below the neutral axis, the picture is much the same except that the stress in now compressive.[12]

Using the apparatus shown in Figure 5.16, Gent and Hamed were able to control the radius of curvature of the peel bend and thus the level of strain experienced by the Mylar. For a gap spacing D between the two sides of the test frame of 3.6 mm or larger, the maximum strain experienced by the Mylar was below the elastic limit,

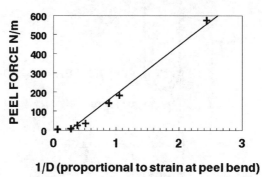

FIGURE 5.19 Peel force versus plate separation for Mylar peel strip in the apparatus shown in Figure 5.16. Note that the strain in the Mylar sample is inversely proportional to the plate separation D.

and the peel force per unit width was effectively constant at close to 4 N/m. For closer spacings, the strain at the surfaces of the Mylar exceeded the elastic limit, and plastic flow set in as illustrated in Figure 5.18. At even higher strains, the peel force shoots up dramatically as the amount of peel force energy dissipated in plastic flow starts to increase. This is shown graphically in Figure 5.19. The data in this figure show a clear break as the strain in the Mylar exceeds the elastic limit. Plasticity effects are now clearly dominating the peeling process. When peeling at low strains, the peel force need only supply enough energy to delaminate the peel strip from the substrate and thus the peel force per unit width can be taken as a true measure of the attraction between the adherend and the substrate. For the data shown in Figure 5.19, the Mylar strips were essentially pressed onto rubber backing sheets attached to the test frame; thus, the level of adhesion is a low 4 N/m or 4 J/m². Note from the figure, however, that without changing the true adhesion strength in the slightest, the apparent adhesion strength can go to nearly 600 J/m² just because of plasticity effects.

As an offshoot of their studies on the effect of plasticity on adhesion strength, Gent and Hamed were able to explain previous results that showed that the peel force went through a maximum as the peel strip thickness increased. This phenomenon can be readily explained in terms of plasticity effects as follows: Assuming strong adhesion of the strip to the substrate, the peel force will go up with increasing peel strip thickness caused by the larger volume of the thicker strips, which require more plastic work to bend and thus a higher peel force. However, this trend cannot continue indefinitely because as the strip gets thicker it also gets stiffer, and a point is eventually reached at which the bend radius increases to the point at which only the outermost regions of the strip are sufficiently strained to go into plastic flow, thus forcing the effective peel force down. As the thickness of the peel strip increases further, the increased stiffness prevents the bending strain from exceeding the yield point in any part of the peel strip; thus, perfectly elastic peeling occurs, and the peel force levels off again.

5.2.2.1 Theory of Elastic-Plastic Peeling for Soft Metals

We conclude our analysis of the quantitative aspects of peel testing with an overview the elastic-plastic analysis of the peel test by Kim and co-workers.[13-18] These authors have been among the most active of the modern investigators who have analyzed the effects of nonlinear mechanical behavior of the peel strip during peel test measurements. Unlike all of the work discussed to this point, which has addressed the problem from the point of view of applied forces and moments, the analysis of plasticity and viscoelasticity effects relies more heavily on energy considerations. Therefore, by way of introduction to the energy approach to peel test analysis, we revisit the problem of peeling a perfectly elastic flexible film off a rigid substrate.

The geometry of the problem is illustrated in Figure 5.20. This figure illustrates a peel force F acting at an angle θ to the substrate and peeling off an increment of the film of length dl. The basic formula of applied work states that the increment of work done peeling the element **dl** with an applied force **F** is dW = **F** · **dl**. Note that dW is the scalar product of two vectors; therefore, it is only that portion of the segment dl that was removed parallel to the direction of F that counts as contributing to dW. Referring to Figure 5.20, we see that the point P1 where the peel starts moves through some arc to the point P2 after the increment of film of length dl is removed. What we need is the distance between P2 and P1 parallel to the force vector **F** as the effective distance covered.

Referring to Figure 5.20, we see that the total distance between P2 and P1 is given by the displacement vector **dr**. The component of **dr** parallel to the applied force is shown as dl′ in the diagram. From the geometry shown, we easily see that dl′ = dl $(1 - \cos \theta)$. Thus, the increment of work done is dW = Fdl$(1 - \cos \theta)$. In peel testing, we like to think in terms of work per unit area of film removed U = dW/dA and peel load per unit width of film P = F/w, where w is the width of the peel strip. In terms of these normalized variables, our peel formula for the perfectly elastic strip becomes U = P$(1 - \cos \theta)$. The peel force must act against the molecular

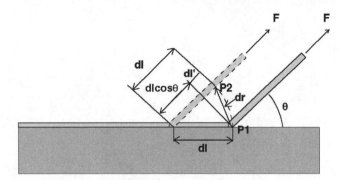

FIGURE 5.20 The energetics of peeling. A peel force F acting at an angle θ to the substrate peels an increment dl of peel strip off the substrate. During this motion, the original contact point P1 moves to P2. Note that, for purposes of computing the work done, it is the distance dl′ = dl$(1 - \cos \theta)$ parallel to the force direction that must be used in computing the work done by the peel force.

forces tending to hold the peel strip to the substrate. The basic assumption of peel testing is that the energy required to remove the coating is a material property of the interfacial region between the coating and the substrate and is commonly designated as γ. In the literature, you will commonly see γ referred to as the surface fracture energy,[19] and it is taken as a truly quantitative measure of the strength of the bonding between the adherend and the substrate. It is this surface fracture energy that the peel load has to overcome to remove the coating. We can therefore write the fundamental formula for perfectly elastic peeling of a film from a rigid substrate as

$$U = P(1 - \cos\theta) = \gamma \tag{5.24}$$

where

U = Work per unit area of peeling
P = Peel force per unit width of peel strip
θ = Peel angle
γ = Surface fracture energy

If all materials deformed according to the laws of linear elasticity, then Equation (5.24) would be all that need be said about the peel test. We know, however, that the majority of materials behave elastically only for small levels of strain, and that the peel test can impose rather large strains on the peel strip material. An elementary calculation can give us some guidance regarding when we need to be concerned about nonelastic behavior when performing the peel test. Figure 5.21 shows the behavior of a peel strip near the delamination point for a 90° peel test off a rigid substrate. We would like to estimate the stress in the peel strip as it approaches the yield point and further relate it to the applied peel load and the thickness of the peel strip. If we know the radius of curvature r of the peel strip at the liftoff point, we can use the beam equation, Equation (5.14), to relate the applied bending moment and the stiffness *EI* of the peel strip as follows:

$$M = \frac{EI}{r} \tag{5.25}$$

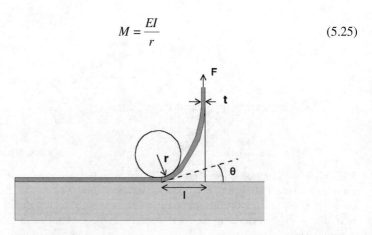

FIGURE 5.21 Geometric parameters of the 90° peel test.

We also see from Figure 5.21 that the applied peel load F acts with a moment arm of length *l* at the liftoff point, so the bending moment is also equal to *Fl*, which we can substitute into Equation (5.25) to obtain

$$Fl = \frac{EI}{r} \qquad (5.26)$$

At this point, we need to estimate how large the radius of curvature r is in terms of the moment arm length l. In general, we do not know these quantities exactly, but we do know that they are roughly the same size. Thus, we write $l = nr$ where n is some constant not too different from 1 or 2. Equation (5.26) can now be written as

$$F = \frac{EI}{nr^2} \qquad (5.27)$$

We now recall that, for simple bending, the strain in the peel strip is related to the radius of curvature by Equation (3.55), which was derived in Chapter 3 as follows:

$$\varepsilon = \frac{x}{r} \qquad (5.28)$$

where

ε = Strain level
x = Distance measured from neutral line
r = Radius of curvature

The maximum strain is attained when x reaches the surface of the peel strip at a distance of $t/2$ from the neutral line where t is the strip thickness as shown in Figure 5.21. Now, we are interested in the strain level at which plastic yielding just sets in, which we designate as ε_y. Because we know that the maximum strain will occur at the strip surface for any given level of bending, we can rewrite Equation (5.28) to express the condition of the onset of yielding in the peel strip as

$$\varepsilon_y = \frac{t}{2r_y} \qquad (5.29)$$

where

ε_y = Strain level at the onset of plastic yielding
r_y = Radius of curvature at the onset of plastic yielding

Equation (5.29) can now be used to eliminate the radius of curvature from Equation (5.27) and arrive at an expression for the peel load at the onset of yielding at the liftoff point

$$F = \frac{4\varepsilon_y^2 EI}{nt^2} \qquad (5.30)$$

We now use Hooke's law to express the yield strain in terms of the yield stress $\varepsilon = \sigma/E$ to obtain

$$F = \frac{4\sigma_y^2 I}{nEt^2} \qquad (5.31)$$

Assuming a rectangular cross section for the peel strip, we have the following standard formula for the moment of inertia I in terms of the strip thickness t and width w:

$$I = \frac{wt^3}{12} \qquad (5.32)$$

Substituting Equation (5.32) into Equation (5.31) and dividing both sides by the peel strip width w,

$$P = \frac{F}{w} = \frac{t\sigma_y^2}{3nE} \qquad (5.33)$$

Equation (5.33) can be solved for the thickness of the peel strip in terms of the peel load:

$$t_c = \frac{3nEP}{\sigma_y^2} \qquad (5.34)$$

where

t_c = Critical peel strip thickness for onset of yielding
E = Modulus of peel strip material
P = Applied peel load per peel strip width
σ_y = Yield stress of peel strip material
n = Numerical factor in range of 1–2

Equation (5.34) is most reliably applied to relatively soft metals with well-defined yielding behavior, such as copper, aluminum, lead, and the like. This relation can be handy in a number of practical situations. Say, for instance, that you are working in the microelectronics industry, and you need to know the adhesion strength of 10-μm copper metal lines on some rigid substrate material such as silicon. The peel test is convenient for this application, but you are concerned about plasticity effects on your results. You perform a few trial peel experiments and note that the observed peel load is 100 N/m (equivalent to 100 J/m²). This is not a very high peel strength, but you need to know whether plasticity effects are important in your peel test measurements. Looking up the modulus and yield stress of copper in some standard materials handbook, you find the following regarding the mechanical properties of reasonably pure annealed copper:

$$E = 110 \text{ GPa} = 110 \times 10^9 \text{ Pa}$$

$$\sigma_y = 70 \text{ MPa} = 70 \times 10^6 \text{ Pa}$$

Substituting these values into Equation (5.34) along with the observed peel strength of 100 Nt/m and taking a conservative estimate of n as equal to 1, you arrive at a value of t_c of approximately 6.7 mm. Thus, to avoid plasticity effects, your peel strips would have to be at least two orders of magnitude thicker than your 10-μm lines. Clearly, to achieve quantitative results you must subject your peel data to a full elastic-plastic analysis.

5.2.3 FULL ELASTIC-PLASTIC ANALYSIS

5.2.3.1 General Equations for Deformation of Peel Strip

In what follows, a fairly detailed outline is given of the elastic-plastic analysis of the peeling process as developed by Kim and co-workers and reported in the general literature.[13–18] From this point on, some fairly heavy sledding through detailed theoretical mechanics-type calculations is involved, although the basic ideas are fairly straightforward. Every effort will be made to be as clear and concise as possible, with a premium given to comprehension as opposed to achieving full generality or maximal terseness.

5.2.3.2 Basic Goal

The object of this exercise is to obtain formulae or workable procedures (numerical or graphical) to analyze peel test data on samples for which plasticity effects cannot be ignored, as in the case of the thin copper strips on silicon discussed in Section 5.2.2.1. We know that if everything remains elastic, then the measured peel force per unit width gives the quantity we are after, which we call the surface fracture energy and, following the general literature, designate it as γ. The main reason for wanting γ is that it should represent more closely the work required to separate the coating from its substrate without confounding effects related to the mechanical properties of the substrate or the coating materials. One further reason for wanting γ is that it should allow one to compute theoretically the stability of a given interface under the loading conditions it will have to endure under service conditions. Thus, the ultimate benefit of determining γ for a given coating/substrate system under study is that it can be used in conjunction with fracture mechanics analyses to determine whether the interface will survive any particular prescribed loading conditions.

5.2.3.3 Analysis Strategy and Assumptions

The analysis strategy is quite simple in principle. We need to estimate how much of the work exerted by the tensile test machine is dissipated by irreversible plasticity effects for every unit length of the peel strip that is removed. Once we know what this work is, we can subtract it and be left with only that fraction of the work that was actually needed to separate the interface, which is our coveted quantity γ. Following the work of Kim and co-workers,[13] we make the following assumptions:

1. We assume the peel strip material closely follows the elastic perfectly plastic stress-strain behavior as depicted in Figure 5.18.
2. All of the nonelastic energy dissipation occurs as a result of plastic bending. This assumption glosses over the plasticity effects caused by the large stresses induced near the delamination point. This is a highly localized region and accounts for a small part of the total energy dissipated. The current analysis effectively lumps this energy into the surface fracture energy γ.
3. The deformation of the peel strip is governed by the equations of the "elastica," which will be derived in Section 5.2.3.4.

The analysis procedure proceeds in the following manner. First, a set of equations that describe the behavior of the peel strip as it is removed from the substrate is derived. These are the equations of the elastica. They relate the bending moment, the tensile stress, and the shear stress in the peel strip to the strip curvature as a function of the angle the strip makes with the plane of the substrate. These equations are needed to provide a relationship between the bending moment of the strip and the curvature, which in turn is needed to compute the energy expended in peeling. These relations are not enough, however, because they only provide three relations among four quantities. To complete the formulation, we further need a set of relations describing the constitutive behavior of the peel strip material (i.e., the stress-strain curve or, what is equivalent, the moment curvature relation). The required constitutive relation is taken directly from the classic plasticity literature.[20] These equations are then manipulated to determine the energy dissipated per unit length advance of the peel strip under conditions of steady-state peeling.

5.2.3.4 Equations of the Elastica

We now need to derive relationships among the various forces that determine the deformation of the peel strip as it is removed from the substrate. Figure 5.22 shows a section of peel strip under a tensile load $\mathbf{F}$. The strip is anchored to the substrate somewhere to the left of the diagram at the point of detachment P0 (not shown). The parameter s measures the distance between P0 and some arbitrary point P1 on the strip. The line L1 is tangent to the strip at P1 and makes an angle ϕ with the horizontal. The peel load $\mathbf{F}$ induces an axial tension $\mathbf{T}(s)$, a normal shear $\mathbf{N}(s)$, and a bending moment $\mathbf{M}(s)$ at every point on the strip; as designated, these quantities depend on the distance s from the detachment point P0. Our task is to derive the

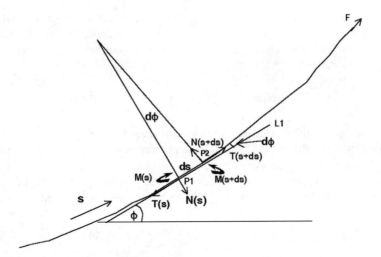

FIGURE 5.22 Force balance diagram for a small element ds of the peel strip. At equilibrium, the bending moment M(s), the normal force N(s), and the tension T(s) must be balanced between both ends of the element ds.

appropriate relationship among these quantities as dictated by the laws of momentum or force equilibrium. As we advance from point P1 by some small distance ds along the strip, we arrive at point P2. The angle ϕ increases by an amount $d\phi$, and the tangent line at P2 is given by L2.

Figure 5.23 gives an expanded detailed picture of the situation. At the point P2, the quantities $\mathbf{T}(s)$, $\mathbf{N}(s)$, and $\mathbf{M}(s)$ become $\mathbf{T}(s + ds)$, $\mathbf{N}(s + ds)$, and $\mathbf{M}(s + ds)$. Equilibrium of the increment of strip ds demands that the axial and normal components of each of these forces balance between the points P1 and P2 as the increment ds tends to zero. We take the right as the positive direction in Figure 5.23 and start with the axial tension $\mathbf{T}$. We note from the figure that $\mathbf{T}(s + ds)$ has a component $T(s + ds)\cos(d\phi)$ along the line L1 and must be balanced by the tension $T(s)$ in the opposite direction. However, looking at the line L1′ in the diagram, which is parallel

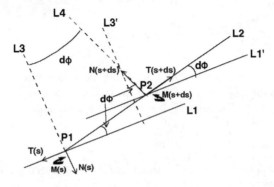

FIGURE 5.23 Greatly expanded view of the peel strip element ds to show more clearly the relationships among the acting forces and moments.

to L1, we notice that the shear force $N(s + ds)$ contributes a negatively directed component along the direction of L1 given by $N(s + ds)\sin(d\phi)$. At equilibrium, the algebraic sum of all these components must come to zero, so we can write our first equation as

$$T(s+ds)\cos(d\phi) - T(s) - N(s+ds)\sin(d\phi) = 0 \qquad (5.35)$$

We proceed in the same way for the normal shear force $N(s)$. This time, we take components along the line L3, which is perpendicular to the peel strip and parallel to the line L3'. Taking the direction toward the center of curvature as positive and noticing that the tension $T(s + ds)$ makes a contribution $T(s + ds)\sin(d\phi)$ along the L3 direction, we derive the following relation for the normal shear force:

$$N(s+ds)\cos(d\phi) - N(s) + T(s+ds)\sin(d\phi) = 0 \qquad (5.36)$$

Last, we need to consider the moment $M(s)$ at point P1. Taking the direction of $M(s + ds)$ as positive, we again balance out the components at point P1 and notice that the normal shear force $N(s + ds)$ contributes a component $N(s + ds)ds$ to the moment at P1; we obtain*

$$M(s+ds) - M(s) + N(s+ds)ds = 0 \qquad (5.37)$$

Now, we divide Equation (5.35) to (5.37) by ds; noting the standard definition of a derivative and the limits $\cos(d\phi) \to 1$ and $\sin(d\phi) \to d\phi$ as $d\phi$ tends to zero, we obtain the following final collection of equations:

$$\frac{dT}{ds} - N(s)\frac{d\phi}{ds} = 0$$

$$\frac{dN}{ds} + T(s)\frac{d\phi}{ds} = 0 \qquad (5.38)$$

$$\frac{dM}{ds} + N(s) = 0$$

The quantity $d\phi/ds$ in Equation (5.38) is the local curvature of the peel strip at some location a distance s along the strip axis from the removal point and is designated as $K(s)$.** Equations (5.38) can now be written as

---

* The reader may note that the tension $T(s + ds)$ also makes a contribution given by $T(s + ds) \sin(d\phi)ds$. However, in the limit of vanishing $d\phi$ and ds, this is a second-order quantity of order $d\phi ds$ and does not contribute to the equation.
** The curvature is also given by $1/R(s)$, where $R(s)$ is the radius of the circle exactly tangent to the peel strip at location s. Thus, a handy set of relations to be cognizant of is $K(s) = d\phi/ds = 1/R(s)$.

$$\text{a) } \frac{dT}{ds} - N(s)K(s) = 0$$

$$\text{b) } \frac{dN}{ds} + T(s)K(s) = 0 \qquad\qquad (5.39)$$

$$\text{c) } \frac{dM}{ds} + N(s) = 0$$

These equations can be simplified significantly by noting that Equation (5.39a) and (5.39b) have the following solutions:

$$T(s) = P\cos(\phi - \theta)$$
$$\qquad\qquad\qquad (5.40)$$
$$N(s) = P\sin(\phi - \theta)$$

where θ is a constant angle, taken to be the peel angle in this case. This can be verified by direct substitution and noting that $K(s) = d\phi/ds$. We can now focus on Equation (5.39c). We need to forge this equation into a single relationship involving only the moment and curvature functions $M(s)$ and $K(s)$, respectively. The first step is to eliminate the shear force function $N(s)$ using Equation (5.39a) to obtain

$$K(s)\frac{dM(s)}{ds} + \frac{dT(s)}{ds} = 0 \qquad\qquad (5.41)$$

Equation (5.41) can now be integrated over ds to give

$$\int KdM + T(s) = C \qquad\qquad (5.42)$$

C is some as yet undetermined constant of integration. We now apply the first of Equations (5.40) to eliminate T in favor of the applied peel load and further apply the standard calculus formula $d(MK) = KdM + MdK$ to the integrand to transform Equation (5.42) into

$$P\cos(\phi - \theta) + MK - \int KdM = C \qquad\qquad (5.43)$$

Equation (5.43) is the basic force balance relation that governs the steady-state equilibrium of the peel strip. The peel force P is measured in newtons, the local bending moment M is measured in newton-meters, and the local curvature K has dimensions of inverse meters. Thus, we see that every term in Equation (5.43) has the dimensions of force and is therefore dimensionally correct. When dealing with peel testing, however, it is always convenient to work in terms of force per unit width of the peel strip. Thus, we divide both sides of Equation (5.43) by the strip width w to obtain

$$p\cos(\phi - \theta) + \underline{M}K - \int Kd\underline{M} = C \tag{5.44}$$

In Equation (5.44), p is a force per unit width, and $\underline{M}$ is the moment per unit width. The next step is to introduce the following set of dimensionless variables:

a) $k = \dfrac{K}{K_0}$ $\left(\text{Normalized bending curvature}\right)$

b) $K_0 = \dfrac{2\sigma_0}{Et}$ $\left(\text{characteristic curvature}\right)$

c) $m = \dfrac{M}{M_0}$ $\left(\text{normalized bending moment}\right)$ $\qquad$ (5.45)

d) $M_0 = \dfrac{\sigma_0 t^2}{4}$ $\left(\text{characterisic bending moment}\right)$

e) $\eta = \dfrac{3p}{M_0 K_0} = \dfrac{6Ep}{t\sigma_0^2}$ $\left(\text{characteristic peel force}\right)$

The parameters in Equations (5.45) are defined as follows:

σ_0 = Yield stress of peel strip material
E = Modulus of peel strip
t = Thickness of peel strip
p = Peel load per unit width of peel strip

The momentum balance equation, Equation (5.44), can now be recast in terms of the dimensionless variables in Equation (5.45) as follows:

$$\frac{\eta}{3}\cos(\phi - \theta) + mk - \int kdm = C \tag{5.46}$$

Equation (5.46) is the working form of the force balance relation for a peel strip where θ is the peel angle. In all that follows, we assume a 90° peel test, so $\theta = \pi/2$, and because $\cos(\phi - \pi/2) = \sin(\phi)$, Equation (5.46) is slightly modified to the following:

$$\frac{\eta}{3}\sin(\phi) + mk - \int kdm = C \quad \text{(90° peel test)} \tag{5.47}$$

Equation (5.47) simply sets the conditions of force balance for the peel strip and by itself is not sufficient to determine the moment or the curvature of the strip as a function of s, the distance from the liftoff point. As is always the case in the mechanics of continuous media, a further constitutive relation is required, which in the case of the peel strip problem is a relation between the bending moment and the curvature. For this relationship, we follow the work of Kim and Kim[21]:

a) $m = \dfrac{2}{3}k \quad 0 \le k \le 1$ (O-A Elastic loading)

b) $m = 1 - \dfrac{1}{3k^2} \quad 1 \le k \le k_B$ (A-B Plastic flow stage)

c) $m = \dfrac{1}{3}\left[3 - \left(\dfrac{1}{k_B^2} + 2k_B \right) + 2k \right] \quad 1 < k_B \le 2$ (B-C Elastic unloading)

d) $m = \dfrac{1}{3}\left[-3 - \dfrac{1}{k_B^2} + \dfrac{8}{\left(k - k_B\right)^2} \right] \quad k_B > 2$ (C-D Reverse plastic loading)

(5.48)

Because elastic and plastic deformation are fundamentally different phenomena, Equations (5.48) have to employ a piecewise representation of the moment curvature constitutive relation. This relation is plotted in Figure 5.24. When the amount of bending of the strip gives rise to a normalized curvature k less than 1, the response of the peel strip is entirely elastic, and the normalized bending moment follows the section marked O-A in the diagram. This will be the case when either the peel strip is quite thick, causing it to bend much less, or the adhesion is low, requiring only a small peel force to effect delamination. As the bending becomes more severe,

MOMENT CURVATURE DIAGRAM
ELASTIC-PLASTIC MATERIAL

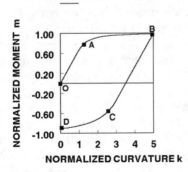

FIGURE 5.24 Moment curvature relation for an elastic-plastic material. This is essentially equivalent to a more usual stress strain plot except that angular units are employed, which are more convenient for examining the peel test where the dominant mode of deformation is bending.

causing k to be greater than 1 but less than some maximum k_B, plastic flow sets in, and the bending moment follows the section marked A-B. For those sections of the peel strip that have passed through the maximum bending, which occurs at the liftoff point, the strip starts to straighten out, and reverse bending occurs. This behavior is governed by the section marked B-C in the diagram. Finally, if the peel load is high enough and the maximum normalized curvature k_B exceeds 2, then reverse plastic unloading also occurs; this is described by Section C-D of the moment curvature relationship.

One final relationship is required to complete the picture so that the surface fracture energy γ can be determined; this is the energy balance relationship. For simple elastic peeling, this is given by Equation (5.24), which for the case of 90° peeling can be written as

$$p = \gamma \qquad\qquad (5.49)$$

where p is the peel load per unit width of the peel strip. Thus, for elastic peeling, energy balance says that the energy expended by the test equipment goes entirely into the work required to cleanly separate the peel strip/substrate interface. However, when dissipative mechanisms such as plasticity are operating, another term is required in Equation (5.49) to account for the energy that is soaked up by such processes, which leads to the following:

$$p = \gamma + \psi \qquad\qquad (5.50)$$

During steady-state peeling, Equation (5.50) states that the energy expended by the test equipment goes not only into separating the interface but also into plastic dissipation, as given by the function Ψ, which is the dissipation rate per unit advance of the peel strip. Our task now becomes the evaluation of Ψ so that the dissipative effects can be subtracted out and the "true" surface fracture energy can be recovered.

Referring to Figure 5.25, we imagine that, during steady-state peeling, every small segment of the peel strip that is originally attached to the substrate goes successively through the zones marked O-A through C-D as the peeling process proceeds. Thus, each segment sees exactly the same load history; we therefore need to examine only one typical case to determine the energy loss per unit length, which is the function Ψ we are after. Thus, referring to the figure, each segment goes through a sequence of loading and unloading steps as follows:

- *Segment O-A: Elastic loading.* As the peel front approaches a segment that is about to be lifted off the substrate, the stress caused by peeling is fairly low; thus, the loading is entirely elastic. The moment curvature behavior for this region is shown by segment O-A in Figure 5.24.
- *Segment A-B: Plastic loading.* As the segment enters the peel bend and starts lifting off, the curvature increases dramatically and so does the elastic strain, which gives rise to plastic flow in the peel strip. The moment curvature behavior for this segment is shown by segment A-B in Figure 5.24.

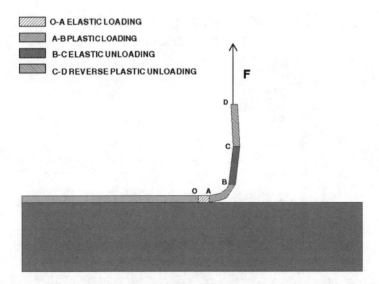

FIGURE 5.25 Schematic of the bending process that occurs on peeling an elastic-plastic material from a rigid substrate. The various stages of the bending process correlate with the similarly labeled sections of the moment curvature diagram in Figure 5.24.

- *Segment B-C: Elastic unloading*. Eventually, the segment passes through the peel bend and starts to straighten again. This gives rise to reverse bending, in which the strains are imposed in the reverse direction and are initially small; thus, the deformation is in the elastic range. This behavior is shown by segment B-C in Figure 5.24.
- *Segment C-D: Plastic reverse unloading*. When the maximum curvature is high enough, reverse plastic bending also occurs, as shown by segment C-D in Figure 5.24.

Note that Figure 5.24 illustrates the general case of fairly high loading, for which the bending is high enough to drive the peel strip through every stage of loading shown in the figure. For cases of low adhesion, the deformation may be entirely elastic and thus confined to segment O-A. As the loading increases, plasticity eventually sets in, and segments A-B and B-C must be included. In general, however the entire moment curvature loop shown in Figure 5.24 must be considered. In all cases, however, the energy dissipated per unit length of peel strip is given by the area enclosed by the moment curvature loop. This can be easily calculated from our model because we know the equations describing the moment curvature loop from Equation (5.48). In what follows, we consider each case in turn, starting from the simplest case of elastic peeling.

5.2.3.5 Case 1: Elastic Peeling

If the normalized curvature k is always less than 1 for the entire peel process, then we have elastic peeling. The moment curvature diagram for this case is illustrated in Figure 5.26. In this case, the typical peel strip segment experiences only elastic

MOMENT CURVATURE DIAGRAM
ELASTIC PEELING

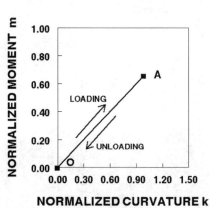

FIGURE 5.26 Moment curvature diagram for the case of totally elastic deformation of the peel strip.

loading on going through the peel bend and elastic unloading on exiting. There is no hysteresis loop in the moment curvature diagram; thus, no energy is dissipated, and the dissipation function Ψ is exactly zero. From Equation (5.50), we see that in this case $p = \gamma$, which is what we knew had to be the case all along for elastic peeling.

5.2.3.6   Case 2: Elastic-Plastic Peeling/Unloading

Things get more interesting as the peel load increases, driving the maximum normalized curvature k above 1. The moment curvature diagram for this case is illustrated in Figure 5.27. As the peel strip segment enters the peel bend, the bending stress exceeds the yield stress of the material, plastic flow sets in, and the material in the bend region starts behaving more like a highly viscous liquid than an elastic solid. All the strain deformation incurred from this point is not recoverable, resulting in permanent deformation of the strip. The moment curvature relation now follows the segment labeled A-B in Figure 5.27. Because the strain incurred is permanent, there is no elastic recovery, and the area under the moment curvature line is proportional to the energy dissipated by the plastic deformation. The contribution to the energy dissipation function Ψ is given by the following integral:

$$\psi = \int_A^B M dK = M_0 K_0 \int_A^B \frac{M}{M_0} d\frac{K}{K_0} = M_0 K_0 \int_A^B m dk \qquad (5.51)$$

As the peel segment exits the bend region, the peel load now acts to straighten it out again, and it is now subjected to a reverse bending load. If the maximum

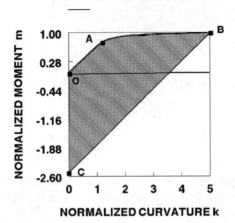

MOMENT CURVATURE DIAGRAM

ELASTIC-PLASTIC PEELING

FIGURE 5.27 Moment curvature diagram for the case of plastic deformation of the peel strip with elastic unloading.

curvature experienced by the strip in the bend region k_B is less than 2, then according to the moment curvature constitutive relations of Equations (5.48), the unloading is elastic in nature and follows a straight line as shown in Figure 5.27. Because the unloading is elastic, it gives back energy to the peel strip, as in the case of perfect elastic peeling shown in Figure 5.26. This elastic unloading energy is given by the area under the elastic unloading line B-C. The sum total of the energy dissipated is thus the integral over the entire transit O-A-B-C as follows:

$$\psi = M_0 K_0 \left[\int_0^1 m_a dk + \int_1^{k_B} m_b dk - \int_0^{k_B} m_c dk \right] \tag{5.52}$$

The integrands in Equation (5.52) are the various sections a, b, and c of the moment curvature relation defined in Equation (5.48). It is readily seen that the dissipation function Ψ is given by $M_0 K_0$ times the shaded area shown in Figure 5.27. A straightforward calculation gives the following result for Ψ:

$$\psi = M_0 K_0 \left(\frac{2}{3k_B} + \frac{k_B^2}{3} - 1 \right) \quad 1 \leq k_B \leq 2 \tag{5.53}$$

5.2.3.7 Case 3: Elastic-Plastic Loading and Unloading

If the maximum curvature k_B is greater than 2, then the full moment curvature relation given in Figure 5.24 must be used. In this case, the reverse bending phase of the peel process also involves plastic flow, and the dissipation function is given by $M_0 K_0$

times the area inside the loop O-A-B-C-D shown in Figure 5.24. The calculation is straightforward, albeit tedious, as above. The final results for the dissipation function for all ranges of the maximum curvature k_B are summarized as follows:

a) $\psi = 0 \quad 0 \le k_B \le 1$

b) $\psi = M_0 K_0 \left(\dfrac{2}{3k_B} + \dfrac{k_B^2}{3} - 1 \right) \quad 1 \le k_B \le 2$ (5.54)

c) $\psi = M_0 K_0 \left(2k_B - 5 + \dfrac{10}{3k_B} \right) \quad k_B \ge 2$

Combining the above expressions for the dissipation function with Equation (5.50) allows us to determine the surface fracture energy for each case:

a) $\gamma = p \quad 0 \le k_B \le 1$

b) $\gamma = p - M_0 K_0 \left(\dfrac{2}{3k_B} + \dfrac{k_B^2}{3} - 1 \right) \quad 1 \le k_B \le 2$

c) $\gamma = p - M_0 K_0 \left(2k_B - 5 + \dfrac{10}{3k_B} \right) \quad k_B \ge 2$ (5.55)

$M_0 K_0 = \dfrac{\sigma_0^2 t}{2E}$

Equations (5.55) give the answer to the problem of computing the fundamental adhesion for the elastic-plastic peel test in terms of the measured peel load and the physical properties of the peel strip. The main problem, however, is that we do not know the maximum curvature k_B. This problem can be addressed in two different ways. The first approach is purely theoretical, involving another calculation that is at least 10 times more tedious than that leading up to Equations (5.55). Kim has carried out this calculation and documented it in a technical report.[22] A number of tricky problems arise in this analysis, not the least of which is that you have to be concerned with the deformation of the substrate to determine the peel angle θ_B. Because the substrate will lift up slightly under the peel force, θ_B will depend on the elastic properties of the substrate in a complicated way. Kim further pointed out that the results are quite sensitive to the precise value of θ_B, so this is not a calculation you can fudge by making coarse approximations. In the end, one ends with a collection of messy nonlinear algebraic equations that must be solved by numerical methods, which is not an undertaking for the fainthearted.

The second approach is by experimentally measuring k_B during steady-state peeling. All things considered, this is quite likely the better approach for those who

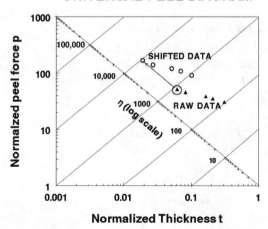

FIGURE 5.28 Universal peel diagram for experimentally determining the surface fracture energy from peel test data by plotting Equation (5.57) using log-log coordinates. After analysis of Kim and Kim.[13]

would seriously like to pursue this type of measurement. One can either measure the maximum curvature of the strip at the peel bend using interferometric methods or optically measure θ_B during steady-state peeling. Either approach should yield reliable values of k_B given the availability of sophisticated computer-controlled metrology equipment.

One final method of estimating γ by purely experimental measurement was mentioned by Kim and Kim*; they called it the universal peel diagram. Figure 5.28 shows such a diagram. Using dimensional analysis arguments, they proposed that for the case of simple elastic-plastic peeling, the problem can be framed in terms of the following dimensionless variables:

a) $\mathbf{p} = \dfrac{p}{\gamma}$ (Normalized peel force)

b) $\mathbf{t} = \dfrac{\sigma_0^2\, t}{6E\gamma}$ (Normalized film thickness)

$$(5.56)$$

c) $\boldsymbol{\eta} = \dfrac{6Ep}{\sigma_0^2\, t}$ (Force/thickness ratio)

d) $\boldsymbol{\eta} = \dfrac{p}{t}$ (Scaling relationship)

---

* See Note 13, Figure 8.

TABLE 5.4
Peel Test Data for Copper on Silicon[a,b]

Copper Thickness t (μm)	Measured Peel Force p = p[c] (g/mm)	Normalized Thickness t (γ = 1)	η (from data)	η (from peel diagram Figure 5.28)
15	52	0.0612	8330	900
20	46	0.0817	5520	500
40	39	0.163	2340	220
51	35.5	0.208	1670	160
75	31	0.306	992	100

[a] Note 13.
[b] $E = 200$ GPa; $\sigma_0 = 70$ MPa for copper.
[c] Note: $p = \mathbf{p}$ assuming $\gamma = 1$.

The parameters in Equations (5.56) are the usual ones defined as follows:

γ = Surface fracture energy
σ_0 = Yield stress of peel strip material
E = Modulus of peel strip
t = Thickness of peel strip
p = Peel load per unit width of peel strip

The argument is as follows: For a given coating substrate system in which the only parameter varied is the coating thickness, the normalized peel force should be some universal function of the normalized thickness, such as $\mathbf{p} = g(\mathbf{t}, a_1, a_2, ..., a_n)$. The a_i are fixed parameters that are particular to the specific system, such as power law hardening parameters and the like. The universal function g can be determined experimentally with the aid of the diagram in Figure 5.28 by measuring the peel force as a function of coating thickness. We first note from Equation (5.56d) that:

$$\log \mathbf{p} = \log \mathbf{t} + \log \mathbf{\eta} \qquad (5.57)$$

Thus, on the log-log plot shown in Figure 5.28, the slanted 45° lines represent lines of constant $\mathbf{\eta}$. The slanted line perpendicular to the constant η lines is obtained from the relationship $\mathbf{p} = 1/\mathbf{t}^{1/2}$, which can be used to determine a log scale for the constant $\mathbf{\eta}$ lines shown in the figure by substituting the $\mathbf{t}$ scale into this equation and plotting the computed $\mathbf{p}$ values on the diagram. The points will all fall on the diagonal line, giving the appropriate log scale for $\mathbf{\eta}$ as shown.

As an example, consider the data in Table 5.4 for a copper film on single-crystal silicon.* Starting with the raw data, we cannot exactly determine the normalized peel force from the normalized thickness because the surface fracture energy γ is

---

* The raw data have been extracted from Note 13, Figure 9 using a ruler and pencil. The accuracy is probably no better than 10 or 20%, but it does not matter because this is only an example.

TABLE 5.5
Summary of Peel Test Data in Figure 5.28

Unshifted Raw Data (γ = 1)			Shifted Data			
t	p = γ	η	t′	p′	η′	γ′
0.0612	52	8330	0.018	155	8600	0.33
0.0817	46	5520	0.026	140	5400	0.33
0.1633	39	2340	0.048	105	2200	0.37
0.2082	35.5	1670	0.062	100	1600	0.35
0.3062	31	992	0.1	92	920	0.34

not known. Therefore, the procedure is as follows: Assume some convenient value for γ and then plot the estimated normalized peel force and thickness on the diagram. For this case, we choose γ = 1 for convenience. The resulting data are shown as the filled triangles in Figure 5.28. Now, if we check, for example, the η value of the point at 15 μm by reading it off the diagonal scale in the diagram, we obtain a value of roughly 900. However, we know that the correct value of η must be 8330 from the experimental data in Table 5.4. Thus, to be consistent we must shift this point parallel to the η axis to the correct value, shown by the arrow in Figure 5.28. Thus, reading from the diagram, new values of **p** and **t** can be estimated as 155 and 0.018, respectively. The new value of η given by the ratio of the shifted values of **p** and **t** is 155/0.018= 8600, which is consistent with the original unshifted data given the crude eyeball estimates of **p** and **t**. Because **p** = p/γ, we can now obtain a corrected estimate for γ as p/**p** = 52/155 = 0.33. The remaining points are shifted in a similar manner. Note the huge difference that plasticity makes. Assuming perfectly elastic behavior of the copper, we would calculate γ = p = 52 gm/mm, whereas the corrected value including plasticity effects gives 0.32, which is only 0.5% of the elastic estimate.

The peel samples used to measure the peel force data in Table 5.4 were prepared in such a manner as to ensure that the interfacial adhesion and thus γ should be the same for each film thickness. Essentially, an initial layer of Cu/Cr was deposited onto a silicon wafer, and coatings of different thickness were obtained by electrodeposition of a further layer of copper. Thus, if we use the corrected values of **p** obtained from the shifted data points in Figure 5.28, then we should obtain a constant value of γ for each point. This is demonstrated in Table 5.5, which summarizes the raw and shifted data in Figure 5.28.

Two points of interest in Table 5.5 are the following: (1) the values of η are consistent between the shifted and unshifted data as they must be, and (2) the corrected value of γ for the shifted data is now constant, which is consistent with the sample preparation discussed above.

5.3 THE SCRATCH/CUT TEST

Alongside the peel test, the scratch test is one of the most popular adhesion tests in regular use in testing laboratories throughout the world. In all likelihood, it outranks the peel test in terms of the number of test units in operation in both industry and academic laboratories as well as industrial manufacturing lines. In a sense, the scratch test is entirely complementary to the peel test in that it applies most satisfactorily to brittle coatings, whereas the peel test is nearly exclusively used for flexible coatings. In addition, the scratch test finds use as a general measure of coating durability, which involves resistance to abrasion as well as resistance to delamination. A typical scratch test unit can be used to estimate the hardness of a coating and surface tribological properties, such as the coefficient of friction. One of the major advantages of the scratch test is its amazing versatility. The following is a sampling of applications from the biomedical and automotive industries:

- Biomedical applications
  - Adhesion of contact lens coatings
  - Mechanical properties of teeth
  - Surface properties of arterial implants
  - Hardness of bone tissue
  - Surface properties of hip prostheses
- Automotive applications
  - Elastic properties of automotive varnishes
  - Wear properties of rubber seals
  - Wear properties of brake pads
  - Hardness of coated tappets and valves
  - Adhesion of bumper paint
  - Scratch resistance of chrome trim

Although the scratch test finds many useful applications, it does suffer significant problems when used strictly for purposes of adhesion measurement. In particular, it is difficult to get fully quantitative data on adhesion strength from the scratch test. A typical scratch test proceeds somewhat as follows:

- A hard stylus is loaded gently onto the surface of the coating to be tested. The stylus is rigidly attached to load cells, which can measure both the normal and tangential force applied to the stylus tip.
- The stylus is slowly dragged over the test surface while the normal load is increased in a linear fashion. The drag and reaction forces are continuously monitored during this process. Modern instrumentation allows these forces to be recorded digitally and saved to a computer file.
- At some point during the test, the increasing load on the stylus causes a failure event of some kind to occur either in the coating or in the underlying substrate. This shows up as a discontinuity in the measured forces on the stylus. The precise nature of the failure event must be evaluated by microscopy either during or after the test. Failure events can also be detected by the acoustic signal they give off.
- As the test continues, a series of failure events will occur, growing more drastic as the normal load increases. The test typically terminates when the stylus has penetrated all the way through the coating and into the substrate.

TABLE 5.6
Factors Affecting the Critical Load for Film Detachment in the Scratch Test

Intrinsic Parameters	Extrinsic Parameters
Loading rate	Substrate properties
Scratching speed	Hardness
Indenter tip radius	Modulus
Indenter wear	Thermal expansion coefficient
Machine factors	Coating properties
	Hardness
	Modulus
	Stress/interface properties
	Thickness
	Friction force/coefficient
	Surface condition/testing environment

From the point of view of adhesion testing, those failure events that cause the coating to separate from the substrate are of most interest. This can arise directly from the load applied to the stylus or as a consequence of residual stresses in the coating. In either case, evaluating the adhesion strength of the coating from the scratch test data is not straightforward and in many cases is not possible. Table 5.6 summarizes the key parameters governing the scratch test as an adhesion measurement tool.

One special case of the scratch test that is amenable to a full quantitative adhesion analysis is the cut test. By this method, a uniform cut is made in the coating down to the substrate, which induces delamination caused by residual stresses in the coating. A review of this technique is outlined in the following section.

5.3.1 THE CUT TEST

Those familiar with the delamination behavior of coatings are aware that film liftoff generally initiates at the edge of the coating or near some internal defect, such as a particulate inclusion, a pit, or a scratch. The basic reason for this is that the stress field near such a defect varies rapidly and can grow quite large near the defect boundary. Elasticity theory in fact predicts that the normal and tangential components of the principal stress tensor become infinite near a sharp edge, although the total strain energy associated with these stresses is finite. This suggests that a simple, straightforward adhesion test would consist of making a simple cut in the interior of the coating and watching whether delamination ensues and if so how far. In what follows, we find that inserting a cut in a coated film can indeed cause delamination, but the analysis involved is surprisingly subtle and complex.

Jensen et al.[23] gave a fracture mechanics analysis of the simple cut test, and Figure 5.29 illustrates the two simplest modes of behavior they investigated. In both of the cases illustrated in the figure, delamination starts from a long cut and then propagates into the bulk of the coating. The only difference between the two cases

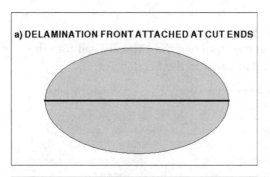

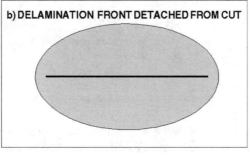

FIGURE 5.29 Elliptically shaped delamination regions caused by a simple cut. (a) The delamination front remains attached to the ends of the cut. (b) The delamination front is detached from the ends of the cut.

is that in Figure 5.29a the delamination front remains attached to the ends of the initial cut. In Figure 5.29b, the delamination front detaches from the cut ends and stands off at a certain distance. Whether or not the delamination front stands off from the ends of the cut has significant consequences for the nature of the delamination process, as discussed in this section.

Some initial insight into the delamination process can be gained by examining the delamination behavior of a simple strip of material coated onto a perfectly rigid substrate. This problem was investigated by Thouless et al.[24] and is illustrated schematically in Figure 5.30. We see from the figure that the strain energy release rate G, which is a measure of the driving force for delamination, is largest at zero crack length and decays asymptotically to the following value:

$$G = \frac{(1-v^2)t\sigma^2}{2E} \tag{5.58}$$

where
G = Strain energy release rate
v = Poisson's ratio of coating
t = Coating thickness
σ = Biaxial stress level in coating
E = Coating modulus

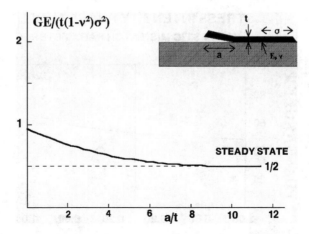

FIGURE 5.30 Plot of normalized strain energy release rate versus normalized thickness for a simple edge delamination process.

Equation (5.58) suggests that we can define a critical level of stress σ_c in the coating and an associated critical strain energy release rate G_c for steady-state delamination as follows:

$$\sigma_c = \left(\frac{2EG_c(\psi)}{(1-v^2)t} \right)^{1/2} \tag{5.59}$$

All parameters are as defined above except for the stress intensity phase angle ψ, which requires some further explanation. Recall from Chapter 4, Figure 4.3, that there are generally three modes of fracture behavior, which are characterized by three stress intensity factors K_1, K_2, and K_3. The K_1 mode is simple tensile opening of the crack front, K_2 corresponds to a shearing motion in the direction of crack propagation, and K_3 represents a side-to-side tearing motion perpendicular to the direction of crack propagation. Equation (5.59) refers to plane strain delamination conditions that involve only the K_1 and K_2 factors, which are defined in terms of the local singular crack tip normal and shear stresses as follows:

$$\text{a) } \sigma_n = \frac{K_1}{\sqrt{2\pi r}} \quad \text{(Normal stress)}$$

$$\tag{5.60}$$

$$\text{b) } \sigma_t = \frac{K_2}{\sqrt{2\pi r}} \quad \text{(Tangential shear stress)}$$

The phase angle is defined in terms of the stress intensity factors by the relation $\psi = \tan^{-1}(K_2/K_1)$. To determine ψ, we need to know the elastic properties of the

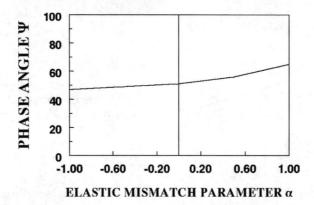

FIGURE 5.31 Behavior of the stress intensity phase angle as a function of the Dundurs' elastic mismatch parameter α.

coating and the substrate that define what are known as the Dundurs' parameters*
of the coating/substrate system:

$$\text{a) } \alpha = \frac{E\big/\!\left(1-v^2\right)-E_s\big/\!\left(1-v_s^2\right)}{E\big/\!\left(1-v^2\right)+E_s\big/\!\left(1-v_s^2\right)}$$

$$\text{b) } \beta = \frac{1}{2}\frac{\mu\left(1-2v_s\right)-\mu_s\left(1-2v\right)}{\mu\left(1-v_s\right)+\mu_s\left(1-v\right)}$$

(5.61)

These parameters characterize the elastic properties of simple bilayer laminates.
The parameter β is somewhat troublesome in that it induces oscillatory stresses into
the problem that are difficult to justify. The effects of this parameter are assumed to
be small in this analysis, leaving us to deal with only the parameter α. Figure 5.31
illustrates the variation of ψ as a function of the elastic mismatch parameter α.

Referring to Equation (5.59), we see that σ_c is the minimum stress level required
to drive steady-state delamination of a coating under plane strain conditions. However, from Figure 5.30 we see that delamination can be initiated at a lower stress,
which however will terminate at some finite length and not further propagate. Let
σ_i be the lowest level of stress in the coating that can initiate a delamination event.
There are then three cases of interest depending on the level of stress in the coating:

- Case I: $0 < \sigma < \sigma_i$. Cut remains stable with no lateral delamination.
- Case II: $\sigma_i < \sigma < \sigma_c$. Limited delamination of coating close to cut.
- Case III: $\sigma > \sigma_c$. Extensive delamination extending out from cut boundary.

---

* We touched on this topic in Chapter 4, Equations (4.40) and Note 11.

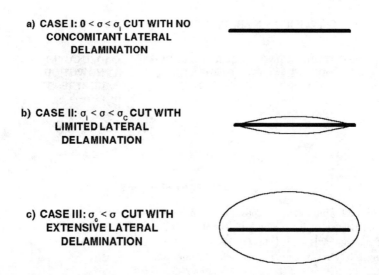

a) CASE I: $0 < \sigma < \sigma_i$ CUT WITH NO
CONCOMITANT LATERAL
DELAMINATION

b) CASE II: $\sigma_i < \sigma < \sigma_c$ CUT WITH
LIMITED LATERAL
DELAMINATION

c) CASE III: $\sigma_c < \sigma$ CUT WITH
EXTENSIVE LATERAL
DELAMINATION

FIGURE 5.32 Three modes of behavior arising from the simple scratch test. (a) Strong adhesion: Scratch remains stable with no further delamination from the edges. (b) Moderate adhesion: Collateral delamination from edges of scratch is observed. However, the delaminated region remains close to the original scratch. (c) Poor adhesion: Extensive collateral delamination observed extending out from the original cut. The delamination front may or may not remain attached to the ends of the original cut.

Figure 5.32 illustrates these delamination scenarios.

5.3.2 SIMPLIFIED ANALYTICAL MODEL FOR CUT TEST

With the above background material in mind, Jensen et al.[23] examined a simplified analytical model for the behavior of the cut test based on a strain energy release rate criterion. For the case of simple two-dimensional plane strain delamination exhibited in Figure 5.30, the strain energy release rate is related to the stress intensity factors K_I and K_{II} by the following standard formula:

$$G = \frac{1}{2}\left(\frac{1-v^2}{E} + \frac{1-v_s^2}{E_s}\right)\left(K_I^2 + K_{II}^2\right) \tag{5.62}$$

v, E = Poisson ratio, modulus of coating
v_s, E_s = Poisson ratio, modulus of substrate

This equation will not suffice for the cut test because there is also a K_{III} factor present. Figure 5.33 illustrates a simplified model for including this effect. The model says that the problem of a long cut in a biaxially stressed coating is the same as the superposition of a biaxially stressed coating without a cut plus a cut coating with no biaxial stress but with the lateral sides of the cut loaded as shown in the figure. The edge of the delamination front is also under stress by both normal and tangential

SIMPLIFIED SUPERPOSITION MODEL FOR CUT TEST

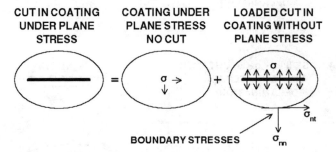

CUT IN COATING   COATING UNDER   LOADED CUT IN
UNDER PLANE   PLANE STRESS   COATING WITHOUT
STRESS   NO CUT   PLANE STRESS

BOUNDARY STRESSES σ_{nn}

FIGURE 5.33 Simplified superposition model for a cut in a stressed coating. The original problem is viewed as the superposition of an uncut film under plane stress plus an unstressed film with a cut that is loaded at the edges only by the original plane stress with further normal and tangential stresses acting at the delamination front.

tractions σ_{nn} and σ_{nt}, as shown in the figure. Under these conditions, the strain energy release rate is related to the stress intensity factors by the following generalization of Equation (5.62):

$$G = \frac{1}{2}\left(\frac{1-v^2}{E} + \frac{1-v_s^2}{E_s} \right)\left(K_I^2 + K_{II}^2 \right) + \frac{1}{4}\left(\frac{1}{\mu} + \frac{1}{\mu_s} \right) K_{III}^2 \qquad (5.63)$$

where μ and μ_s are the shear moduli of the coating and substrate, respectively. The shearing traction σ_{nt} at the delamination front introduces a tearing component and thus the appearance of the Mode III stress intensity factor. Thus, it is natural to assume that the K_I and K_{II} factors are associated with the normal traction σ_{nn}, and K_{III} is associated with σ_{nt}. The basic problem, however, is to be able to calculate the shape and extent of the delamination front knowing the tractions σ_{nn} and σ_{nt}. To do this, Jensen et al. assumed that the following condition must be satisfied at the delamination front, which is essentially a generalization of Equation (5.59):

$$\sigma_{nn}^2 + \frac{2\lambda}{1-v}\sigma_{nt}^2 = \sigma_c^2 \qquad (5.64)$$

This equation essentially states that, at the delamination boundary, the driving force for delamination given by the left-hand side of Equation (5.64) is exactly balanced by the strain energy required for delamination, which is proportional to σ_c^2. To proceed, one must now calculate the stress components σ_{nn} and σ_{nt}. This will generally require the use of powerful numerical methods, such as finite element analysis. However, an elementary solution to the problem can be found by limiting the analysis to the local region near the end of the cut where the delamination front is pinned to the end. The model is that of a simple wedge region, as shown in Figure 5.34. The stresses are easily computed for this simple two-dimensional model using

**SIMPLIFIED WEDGE MODEL OF CUT END
IN POLAR COORDINATES**

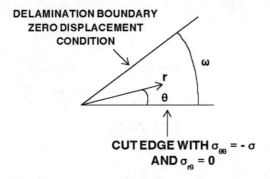

FIGURE 5.34 Further simplified model of cut test in which the analysis is confined to the end of the cut assuming that the delamination front remains attached at the cut end.

methods derived from complex variable theory. Shifting to polar coordinates with the origin at the left end of the cut, the stresses at the delamination front are:

$$\text{a) } \sigma_{\theta\theta} = \frac{-2\sigma \cos 2\omega}{1 - v + (1 + v)\cos 2\omega}$$

$$\text{b) } \sigma_{r\theta} = \frac{-\sigma(1 - v)\sin 2\omega}{1 - v + (1 + v)\cos 2\omega} \qquad (5.65)$$

$$\text{c) } \sigma_{rr} = \sigma \frac{1 - v - (1 + v)\cos 2\omega}{1 - v + (1 + v)\cos 2\omega}$$

The delamination condition can now be derived by substituting the stress components in Equation (5.65) into Equation (5.64), recognizing that $\sigma_{\theta\theta} = \sigma_{nn}$ and $\sigma_{r\theta} = \sigma_{nt}$, to arrive at the following condition for critical stability at the delamination front:

$$\text{a) } F(\omega, \lambda) = (\sigma_c / \sigma)^2 \qquad (5.66)$$

where

$$\text{b) } F(\omega, \lambda) = \frac{4\cos^2 2\omega + \lambda(1 - v)\sin^2 2\omega}{\left[1 - v + (1 + v)\cos 2\omega\right]^2}$$

A plot of the function $F(\omega, \lambda)$ is shown in Figure 5.35 for various values of the parameter λ. A number of things are noteworthy from this diagram. First, referring

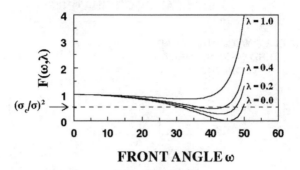

FIGURE 5.35 Stability condition for delamination of the cut model in Figure 5.34. The function $F(\omega,\lambda)$ is given by Equation (5.66).

to Equation (5.64), we see that the parameter λ controls how much the tearing component σ_{nt} of the stress field (and thus the K_{III} stress intensity) contributes to the delamination criterion. In particular, for $\lambda = 0$ there is no Mode III contribution, and Mode III becomes progressively more dominant as λ increases. The dashed line represents a fixed value of the quantity $(\sigma_c/\sigma)^2$, and where it intersects the curve $F(\omega,\lambda)$ represents a value of the delamination front angle ω, satisfying Equation (5.66). It should be noted that there are typically two solutions for a given fixed value of λ; however, the larger solution represents an unstable nonphysical state and is therefore discarded.

For the case $\lambda = 0$, we note that there is always a physical solution to Equation (5.64), and we can thus conclude that for a delamination front that is pinned at the cut end, the contribution of Mode III fracture is likely to be small. However, for the case $\lambda = 0.2$, we see that there may or may not be a solution to Equation (5.64), depending on the value $(\sigma_c/\sigma)^2$. If the film stress is too high, thus making $(\sigma_c/\sigma)^2$ small, there will be no solution to Equation (5.64), indicating that the delamination front cannot be pinned to the end of the cut (Figure 5.29b). Thus, a nonpinned delamination front indicates the presence of Mode III behavior plus relatively high film stresses.

The work of Jensen et al. suggested a further way to pin down the Mode III contribution to the delamination behavior. If one has a coating that delaminates in the presence of a circular cut and a straight cut simultaneously as shown in Figure 5.36, then one can proceed as follows to determine a value for the parameter λ. For the circular cut test, there can be no Mode III behavior because $\sigma_{nt} = 0$ by symmetry. An initial circular cut of radius R_1 delaminates further out to a radius R_2. This problem can be worked out in detail, and it is found that the ratio R_2/R_1 satisfies the following equation:

$$\frac{\sigma}{\sigma_c} = \frac{1}{2}\left[(1-v)\frac{R_2}{R_1} + 1 + v\right] \tag{5.67}$$

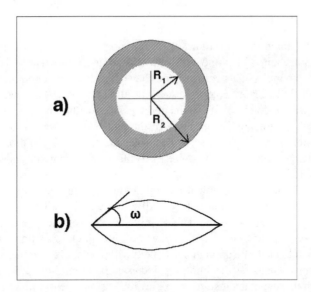

FIGURE 5.36 Coating with both a straight and circular cut. Data from the circular cut can be used to determine the ratio (σ/σ_c) in Equation (5.68), and this along with the measured cut angle ω and the Poisson ratio of the coating will determine the parameter λ in Figure 5.35 and thus the amount of Mode III tearing present at the delamination front.

This relation provides a convenient way of getting at the critical ratio $(\sigma_c/\sigma)^2$ knowing only the Poisson ratio of the coating and by measuring the radii R_2 and R_1. For most polymer coatings, for instance, ν is close to 1/3, and thus it suffices only to measure the radii. With these data in hand, one can now solve Equation (5.66) for λ as follows:

$$\lambda = \frac{\left(\sigma_c/\sigma\right)^2\left[1-\nu+\left(1+\nu\right)\cos 2\omega\right]^2 - 4\cos^2 2\omega}{\left(1-\nu\right)\sin^2\left(2\omega\right)} \tag{5.68}$$

It now remains only to measure the angle ω, and Equation (5.68) can be used to determine λ and thus the importance of Mode III delamination directly. As a hypothetical example, consider a coating as depicted in Figure 5.36 with both a circular and a straight cut, both showing peripheral delamination. If the radii R_2 and R_1 come out to 0.012 and 0.01 m, respectively, then Equation (5.67) predicts that $(\sigma_c/\sigma)^2$ should equal 0.76. If we further assume that the angle ω is measured as 30° degrees (0.52 rad), then Equation (5.68) gives λ as 0.69. This would indicate that, for this case, there is a strong Mode III component involved in the delamination process. What is remarkable about this situation is that we learn something quite fundamental about the delamination behavior of this system on the basis of a handful of simple geometric measurements. If we were to go further and measure the film stress along with the film modulus and thickness, then we could determine the critical strain energy release rate and the critical stress σ_c.

5.4 THE PULL TEST

The peel and scratch tests are quite likely the most popular experiments for determining the adhesion of flexible and brittle coatings, respectively. The pull test, although not as popular as its cousins, does have the advantage of equal applicability to both flexible and brittle coatings. As with all adhesion tests, however, the pull test has its own catalogue of strengths and weaknesses that must be understood to be effectively implemented. The nature of the pull test can best be illustrated by a specific example. For this purpose, we consider the adhesion of a pin soldered to a ceramic substrate because this is quite likely the most important and widely used application of the pull test.

Figure 5.37 illustrates a typical configuration. The figure attempts to illustrate that the major underlying feature of the pull test is that the failure load depends on the nature and distribution of the defects in the solder and the supporting ceramic. Despite our best efforts, there will always be a distribution of flaws present, and when the pin is loaded, the applied stress field will unerringly find the weakest flaw, which will turn out to be the initiating locus of failure, which causes the pin to separate from the substrate at some stress level σ_F. This type of failure phenomenon is well known in the world of mechanical testing, and the probability model for failure has been shown to follow Weibull statistics.[25]

As an example of how a typical Weibull analysis works, consider that we have a large number of pins soldered to a ceramic substrate similar to the situation depicted in Figure 5.37. These might typically be pins soldered to a multichip module that will be used in a high-end computer of some type. The number of pins on such modules can range from hundreds to a thousand or more. Such modules also tend to be very expensive, so it is critical to know how well the pins are attached to the ceramic substrate for quality control purposes. We take our module to the lab, mount

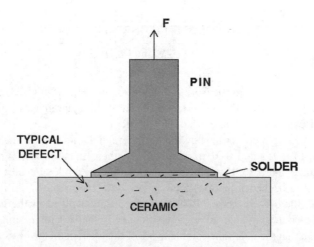

FIGURE 5.37 Typical pull test configuration. Note that even in a well-formed joint there will be some distribution of microscopic defects, any of which could be the site of a fracture failure in the event that an excessive load is applied to the pin.

TABLE 5.7
Hypothetical Distribution of Failure Strengths
for a Pin Pull Experiment[a]

18.940	18.906	19.021	18.795	18.779	19.165	19.094	18.755
18.833	19.394	19.246	19.399	18.505	18.643	18.499	18.778
19.142	19.195	19.267	18.836	19.075	19.122	19.103	18.567
18.727	18.377	19.197	18.784	19.350	19.312	19.003	18.770
19.175	19.372	19.232	19.395	19.437	18.790	18.676	19.011

[a] All stresses in megapascals (MPa).

it in a tensile test unit, and carefully pull off a large number of pins, recording the load at which each pin failed. As a matter of procedure, we censor the data to include only those fails that occur in the solder or at one of the solder's interfaces with either the pin or the underlying ceramic. Thus, cases for which either the pin failed or there was pullout of the ceramic are discarded, so we will be evaluating in essence the strength of the solder bond. The raw data gathered might look like that given in the Table 5.7.

The data in Table 5.7 have been adapted from an actual experiment, the details of which are unimportant except for the fact that they are expected to follow Weibull statistics. According to the Weibull model, the probability that a given pin will fail in a given range $\sigma + d\sigma$ is given by

$$p(\sigma) = m \left(\frac{\sigma - \sigma_u}{\sigma_0} \right)^{m-1} \exp\left[-\left(\frac{\sigma - \sigma_u}{\sigma_0} \right)^m \right] \tag{5.69}$$

where

σ_u = Stress below which failure cannot occur, typically taken as 0
σ_0 = Scale factor
m = Weibull modulus

The probability $p(\sigma)$ can be estimated from Table 5.7 in the following manner: First, select an appropriate value for the increment $d\sigma$, say 0.1. Assume we want to know $p(\sigma)$ near 19 MPa. We need to count the number of times σ falls in the range 19 and 19.1. From the table, that number is 5. We then divide 5 by the total number of data points (40) to obtain 0.125. Finally, we take σ to be at the center of the range 19 to 19.1 and therefore conclude that p(19.05) = 0.125. All other relevant values of $p(\sigma)$ are obtained in a similar manner. In essence, we first bin the data into groups falling within some appropriately chosen range and then count the number of points falling into each bin. Dividing by the total number of points gives an estimate of the probability that the true failure stress will fall within a given bin. Taking the stress to be at the center of the range of the bin improves the estimate by removing any bias toward either the upper or lower bounds.

In addition to the probability density $p(\sigma)$, it is important to know the cumulative probability $P(\sigma)$, which is defined to be the probability that a typical pin will fail at a stress level at or below σ. $P(\sigma)$ is simply the integral of $p(\sigma)$ from 0 to σ as follows:

$$P(\sigma) = \int_0^\sigma p(\sigma)d\sigma = 1 - \exp\left[-\left(\frac{\sigma - \sigma_u}{\sigma_0}\right)^m\right] \tag{5.70}$$

$P(\sigma)$ is easily estimated from binned data by simply adding all the probabilities for every bin up to and including the one containing σ. It is also important to know the average stress level at which failure will occur and the variance about this average. These quantities are given by the following formulae:

$$\text{a) } \text{EXP}\left(\frac{\sigma - \sigma_u}{\sigma_0}\right) = \Gamma\left(1 + \frac{1}{m}\right)$$

$$\text{b) } \text{VAR}\left(\frac{\sigma - \sigma_u}{\sigma_0}\right) = \Gamma\left(1 + \frac{2}{m}\right) - \Gamma^2\left(1 + \frac{1}{m}\right) \tag{5.71}$$

Equation (5.71a) gives the average or expected value of the normalized failure stress $(\sigma - \sigma_u)/\sigma_0$, and Equation (5.71b) gives the variance or the variation of this quantity. The variance is simply a measure of how much the actual data vary from the average value. It is more common to use the standard deviation in this regard, which is just the square root of the variance. Readers not familiar with statistics can find all this information in any textbook. The book by Meyer cited in Note 25 is highly recommended because of its focus on engineering problems. The quantity Γ in Equation (5.71) is the gamma function, which can be looked up in common reference tables.[26] Now, with all this statistical lore at our disposal, we can analyze the data in Table 5.7 as follows.

Table 5.8 sorts the data of Table 5.7 into nine different bins, each covering a range of 0.106 MPa over the entire range of the raw data from 18.377 to 19.447 MPa. The number of data points in each bin is tabulated in Column 4, and the estimated probability density and cumulative probability are given in Columns 5 and 6, respectively.[27] The task now is to evaluate the parameters in Equation (5.70). First, the shift factor σ_u is set to 0, leaving the Weibull modulus m and the scale factor σ_0 to be determined. The process is well defined if somewhat arcane and involves the use of logarithms to eliminate the exponential function and the power law dependence in the argument of the exponential. This process is commonly referred to as logging the data into submission.

Some preliminary algebra is required to eliminate the negative signs, which will cause problems when taking the logarithms. Subtracting 1 from each side of Equation (5.70), multiplying each side by -1, and taking the inverse of each side transforms the equation to the following:

TABLE 5.8
Weibull Analysis of Data in Table 5.6

Lower Bin Value (MPa)	Upper Bin Value (MPa)	Central Bin Value (MPa)	No. of Data Points in Bin	Probability of Central Bin Value	Cumulative Probability
18.377	18.483	18.430	1.000	.025	.025
18.499	18.605	18.552	3.000	.075	.100
18.643	18.749	18.696	3.000	.075	.175
18.755	18.861	18.808	9.000	.225	.400
18.906	19.012	18.959	4.000	.100	.500
19.021	19.127	19.074	5.000	.125	.625
19.142	19.248	19.195	7.000	.175	.800
19.267	19.373	19.320	4.000	.100	.900
19.394	19.500	19.447	4.000	.100	.999

$$\frac{1}{1-P(\sigma)} = \exp\left[\left(\frac{\sigma}{\sigma_0}\right)^m\right] \tag{5.72}$$

The job is completed by taking the natural logarithm of each side of the above equation two times to obtain

$$\ln\ln\left(\frac{1}{1-P(\sigma)}\right) = m\ln\sigma - m\ln\sigma_0 \tag{5.73}$$

Equation (5.73) tells us that if we take the double logarithm of the reciprocal of 1 minus the cumulative probability given in Column 6 of Table 5.8 and plot it against the logarithm of the failure stress given in Column 3, we should get a straight line with a slope that is the Weibull modulus m and intercept $(-m\ln\sigma_0)$. Figure 5.38 shows the resulting plot, which is indeed close to a straight line. The estimated value of m is 90.61, and σ_0 is 19.07. From Equations (5.71), we estimate the average failure stress as $<\sigma_F> = 18.95$ MPa and the deviation as 0.4 MPa, remembering that the deviation is the square root of the variance. We would thus conclude that the expected adhesion strength of our hypothetical pin population is 18.95 MPa. If we further assume that the strength should never be more than three deviations away from the average, then the strength should never get lower than 17.75 MPa or greater than 20.15 MPa.

5.5 SUMMARY

The primary intent of this chapter is to provide a more intuitive feel for the nature of applied adhesion testing through the use of specific examples that are chronicled in the literature. The collection of examples has of necessity been highly selective.

WEIBULL PLOT OF PIN FAILURE DATA

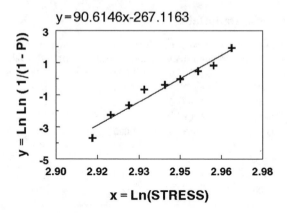

FIGURE 5.38 Typical Weibull plot of pin pull data used to estimate the pin adhesion to the underlying substrate.

Because of its great popularity and applicability in the realm of flexible coatings, the peel test has been singled out for extensive treatment. Starting from the earliest elastic analysis of Spies in the 1950s and working up to the highly detailed elastic-plastic investigations of Kim and co-workers, the peel test clearly demonstrates the dependence of adhesion testing on knowing the continuum behavior of the materials involved. Moving on to the scratch test, the number of examples that can be treated in a fully quantitative manner is much more limited than for the peel test because of the exceeding complexity of the scratch process. The action of the scratch indenter in penetrating and plowing aside the coating material all but excludes any attempt at a linear elastic analysis. The early analysis of Benjamin and Weaver[28–31] has been shown to be of very limited applicability to true quantitative adhesion analysis and is therefore not covered. However, a more recent treatment of the related cut test is included because it demonstrates the application of modern fracture mechanics methods to a very difficult adhesion measurement problem. Finally, a fairly detailed example of the analysis of pull test data is given because it nicely demonstrates the application of statistical methods to the analysis of adhesion data.

Notes

1. One of the earliest investigations into the energy dissipative processes occurring during the peel test is in the following article by A.N. Gent and G.R. Hamed, "Peel Mechanics for an Elastic-Plastic Adherend," *Journal of Applied Polymer Science*, 21, 2817 (1977).
2. "An Experimental Partitioning of the Mechanical Energy Expended During Peel Testing," R.J. Farris and J. Goldfarb, in *Adhesion Measurement of Films and Coatings*, K.L. Mittal, Ed. (VSP, Utrecht, The Netherlands, 1995), p. 265.
3. Ibid.; adapted from Table 3 on page 270.

4. See, for instance, the following overview article: "Contact Angle, Wetting and Adhesion: A Critical Review," R.J. Good, in *Contact Angle Wettability and Adhesion*, K.L. Mittal, Ed. (VSP, Utretcht, The Netherlands, 1993), p. 3. This volume is a Festschrift in honor of Prof. Good and contains 60 more articles dealing with various aspects of the contact angle measurement technique.

5. "The Peeling Test on Redux-Bonded Joints," G.J. Spies, *Aircraft Engineering*, 25, 64 (1953).

6. "Theory of Peeling Through a Hookean Solid," J.J. Bikerman, *Journal of Applied Physics*, 28, 1484 (1957).

7. "Experiments on Peeling," J.J. Bikerman, *Journal of Applied Polymer Science*, 2, 216 (1959).

8. "Theory and Analysis of Peel Adhesion: Mechanisms and Mechanics," D.H. Kaelble, *Transactions of the Society of Rheology*, 3, 161 (1959).

9. "Peel testing of Adhesive Bonded Metal," S. Yurenka, *Journal of Applied Polymer Science*, VI-20, 136 (1962).

10. "Peel Adhesion. II. A Theoretical Analysis," J.L. Gardon, *Journal of Applied Polymer Science*, 7, 643 (1963).

11. "Peel Mechanics for an Elastic-Plastic Adherend," A.N. Gent and G.R. Hamed, *Journal of Applied Polymer Science*, 21, 2817 (1977).

12. This is not quite accurate; because of what is called the Baushinger effect, the stress strain diagram is not symmetric as shown in Figure 5.18. This little technical difficulty does not change the general conclusions, however. Further discussion of the Baushinger effect may be found in the volume by I.M. Ward, *Mechanical Properties of Solid Polymers*, 2nd ed. (John Wiley and Sons, New York, 1983), see Chapter 11.4.6.

13. "Elasto-Plastic Analysis of the Peel Test for Thin Film Adhesion," K.-S. Kim and J. Kim, *Journal of Engineering Materials and Technology*, 110, 266 (1988).

14. "Mechanics of the Peel Test for Thin Film Adhesion," K.S. Kim, *Material Research Society Symposium Proceedings*, 119, 31 (1988).

15. "Elastoplastic Analysis of the Peel Test," K.S. Kim and N. Aravas, *International Journal of Solids Structures*, 24, 417 (1988).

16. "Elasto-Plastic Analysis of the Peel Test for Thin Film Adhesion," K.S. Kim and J. Kim, *Journal of Engineering Material Technology*, 110, 266 (1988).

17. "On the Mechanics of Adhesion Testing of Flexible Films," N. Aravas, K.S. Kim, and M.J. Loukis, *Materials Science and Engineering*, A107, 159 (1989).

18. "Mechanical Effects in Peel Adhesion Test," J. Kim, K.S. Kim, and Y.H. Kim, *Journal of Adhesion Science and Technology*, 3, 175 (1989).

19. The term *surface fracture energy* was coined by the fracture mechanics investigators, who think of delamination as a type of fracture process. A surface chemist would prefer the term *work of adhesion* as related to the surface energetics of contact angle phenomena. In principle, the two should describe precisely the same phenomenon. In practice, this is not so because of complications introduced by the very high stress fields associated with mechanical fracture processes, which do not enter into purely surface energetic experiments such as contact angle measurements.

20. We follow again the analysis of Kim and co-workers and use the results of P.G. Hodge, *Plastic Analysis of Structures*, McGraw-Hill, New York, 1959.

21. Note 15 above, Equations (14)–(17). Kim and Aravas most likely adapted their elastic plastic constitutive relations from the text by P.G. Hodge mentioned in Note 20.

22. K.S. Kim, University of Illinois, TAM Technical Report No. 472, IBM Test No. 441614. It is doubtful that the general reader will have much luck getting a copy of this report directly from the University of Illinois. A better approach might be to contact Prof. Kim directly at the following address: Prof. K.S. Kim, Department of Mechanics, Division of Engineering, Box D, Brown University, Providence, RI 02912; e-mail Kyung-Suk_Kim@brown.edu.

23. "Decohesion of a Cut Prestressed Film on a Substrate," H.M. Jensen, J.W. Hutchinson, and K.-S. Kim, *International Journal of Solids Structures*, 26, 1099 (1990).

24. "Delamination From Surface Cracks in Composite Materials," M.D. Thouless, H.C. Cao, and P.A. Mataga, *Journal of Materials Science*, 24, 1406 (1989).

25. An excellent treatment of Weibull statistics as applied to problems of mechanical failure can be found in the following article: "Probability of Failure Models in Finite Element Analysis of Brittle Materials," C. Georgiadis, *Composites and Structures*, 18, 537 (1984). A mathematical overview of the Weibull distribution can be found in *Data Analysis for Scientists and Engineers*, S. Meyer (John Wiley and Sons, New York, 1975), p. 284. The original work of Weibull himself goes back to 1939. The most accessible of the original papers is quite likely the following: W. Weibull, *Journal of Applied Mechanics,* 18, 293 (1951).

26. Readers who own a standard scientific calculator may be pleased to find that the gamma function is available on their machine. It turns out that, for positive integer values of the variable x, the gamma function is directly related to the factorial function through the following identity: $\Gamma(x + 1) = x!$ All scientific calculators include the factorial function, and most compute it by computing $\Gamma(x + 1)$, which means they will give values of the factorial function even for non-integer input. Thus, to compute $\Gamma(x)$ for any value of x, simply enter x from the keyboard, subtract 1, and then hit the factorial function key. To test it out, try the factorial of -0.5, which should be $\Gamma(0.5)$, which is exactly the square root of $\pi = 1.77245385$. If it works, you are in business. Just one of the nice things to know in life.

27. Note that there is clearly a certain level of subjectivity involved in binning data of this type, and that Table 5.8 is not fully unique. For instance, one could have used 8 bins or 10 bins instead of 9, and the numbers would come out somewhat differently. With only a finite amount of data, the results can only approximate the true probabilities. However, for any reasonable binning scheme the probabilities in Table 5.8 will represent an accurate estimate of the true probabilities to within expected error.

28. "Measurement of Adhesion of Thin Film," P. Benjamin and C. Weaver, *Proceedings Royal Society*, 254A, 163 (1960).

29. "Adhesion of Metal Films to Glass," P. Benjamin and C. Weaver, *Proceedings Royal Society*, 254A, 177 (1960).

30. "The Adhesion of Evaporated Metal Film on Glass," P. Benjamin and C. Weaver, *Proceedings Royal Society*, 261A, 516 (1961).

31. "The Adhesion of Metals to Crystal Faces," P. Benjamin and C. Weaver, *Proceedings Royal Society*, 274A, 267 (1963).

6 Adhesion Aspects of Coating and Thin Film Stresses

A deep, intuitive appreciation of the inherent cussedness of materials is one of the most valuable accomplishments an engineer can have. No purely intellectual quality is really a substitute for this.

J. E. Gordon, *Structures or Why Things Don't Fall Down*

6.1 INTRODUCTION

When trying to deal with problems of film or coating adhesion in a fully quantitative manner, the issue of residual or intrinsic stress in the material inevitably arises. In some cases, these stresses can completely dominate the problem of adhesion strength and stability. Thus, it is important to understand the nature and genesis of these stresses and how they figure into the problem of adhesion measurement. First, however, a few words are in order regarding what is meant precisely by residual or intrinsic stress.

Nearly every material object with which we come in contact contains some level of intrinsic stress that arises from its history and how it was formed or fabricated. A common example is glassware. Did you ever take a water glass and try to fill it with very hot water from a tap or kettle only to have it shatter in your hand? This is because the glass had developed a high level of intrinsic stress at the time of its manufacture. In the fabrication of common inexpensive glassware, a molten slug of liquid glass is injected into a mold and then rapidly cooled to room temperature. The material at the surface solidifies faster than that in the interior, setting up a shrinkage difference and thus a residual stress in the final artifact. This residual stress can be quite high but not enough to fracture the final piece. However, a water glass made in this fashion if filled with very hot water (or very cold water for that matter) will experience a large thermal gradient between the inner and outer surfaces, thus adding an additional thermal stress to the already existing residual stress, which can then cause the glass to shatter.

One way to prevent the buildup of residual stress in a glass object is to put it in an oven heated to near the glass transition temperature of the material and leave it there for an extended period of time. Near the glass transition temperature, the glass molecules have sufficient mobility that allows them to reorient and thus relax out the stress field originally set up because of too rapid cooling. All laboratory

249

glassware is subject to this annealing process to relieve any residual stress incurred during the fabrication process. The result is glassware that can handle very hot liquids and be placed directly on a hot plate without shattering because all the residual stress has been removed and the stresses caused by subsequent thermal gradients are not sufficient to cause fracture themselves.

Of more interest to the problem of coating adhesion, however, is the case of high-temperature polymer materials that are spin coated onto silicon wafers or ceramic substrates and then cured at an elevated temperature. We saw in Chapter 3, Equation (3.50), that on cooling to room temperature quite large thermal expansion mismatch stresses can be developed in the coating, approaching nearly 50% of the ultimate tensile strength of the material. Such stresses can cause spontaneous delamination from the substrate if the level of adhesion is not strong enough. In addition to thermal expansion mismatch stresses, there are stresses set up because of shrinkage associated with the curing chemistry of the coating and shrinkages caused by solvent evaporation. All of these stresses get frozen into the material and remain an invisible but potentially destructive latent property of the coating.

Metal coatings stand in a class by themselves. In the case of metals, in addition to the potential for developing thermal expansion mismatch stresses, there are also a number of other mechanisms available for developing high levels of intrinsic stress. Metal coatings can be applied in a number of ways, with vapor deposition one of the favorites. Because metal coatings are often polycrystalline, one can have stresses develop because of lattice mismatch between the applied coating and the underlying substrate. This happens when the crystal structure of the substrate is incommensurate with the way the vapor-deposited metal would like to crystallize. The coating material is thus forced to accommodate itself to the underlying lattice structure at the interface, giving rise to strains that then induce significant levels of stress. Also, even assuming there is no lattice mismatch between the coating and the substrate, the polycrystalline nature of the metal coating can give rise to significant levels of stress. During vapor deposition, many small islands are formed at the earliest stage of the deposition process. Each of these islands is a small microcrystal that continues to grow as more material is added during the deposition process. Further, each microcrystal has its own orientation determined by more or less random effects. As the film grows, the microcrystals start to coalesce, and a complex pattern of grain boundaries is formed as the different microcrystals merge into each other with their lattice structures misaligned with respect to one another at random angles. Stresses now develop in a similar manner to the lattice mismatch situation only now they are spread throughout the volume of the developing coating.

On top of the lattice mismatch and grain boundary effects, metal coatings can develop excessive stresses caused by the inclusion of impurities. What happens is that an atom that is too large tries to fit itself into the existing crystal lattice, causing a tight squeeze and thus a local compressive stress to develop. The existence of a void or lattice vacancy can have the opposite effect and therefore set up a local tensile stress. What makes the situation complicated for metal coatings is that all of these effects depend sensitively on the details of the deposition process, including the following parameters:

- Substrate temperature
- Deposition rate
- Deposition process
  - Vapor deposition
  - Sputtering
  - Chemical vapor deposition (CVD)
- Ambient pressure, composition
- Presence of impurities

The above list is a sample of the factors that can have a significant effect on the level of intrinsic stress in a metal coating. However, regardless of how intrinsic stresses find their way into a coating, they all share the common property that they tend to be invisible and go undetected until such time as they reach a critical level that causes failure in the form of either delamination or fracture. The process engineer is therefore faced with the problem of detecting and measuring these stresses so they can be accounted for and controlled before they can cause trouble. The following sections give details of several of the more common methods of measuring coating and film stresses.

6.2 GENERAL MEASUREMENT METHODS FOR THIN FILMS AND COATINGS

The classification of stress measurement methods breaks down in a natural way into two separate classes, one dealing with amorphous materials and the other with crystalline substances. For crystalline materials, the powerful x-ray diffraction method can be employed, which is not an option for amorphous coatings.

6.2.1 CANTILEVERED BEAM METHOD

As pointed out in Chapter 3, Section 3.4, the theory of the bending of simple beams is quite likely the most useful application of the continuum theory of elastic bodies, and the measurement of thin film stresses is one of the most prolific of all the applications. For this problem, we can call on the cantilevered beam, discussed in some detail in Chapter 3.

The basic principles are simple. If one coats a cantilevered beam with a thin film that develops an intrinsic stress, then the film will exert a bending moment on the beam and cause it to deflect as shown in Figure 6.1. We can in fact use the elementary principles of beam bending discussed in Chapter 3, Section 3.4, to calculate just how much deflection we can expect for a given film thickness with average intrinsic stress σ. This problem was first worked out by Stoney[1] and was further refined by Brenner and Senderoff[2] and Davidenkov.[3] It is instructive to derive this formula because one needs only elementary beam theory, and the final result is widely used in the literature on thin film stresses. We start with the standard formula of elementary beam mechanics that was derived as Equation (3.59) in Chapter 3:

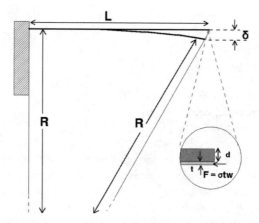

FIGURE 6.1 Cantilevered beam of thickness d and width w with a stressed coating of thickness t. The coating supports a uniform stress level σ that gives rise to an in-plane force F = tw, which causes the beam to deflect an amount δ.

$$M = \frac{EI}{R}$$

$$I = \int\limits_{\Delta A} x^2 \, dA \tag{6.1}$$

where

      M = Bending moment acting on beam because of film stress
      E = Beam modulus
      R = Radius of curvature caused by bending (See Figure 6.1)
      I = Bending moment of inertia of beam (= $wt^3/12$ for beam of rectangular cross section, w = beam width, t = beam thickness)

We estimate the bending moment that the coating exerts on the beam as follows: Referring to Figure 6.2, we see that in the case of a tensile stress the film wants to contract, but it cannot because it is firmly adhered to the relatively massive beam.

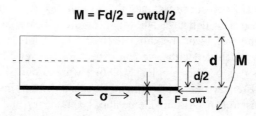

FIGURE 6.2 Detailed elaboration of the forces and moments acting on the cantilevered beam depicted in Figure 6.1. M is the bending moment acting on the beam.

Because of this, the film will exert a force F parallel to the beam of magnitude σwt, which is just the stress acting at the end times the cross-sectional area of the film. This force in turn induces a bending moment M acting at the end of the beam and tending to bend it downward. Because the film is very thin compared to the thickness of the beam, the neutral line of bending is very close to the center line of the beam, so the force exerted by the film acts at a distance $d/2$ to the neutral line, giving a net bending moment $M = \sigma wtd/2$. Inserting this expression for the moment into Equation (6.1) and noting that the moment of inertia of the beam is $I = wt^3/12$, we get the following expression for the stress in the film as a function of the radius of curvature of bending of the beam:

$$\sigma = \frac{Ed^2}{6Rt} \qquad (6.2)$$

This is the original expression for the film stress derived by Stoney.[1] However, because for slender beams the stresses will be largely in the plane of the beam (i.e., plane stress conditions), it is more accurate to use the plane stress expression for the beam modulus $E/(1-v)$, where v is the beam Poisson ratio, which when inserted into Equation (6.2) yields the commonly used expression for the film stress:

$$\sigma = \frac{Ed^2}{6(1-v)Rt} \qquad (6.3)$$

Equation (6.3) is handy when the radius of curvature of the beam is measured directly. In many cases, however, it is the end deflection of the beam that is measured. In such cases, the radius of curvature R appearing in Equation (6.3) must be replaced by the end deflection of the beam δ, for which there is a simple if somewhat subtle relationship with the radius of curvature. This relationship can most easily be derived by referring to Figure 6.3, which for the sake of clarity is a simplified version of Figure 6.1. For small deflections (i.e., $\delta \ll L \ll R$), we can relate the three sides of the right triangle appearing in the figure by the standard Pythagorean theorem:

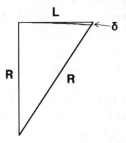

FIGURE 6.3 Diagram revealing the relationship among the radius of curvature R, the length L, and the deflection δ of the beam in Figure 6.1.

$$L^2 + R^2 = \left(R + \delta\right)^2 = R^2 + 2R\delta + \delta^2$$

Retaining only lowest order terms in δ

$$L^2 = 2R\delta \tag{6.4}$$

$$\frac{1}{R} = \frac{2\delta}{L^2}$$

The final expression for $1/R$ derived above can be substituted into Equation (6.3) to give a second version of the Stoney relation as follows:

$$\sigma = \frac{E\delta d^2}{3(1-v)L^2 t} \tag{6.5}$$

Equation (6.5) is quite likely the most popular version of the Stoney relation because it applies directly to the cantilevered beam configuration for which there are several methods available for measuring the beam deflection. One should be aware, however, that Equation (6.5) is an approximation valid for thin coatings on slender beams, which suffer only small deflections. Using the more exact theory of plate deflection, Timoshenko[4] computed correction factors to Equation (6.5) that need to be applied if the conditions of thin beams and small deflections are not met. As a guideline, Timoshenko derived the relation that $w/L < (d/2\delta)^{1/2}$ should hold for Equation (6.5) to be a good approximation. For very stiff coatings such as diamond, the elastic properties of the coating cannot be ignored, and corrections must be made to Equation (6.5). Brenner and Senderoff[5] developed the appropriate correction to Equation (6.5) for this case:

$$\sigma = \sigma_0 \left(1 + 4\frac{(1-v_s)E_f}{(1-v_f)E_s}\frac{t}{d} - \frac{t}{d}\right) \tag{6.5a}$$

where
  σ_0 = Uncorrected Stoney stress Equation (6.5)
  v_f, v_s = Poisson ratios of coating and substrate, respectively
  E_f, E_s = Modulus of coating and substrate, respectively
  t, d = thickness of coating and substrate, respectively

Any number of experimental setups have been built that utilize Equation (6.5) to measure the stress buildup in thin coatings. One of the earliest and best documented was reported on by Klockholm.[6] A schematic of this apparatus is shown in Figure 6.4. The figure illustrates the rough layout of the experiment. Two cantilevered beams and a quartz crystal thickness monitor are suspended above a shuttered metal evaporation source, all of which are enclosed within a vacuum system that can be evacuated to pressures low enough for metal deposition. As shown in the diagram,

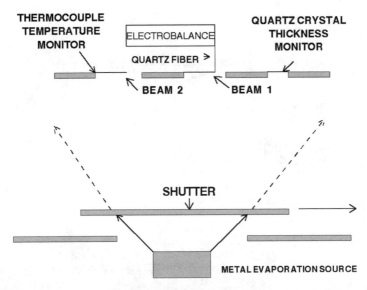

FIGURE 6.4 Schematic drawing of an apparatus designed to use the cantilevered beam method for measuring the intrinsic stress of thin metal films. See the work of Klockholm.[6]

the central cantilevered beam is connected to an electrobalance, which is connected to the end of the beam by a thin glass fiber and thus capable of continuously monitoring the deflection of the end of the beam. The second beam is attached to a thermocouple and is used to monitor the temperature of the metal substrate system continuously during the metal deposition process. The quartz crystal monitor measures the thickness of the metal layer during deposition. Klockholm used glass substrates that were 7 cm long (L), 1.5 cm wide (w), and 0.035 cm thick (d). The maximum deflection reported by Klockholm was $\delta = 0.02$ cm, which is well within the guideline of w/L < $(d/2\ \delta)^{1/2}$ given by Timoshenko.

Klokholm and Berry[7] used this apparatus to measure the stress of 15 different metals evaporated as thin films onto glass substrates as described above. Figure 6.5 shows sample plots of their data for three different metals. The authors pointed out that the measurements at low film thicknesses were subject to a variety of spurious effects and should thus be taken with caution. Table 6.1 summarizes their results for the 15 different metals measured at a thickness of 1000 Å.

As pointed out by Klokholm and Berry, there is really no way that one can predict *a priori* what the stress will be in an evaporated metal film. The Table 6.1 shows little in the way of useful correlation of the stress level with any of the other properties listed. The only obvious correlation is that the refractory metals such as molybdenum with a high melting point and shear modulus also have a very high stress level. What is striking about this table is that, with the exception of titanium, all of these metal films exhibit an enormous level of intrinsic stress. The average stress for all 15 metals is near 0.78 GPa. To put this number in perspective, consider a three-story brick building. The average compressive stress on the very bottom brick loaded by every brick all the way to the top will come to something like 2×10^{-5} GPa,

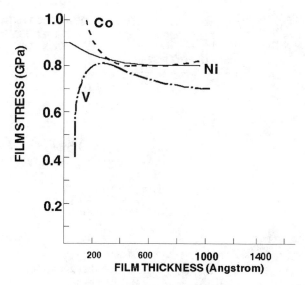

FIGURE 6.5 Intrinsic stress versus film thickness for three different metal thin films. Replotted from data of Klockholm and Berry.[7]

TABLE 6.1
Stresses of 1000-Å Thin Films for 15 Different Elements[a]

Atomic Number	Element	Stress (GPa)	Melting Point (degrees Kelvin)	Shear Modulus (GPa)
22	Ti	0	1930	41
23	V	0.7	2130	47
24	Cr	0.85	2120	90
25	Mn	0.98	1520	78
26	Fe	1.1	1810	84
27	Co	0.84	1770	78
28	Ni	0.8	1710	78
29	Cu	0.06	1360	46
40	Zr	0.7	2135	37
41	Nb	1.05	2740	37
42	Mo	1.08	2900	120
46	Pd	0.6	1830	46
47	Ag	0.02	1235	28
57	La	0.3	1190	15
79	Au	2.6	1340	28

[a] Retabulated and modified from work of Klokholm and Berry.[7]

which is 100 times smaller than the table's smallest nonzero stress, which belongs to the silver film. Intrinsic stress levels in thin metal films can be impressive and lead to rather dramatic failure phenomena.[8]

6.2.2 Variations on Bending Beam Approach

The work of Klockholm and Berry is illustrative of the type of results one can obtain from the bending beam approach to intrinsic stress measurements. This method can be applied to a wide variety of coating materials under a range of environments from ultrahigh vacuum to conditions of high pressure, humidity, and chemically reactive ambients. The type of measurement chamber used will obviously vary from situation to situation. In addition, there are at least two other methods for measuring the beam deflection other than the electromechanical technique. The two most popular approaches employ optical methods and the measurement of differential capacitance. Further details on these techniques are discussed next.

6.2.3 Optical Measurement of Deflection

6.2.3.1 Microscopy

There are essentially three different optical methods for measuring the deflection of a beam in a very sensitive noncontact manner. These methods are the long-distance microscope, optical triangulation, and optical interferometry. The long-distance microscope is the most straightforward method. One possible configuration is illustrated in Figure 6.6. Using the microscope to measure the deflection of the beam has the obvious advantages of simplicity and reliability. One can either measure the deflection by direct observation using a reticle built into the optics or, with more modern equipment, detect the image with a charged coupled device (CCD) and have it transmitted to a computer screen. The main drawback with this approach is that a direct and unobscured optical path must exist between the sample and the viewing lens. This will impose restrictions on the type of sample chamber that can be used. In addition, use of light microscopy will require that the deflection be at least 1 μm or larger to have sufficient resolution. This constraint in turn limits the sensitivity of the stress measurement. Nevertheless, optical microscopy has been used successfully by a number of workers[9] and will quite likely be the most cost-effective approach for many applications.

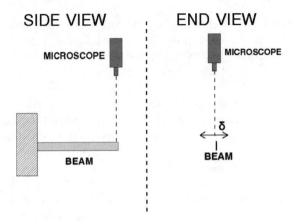

FIGURE 6.6 Measuring the deflection of a cantilevered beam using a long-range microscope.

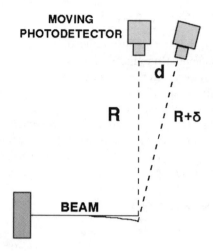

FIGURE 6.7 Measurement of the deflection of a cantilevered beam by triangulation using a laser beam and photodetector.

6.2.3.2 Laser Beam Deflection

Another approach to detecting the deflection of a stressed beam is through simple triangulation methods. Figure 6.7 shows one possible experimental configuration. A beam of light directed at the end of the beam is detected by a photodetector at a carefully measured distance R. After a stressed film is deposited on the beam, it deflects a small distance δ downward. The reflected beam is now deflected to the right, and the photodetector must be moved a distance d to again intercept the beam at maximum intensity. The situation is now the same as shown in Figure 6.1, with L replaced by d, and Equation (6.4) can be used to estimate directly the deflection δ in terms of R and d as $\delta = d^2/2R$. A variation of this approach using commercially available equipment was reported by Wojciechowski[10] and by Bell and Glocker.[11,12]

6.2.3.3 Laser Interferometry

Perhaps the most sensitive method for detecting small displacements is optical interferometry. Figure 6.8 shows a possible setup for using laser interferometry to measure beam deflections. Half of the laser beam is reflected off the beam splitter and then reflected back off the sample beam. The other half passes through the splitter and is reflected back by a mirror that is held a fixed distance away. The two beams then recombine again at the beam splitter and superimpose as they are reflected back to the detector. If the two beams reflecting off the fixed mirror and the sample beam are exactly in phase, a single spot is recorded at the detector. As the sample beam deflects, the beams get out of phase, and interference fringes appear at the detector. The spacing between the fringes can be used to sensitively determine the beam deflection. Deflections on the order of the wavelength of the laser light are easily detected by this method, making it one of the most sensitive techniques for this type of measurement. A specific example of this method has been given by Ennos.[13]

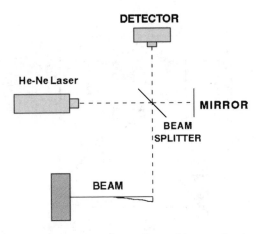

FIGURE 6.8 Measuring the deflection of a cantilevered beam using laser interferometry.

6.2.3.4 Capacitive Measurement of Deflection

If either the substrate or the coating is a conductive material, then the deflecting beam can be considered as one electrode of a small capacitor as shown in Figure 6.9. Placing a fixed electrode near the end of the deflecting beam creates a capacitor with a value that can be sensitively measured through the use of a capacitance bridge circuit as shown in the diagram. The bridge circuit works by sensitively measuring the voltage difference between Nodes 1 and 2 in the circuit. To determine the unknown impedance Z_x, the variable impedance Z_4 is carefully adjusted until the voltage difference between Nodes 1 and 2 is zero, as detected by the voltmeter D. Under this condition, the bridge circuit is said to be in balance and therefore satisfies the well-known balance condition $Z_1 Z_4 = Z_2 Z_x$.[14] If the impedances Z_1 and Z_2 are purely resistive and given by the values R_1 and R_2, respectively, then the capacitance C_x of the sample arm of the bridge is simply related to the adjustable capacitive component of the impedance Z_4 (C_4) by the following simple equation: $C_x = C_4 R_2 / R_1$. Because the capacitance bridge is essentially a differential measurement device, it is very sensitive to small changes in the sample capacitance. As the beam deflects, the capacitance between it and the fixed electrode changes slightly, which throws the bridge out of balance. This is detected by the voltmeter, and the appropriate adjustment must be made to the adjustable capacitor C_4 to bring the bridge back into balance. In this way, the change in capacitance of the sample ΔC_x can be measured. The deflection of the beam is then related to the change in capacitance by the following relation derived by Wilcock and Campbell[15]:

$$\delta = \frac{3wL^2(L-X)}{4\pi C^2(L^2 + XL + X^2)} \Delta C \tag{6.6}$$

The geometric dimensions δ, L, and X are as shown in Figure 6.9, w is the beam width, and C and ΔC are the sample capacitance and change in capacitance, respectively.

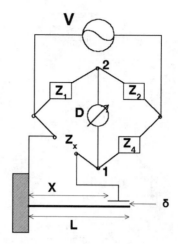

FIGURE 6.9 Measuring the deflection of a cantilevered beam by a capacitance measurement.

The sensitivity of the capacitive technique is such that it gives a resolution of the beam deflection comparable to that of optical interferometry.

6.2.3.5 Stress Measurement by Vibrational Resonance

Up to this point, every measurement technique has relied ultimately on the measurement of a deflection of some sort. Using optical methods, deflections can be measured quite accurately and reliably. However, to derive the stress that is giving rise to the deflection one must at a minimum know the modulus of the material in question. This causes a practical problem because in general practice we are quite lucky if we know the modulus of any material to within 10%, not to mention the Poisson ratio, which most often is simply a guess. In some cases, one may be lucky enough to be working with well-characterized materials such as silica glass or single-crystal silicon, in which case the elastic constants may be known to 1% or better, but this is still not nearly as accurate as one can measure the deflection. Thus, when using the Stoney equation, Equation (6.5), to evaluate the stress in a coating, the level of accuracy achievable is limited by how well we know the elastic properties of the substrate material used.

However, there is a method of stress measurement that avoids this drawback; it is called the vibrational resonance technique and involves the measurement of a frequency, or its inverse time, to evaluate the stress in a thin film without prior knowledge of its elastic properties. The basic principle, first worked out by Raleigh,[16] is most easily understood by considering the problem of a vibrating string as shown in Figure 6.10. The essence of the string is that it cannot support a vertical load unless it is under tension. The string thus effectively has no bending modulus or stiffness and would hang limp unless under tension. Because of this, the elastic modulus of the string material does not enter the problem of the vibrating string. The modulus of the string materials acts only in the axial direction, where it resists the tensile load F, thus setting up a tension that gives the string its ability to resist

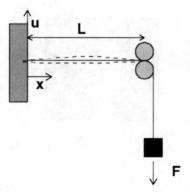

FIGURE 6.10 Thin string under tension. Without the action of load F, the string has essentially zero stiffness and cannot vibrate. The tension induced by the load imparts a definite stiffness to the string, which can now vibrate like an elastic solid.

vertical loads. A straightforward application of the momentum balance law leads to the following well-known equation for a vibrating string:

$$F\frac{\partial^2 u}{\partial x^2} = \rho_l \frac{\partial^2 u}{\partial t^2} \tag{6.7}$$

where

u = Vertical displacement of string
x = Coordinate with origin at left support structure
t = Time
F = Tensile load on string (N)
ρ_l = Linear mass density of string material (kg/m)

For the configuration depicted in Figure 6.10, for which we have a string of length L clamped at both ends, Equation (6.7) can be readily solved by the method of separation of variables, leading to the following expression for the deflection u as a function of x and t:

$$u(x,t) = A \sin\left(\frac{n\pi}{L}x\right)\sin(\omega t) \quad n = 1,2,3\ldots$$

$$\tag{6.8}$$

$$\omega = \frac{n\pi}{L}\sqrt{\frac{F}{\rho_l}}$$

What is most interesting about Equation (6.8) is the fact that the frequency of vibration depends on the string tension and density plus a factor involving the mode number and string length. Assuming the string is vibrating in its lowest frequency state, given by $n = 1$, then from Equation (6.8) we can find the tension in the string in terms of its density and the frequency of vibration as follows:

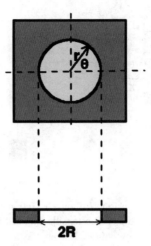

FIGURE 6.11 Membrane in tension formed by etching away a circular section of the substrate supporting a uniform blanket coating.

$$F = \left(\frac{L}{\pi}\right)^2 \rho\omega^2 \tag{6.9}$$

If we know the density of the string material and the cross-sectional area of the string, then using Equation (6.9) allows us to calculate the axial stress in the string simply by measuring the frequency of the fundamental mode of vibration.

The situation is much the same for the case of thin films. Instead of a tensioned string, we have a thin membrane in tension supported by a rigid frame of some sort. The head of a bongo drum is a simple example. A more interesting example is a coating on a rigid substrate, such as a paint coating on a steel plate. To measure the stress in the coating, one first etches away a portion of the substrate, leaving a rigid frame to support the coating, which is now a membrane in tension similar to the membrane of the bongo drum. Measuring the frequency of vibration of the membrane should reveal the stress in the original coating. Figure 6.11 illustrates a simple example. The physical nature of the problem is identical to the string problem except that we now have to work in two dimensions instead of one. Instead of Equation (6.7), we have the following equation for the suspended membrane:*

$$\sigma\nabla^2 u = \rho_a \frac{\partial^2 u}{\partial t^2} \tag{6.10}$$

---

* *Note:* This equation follows directly from the field equations of continuum theory given in Table 3.1. In the displacement formulation, the solution is given by Equation (3.109a). For the membrane problem the shear modulus μ must be replaced by the membrane stress σ. Further, for transverse vibrations $\overline{\nabla}\cdot\mathbf{u} = 0$ and thus Equation (6.10) follows. See also development leading to Equation (6.28).

where

σ = Biaxial stress in membrane

ρ_a = Density of membrane material (kg/m²)

u = Vertical displacement

In Equation (6.10), we use the general coordinate free symbol for the Laplace operator ∇^2 because we have not yet committed to a particular coordinate system.* For the disk-shaped membrane shown in Figure 6.11, it is highly advantageous to use cylindrical coordinates (r, θ, z). However, because the problem is essentially two dimensional (i.e., u does not depend on z), we can ignore the dependence in the z direction and rewrite Equation (6.10) as follows:

$$\frac{1}{r}\frac{\partial}{\partial r}\left(r\frac{\partial u}{\partial r}\right) + \frac{1}{r^2}\frac{\partial^2 u}{\partial \theta^2} = \frac{\rho_a}{\sigma}\frac{\partial^2 u}{\partial t^2} \qquad (6.11)$$

This equation can also be solved by the separation of variables method as was done for Equation (6.9) with the modification that the vertical displacement of the membrane now depends on three variables (r, θ, t). A general solution to Equation (6.11) has the following form:

$$u(r,\theta,t) = A\, J_n(\omega k r)\cos(n\theta)\sin(\omega t) \qquad (6.12)$$

where

u = Vertical displacement of membrane

r = Radial coordinate

θ = Angular coordinate

t = Time

n = Mode number

ω = Angular frequency

$k = (\rho/\sigma)^{1/2}$

ρ = Membrane density

σ = Membrane biaxial stress

The only unusual aspect of Equation (6.12) is the appearance of the function $J_n(x)$, which is the Bessel function of order n that appears in the solution of innumerable problems that display cylindrical symmetry.[17] The fact that the membrane is constrained to adhere perfectly to the supporting substrate at the boundary where $r = R$ forces the condition that $J_n(\omega k R) = 0$ at all times. This constraint then imposes the condition that $\omega k R = \lambda_{nm}$, where λ_{nm} is the mth zero of the nth-order Bessel function.[18] Because $k = (\rho/\sigma)^{1/2}$, this condition directly implies the following relation between the stress in the membrane and the frequency of oscillation:

---

* The precise form of the Laplace operator for a variety of useful coordinate systems is given in Table A.2 of Appendix A.

$$\sigma = \frac{\rho R^2 \omega_{nm}^2}{\lambda_{nm}^2} \tag{6.13}$$

Equation (6.13) demonstrates that the stress in a thin film can be measured knowing its density and a frequency of vibration without reference to any elastic properties. Ku et al.[19] used this formula to measure the stress in thin coatings on top of thin metal membranes. They first inverted Equation (6.13) to find the frequency of oscillation in terms of the stress and the film density:

$$\omega = \frac{C}{R} \left(\frac{\sigma}{\rho} \right)^{1/2} \tag{6.14}$$

The details of the mode numbers are now lumped into the factor C. For a given mode of vibration, Equation (6.13) can be used to determine the stress in the substrate metal membrane. If now a thin coating is deposited on top of the metal membrane, then both the stress and density of the now-composite bilayer laminate will shift. Assuming that the stress and density of the laminate structure will be additive in the stress and density of the component layers, Equation (6.14) can be rewritten as

$$\omega_l = \frac{C}{R} \left(\frac{\sigma_s t_s + \sigma_f t_f}{\rho_s t_s + \rho_f t_f} \right)^{1/2} \tag{6.15}$$

where
$\quad \omega_l$ = Vibration frequency of laminate
$\quad \sigma_s, \sigma_f$ = Stress in substrate and deposited film, respectively
$\quad \rho_s, \rho_f$ = Density of substrate and deposited film, respectively
$\quad t_s, t_f$ = Thickness of substrate and deposited film, respectively
$\quad R$ = Membrane radius
$\quad C$ = Constant related to mode number

Equation (6.15) may be combined with Equation (6.14) to yield the following simple relation for the stress in the deposited film in terms of known or measured quantities:

$$\frac{\sigma_f}{\sigma_s} = \left(\frac{\omega_l}{\omega_s} \right)^2 \left(\frac{t_s}{t_f} + \frac{\rho_f}{\rho_s} \right) - \frac{t_s}{t_f} \tag{6.16}$$

All quantities are as in Equation (6.15) except that ω_l is the vibration frequency of the laminate membrane, and ω_s is the frequency of the uncoated substrate. Again, the main advantage of using Equation (6.16) is that the elastic properties of the substrate or the coating need not be known.

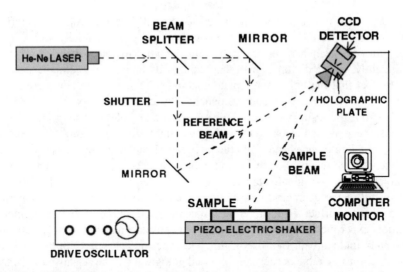

FIGURE 6.12 Simplified schematic of the laser holographic method for measuring the mode and frequency of vibration of a suspended membrane.

6.2.3.6 Holography of Suspended Membrane

A very fruitful approach to the suspended membrane method was given by Farris[20] and co-workers. The reader may have noticed that in the treatment given above the question of which mode of vibration is analyzed has been pushed to the background. Most work is done at the fundamental mode, which corresponds to the lowest frequency of vibration. Many other modes of vibration are available, however, and it is not always easy to sort them out just on the basis of vibration frequency. A clever way of attacking this problem is through the use of laser holography, which allows one to visualize directly the mode of vibration by observing the interference pattern created when the vibrating membrane is observed through a hologram taken while the membrane is at rest.[21]

Figure 6.12 exhibits a simple schematic diagram of one possible experimental setup for implementing this technique. The basic procedure is as follows. First, a hologram of the membrane surface is taken at rest using standard methods of laser holography. By this method, a plate of holographic film is exposed by two coherent beams of light using a beam splitter and mirrors as shown in the diagram. The reference beam comes directly from the laser via the beam splitter and a mirror and serves to define a clean reference signal for the phase of the laser illumination. The second beam is reflected off the surface of the membrane under investigation and contains all the phase information required to define that surface. The two beams interfere in the plane of the holographic film* where the hologram is created. After a brief exposure, the reference beam is shut off by a shutter, and the hologram is fixed in the film. The holographic plate now contains all the geometric information

---

* In practice, this is a special thermoplastic recording film made of material that can be erased and rerecorded many times. Standard holographic plates do the job but are awkward to use.

necessary to reconstruct the sample surface visually. Aside from the imprinted holographic interference pattern, the holographic plate is essentially a clear transparent film that can be seen through from behind with the naked eye or, more conveniently, with a digital camera that displays its output on a computer screen.

The fact that the holographic plate contains all visual information concerning the sample surface is easily and dramatically tested by first blocking off all light coming from the sample surface. This causes the sample to disappear, as would normally be expected. However, leaving the light from the sample blocked but turning on the reference beam from the laser causes the image of the sample surface to miraculously* reappear on the output monitor in precisely the same state as when the hologram was originally taken. This clearly demonstrates that the hologram contains all of the geometric information required to reconstruct the configuration of the sample surface in its reference state. For the purposes of the membrane vibration experiment, the reference state is exposed when the sample surface is completely undisturbed. Because of this, the experimental setup must be mounted on a vibration-free table because the exposed hologram will be sensitive to displacements on the order of a micrometer.

Having exposed the holographic plate and turned off the reference beam, the sample surface can now be viewed directly on the output monitor as it is illuminated by the sample beam from the laser. As long as everything remains quiescent, the sample surface simply appears undisturbed as it was when the hologram was taken. However, even the smallest disturbance of the sample surface will cause the phase of the sample beam to shift, which will cause it to interfere with the recorded pattern in the holographic plate, giving rise to an interference pattern that essentially amounts to a contour map of the surface displacement. In this experiment, the sample is vibrated by a piezoelectric shaker connected to a drive oscillator, which provides a single-frequency output. A confused pattern is observed on the output monitor as the frequency is varied over some fixed range. However, as the drive frequency approaches one of the normal modes of vibration of the membrane, a distinct modal pattern appears on the screen, indicating that the driving signal is at one of the fundamental frequencies that would be calculated by Equation (6.14). The image appearing on the screen is what is called a time-averaged interference pattern in which the maxima and minima of the excited vibration mode remain fixed in time and thus give rise to the observed interference pattern. Figure 6.13 gives a schematic representation of a few of the lower-frequency modal patterns. Knowing the modal pattern essentially solves the problem of fixing the numbers n and m that appear in Equation (6.13). Knowing n and m, one can look up the mode number λ_{nm} in a table, and the remaining quantities in Equation (6.13) are either known or measured.

The holographic measurement method has a number of other less-obvious advantages that make it superior to other approaches. In particular, because from simple physical reasoning we know that the density of the membrane material and the residual stress in the membrane do not change with the vibration frequency, it is

---

* Not to sound too "paranormal" about it, but working with holograms can be an eerie experience. I still find it miraculous that the weird pattern of fringes in the plate manages to capture the full three-dimensional image of the object.

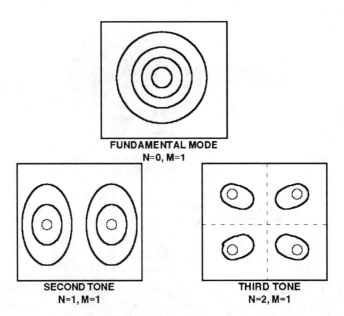

FUNDAMENTAL MODE
N=0, M=1

SECOND TONE
N=1, M=1

THIRD TONE
N=2, M=1

FIGURE 6.13 First three modes of a vibrating membrane. The shape of the interference pattern allows inference of the mode numbers n and m and thus determines λ_{nm}, which is needed in Equation (6.14) to determine the stress in the membrane.

clear that Equation (6.13) must give the same value for the stress for each of the different normal modes. This implies that the normal mode frequencies ω_{nm} and the mode numbers λ_{nm} must shift so that the ratio $(\omega_{nm}/\lambda_{nm})^2$ is constant. This can be verified by performing measurements at several different modal frequencies and provides a valuable check on the validity of the data.

Another valuable feature of the holographic interferometry approach arises from the visual nature of the experiment. In particular, one easily detects the presence of any anisotropic behavior of the residual stress by direct inspection. At the fundamental mode, for example, if the in-plane stress is truly isotropic, then the modal pattern will be one of concentric circles for a disk-shaped membrane. However, if anisotropy is present, then the circles flatten out into ellipses, and the minor and major axes of the ellipses indicate the principle stress directions in the film.

As a final example of the usefulness of the suspended membrane technique, Farris and co-workers demonstrated a clever way to measure the Poisson ratio of a thin film by measuring the stress in both a complete film and one cut down to a thin strip. Figure 6.14 illustrates the sample configurations. For the sake of argument, assume that Figure 6.14a represents a film coated at a high temperature and then allowed to cool to ambient conditions. From Equation (3.50) in Chapter 3, we know that the film will develop a biaxial stress given by the following simple formula:

$$\sigma_{bi} = -\frac{E\Delta\alpha\Delta T}{1-\nu} \tag{6.17}$$

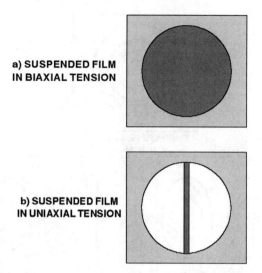

a) SUSPENDED FILM
IN BIAXIAL TENSION

b) SUSPENDED FILM
IN UNIAXIAL TENSION

FIGURE 6.14 Suspended membrane samples for a thin film on a rigid substrate. (a) Full film under biaxial tension. (b) Thin slice of sample a, which is now under uniaxial tension. Comparing the tension in these samples allows determination of the Poisson ratio of the film.

where

σ_{bi} = Biaxial stress developed in film

$\Delta\alpha$ = Thermal expansion mismatch between film and substrate

ΔT = Temperature drop

E = Film modulus

ν = Poisson ratio of film

Starting with the full film in Figure 6.14a, we now cut it down to a narrow strip as shown in Figure 6.14b. The biaxial stress now relaxes down to a uniaxial stress given by

$$\sigma_{uni} = -E\Delta\alpha\Delta T \tag{6.18}$$

Because both σ_{bi} and σ_{uni} can be easily measured by the vibrating membrane method, we can take the ratio of Equations (6.17) and (6.18) to arrive at the following elegant formula for the Poisson ratio:

$$\nu = 1 - \frac{\sigma_{uni}}{\sigma_{bi}} \tag{6.19}$$

As an example, Farris and co-workers carried out this measurement on a polyimide film. The biaxial stress was measured as 7.41 MPa, and the uniaxial stress was 4.28 MPa. Using Equation (6.19), we readily estimate the Poisson ratio as 0.42. As a parting word before moving on to other stress measurement methods, it should

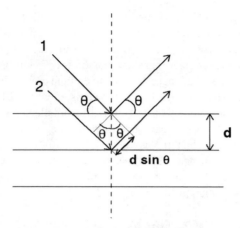

FIGURE 6.15 Schematic diagram of the classic Bragg diffraction of x-rays by crystal lattice planes. The phase difference between the waves reflected by Layers 1 and 2 is (d sinθ). A sharp diffraction peak will be observed at angles θ, for which (d sinθ) is an integral multiple of the x-ray wavelength. Measuring θ allows computation of the lattice spacing d from Equation (6.20).

be pointed out that obtaining reliable data on the Poisson ratio of thin films is not an easy matter.* Thus, Equation (6.19) fills an important gap in our ability to obtain reliable elastic constant data on thin films.

6.2.4 X-RAY MEASUREMENTS

With the x-ray method, we return to measuring the stress in a solid by first accurately measuring the deflection of crystal planes caused by the local stress field. It should be noted that there are two separate types of experiments that use x-rays to estimate the stress in a coating. One method involves using x-rays to detect the deflection of a crystalline substrate on which a coating is deposited. Knowing the deflection of the substrate, the stress in the coating is then deduced using the Stoney equation as outlined above. An example of this approach was given by Hearn.[22] I do not go into details of this approach as it is simply a variation on measuring the deflection of a substrate and then inferring the stress in the coating from the Stoney relation.

The second approach is to measure directly via x-ray diffraction the distortion of the crystal lattice of the coating or film under consideration and from this deformation calculate the associated stress field. A comprehensive review of this method was given by François et al.[23] This technique depends directly on Bragg's law for the diffraction of x-rays from crystal lattice planes. Figure 6.15 illustrates the classic derivation of this principle for a simple two-dimensional array of lattice planes.[24] The diagram shows two rays from an x-ray source; both are completely in phase as they arrive at a set of atomic lattice planes, which are schematically illustrated by

---

* Those who routinely work with material data sheets from polymer material vendors know they are lucky even to obtain data on the Young's modulus of most materials. The Poisson ratio is unlikely even to be mentioned, much less specified.

a set of parallel lines. Ray 1 is assumed to be reflected off the top plane and Ray 2 off the next plane down. By the well-known rule that the angle of reflection is equal to the angle of incidence, we see that both rays are reflected at the incidence angle θ.

From the diagram, it is clear that the reflected Ray 2 will be out of phase with Ray 1 because it must travel a longer distance to exit the crystal. Inspection of Figure 6.15 shows that the extra distance Ray 2 travels is equal to (2d sin θ), where d is the spacing between the lattice planes. However, from elementary wave theory, we know that two sinusoidal waves can be shifted an integral number of wavelengths and still remain in phase. Thus, if λ is the wavelength of the incident radiation, the reflected waves will be in phase and thus superimpose constructively if the extra distance traveled by Ray 2 is an integral multiple of λ, which immediately leads to the well-known Bragg relationship for x-ray diffraction:

$$2d \sin \theta = n\lambda \quad n = 1, 2, 3, \dots \tag{6.20}$$

where

d = Lattice spacing
λ = X-ray wavelength
θ = Angle of incidence

The x-ray measurement is carried out on a device called a goniometer, which essentially allows careful measurement of the angle θ. What Equation (6.20) implies is that if θ satisfies the Bragg relationship, then there will be a high relative intensity of reflected radiation at that angle because the reinforcing effect of many scattered waves adding up constructively to give a sharp diffraction peak. What is most important from the point of view of estimating the stress in a crystal is the fact that Equation (6.20) allows measurement of the lattice spacing d. Thus, in principle one takes a single crystal of a material known to be in a zero-stress state and performs an x-ray experiment to determine the lattice spacing d. Stressing the crystal causes the lattice planes either to separate because of tension or close together because of compression, and this new lattice spacing can also be measured. Assuming d_0 is the lattice spacing in the unstressed reference state and d is the new spacing when stressed, we can immediately estimate the strain in the direction normal to the relevant lattice planes as

$$\varepsilon = \frac{d - d_0}{d_0} = \frac{\sin \theta_0}{\sin \theta} - 1 \tag{6.21}$$

As with all of the methods that directly measure a displacement, however, one now needs information on the crystal elastic constants to calculate the stress from the measured strain. Thus, in principle at least, the problem of measuring the stress in a solid by the x-ray technique is fairly straightforward. The method in effect cleverly uses the regularly spaced lattice planes as a sort of built-in atomic-level strain gauge. However, as with nearly everything else in the real world, reducing the elementary theory to practice raises a number of practical problems. The first issue of course is that one is confined to investigating crystalline materials. More

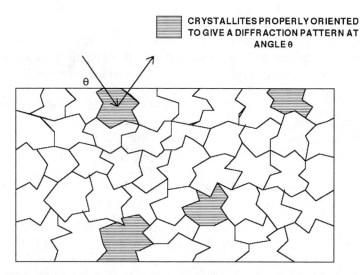

FIGURE 6.16 Schematic illustration of a polycrystaline solid. The shaded regions indicate crystallites with the same lattice orientation and thus will diffract in unison at some angle θ.

serious, though, is the fact that even when dealing with materials that crystallize one rarely is presented with nice, uniform single crystals. Rather, just about everything of practical use comes in the form of a polycrystalline material consisting of some random packing of an enormous number of small single crystals, each with its own different orientation.

Figure 6.16 gives a schematic illustration of what occurs in most metals. Instead of a single crystal, we have an aggregate of small, close-packed crystallites ordered in random directions. Thus, if the x-ray beam illuminates some well-defined volume of the material, only a fraction of the crystallites in that volume will have the proper orientation to contribute to a diffraction peak.

A second issue is the size of the illuminated volume. As with any other form of electromagnetic radiation, x-rays are attenuated exponentially with penetration depth. With metals, for instance, the intensity of the incident beam is decreased by 95% within the first 5 to 100 μm from the surface. A variety of corrections must be applied to account for this problem when dealing with actual metal parts. However, this is not the only problem facing practical use of the x-ray method. The following is a short list of difficulties that can be encountered:

- *Problem of large crystallites*: In certain cases, such as when examining welds, a number of large crystallites are formed that dominate the illuminated volume and distort the normal averaging process. Special precautions have to be taken to overcome this problem, such as enlarging the illuminated volume and/or oscillating of the specimen.
- *Problem of stress and composition gradients*: In some samples, the stress may vary sharply with depth from the surface or the sample composition

may vary as well. In either case, interpretation of the diffraction data in such situations can be greatly complicated.

- *Problem of multiphase materials*: When there is more than one phase present, there can be large anisotropy in the local elastic behavior of the material. The x-ray analysis will be strongly affected by this behavior, and appropriate adjustments must be applied.
- *Problem with very thin films*: When examining coatings, there must be sufficient thickness present to give a reasonable diffraction pattern.
- *Problem of low crystallographic symmetry*: In materials with low symmetry such as rhombohedral or triclinic crystals, the number of diffraction peaks increases, and their intensity drops. In addition, their separation decreases, making it difficult to resolve specific diffracting planes. On top of this, the elastic constants are more numerous and may not be well known.

One or more of the above problems may surface in any particular attempt to use the x-ray technique to evaluate the stress in a given sample. A number of advanced methods have been developed to overcome these problems, however. The serious investigator needs to consult the relevant literature to come up with an appropriate strategy. The treatment given in Note 22 is a good place to start.

6.2.5 ULTRASONICS

One can also probe the stress distribution in a solid using sound waves. As with the vibrating membrane method, this approach is one that relies on the ability to measure time and time differences accurately. Again, to obtain some understanding of how this method works, we need to go back to the fundamental equations of continuum theory developed in Chapter 3. When solving the field equations for the displacement vector **u** that describes the deformation of the solid in response to applied loads, we came up with Equation (3.109b), which holds for homogeneous, isotropic solids in the elastic limit:

$$\mu\nabla^2\mathbf{u}+\left(\mu+\lambda\right)\nabla\left(\nabla\cdot\mathbf{u}\right)+\mathbf{f}_b =\rho\frac{d^2\mathbf{u}}{dt^2} \tag{6.22}$$

where

$\mathbf{u}$ = Displacement field vector (m)
μ = Shear modulus (GPa)
ρ = Material density (kg/m^3)
$\mathbf{f}_b$ = Body force density (i.e., gravity) (N/m^3)
λ = Lame elastic constant = $\nu E/[(1 + \nu)(1 - 2\nu)]$ (N/m^2)
ν = Poisson ratio, E = Young's modulus (N/m^2)

For present purposes, we assume that body forces $\mathbf{f}_b$ such as gravitation can be neglected, which leads to the following significantly simpler relationship:

$$\mu\nabla^2 u + (\mu+\lambda)\nabla(\nabla\cdot\mathbf{u}) = \rho\frac{d^2 u}{dt^2} \tag{6.23}$$

The derivation of Equation (6.23) is rather cumbersome, as those who followed the development in Chapter 3 may be aware. However, Feynman et al.[25] pointed out that there is a neat way to demonstrate that the form of Equation (6.23) is indeed correct; it relies on some interesting physical reasoning plus some fairly simple vector calculus. The first thing one notices is that Equation (6.23) looks like Newton's second law, F = ma, where the right-hand side represents the mass times acceleration part. This means that the left-hand side represents the force on the material caused by the displacement field, and for reasons of mirror symmetry this force can depend only on second derivatives of the displacement.

It is easier to visualize this in the case of one dimension, for which the equation must stay the same if x goes to −x, which is true for second derivatives like $\partial^2 u/\partial x^2$ but not for first-order terms like $\partial u/\partial x$. The same reasoning must hold true in three dimensions, but now we must consider second-order vector derivatives. Looking through the mathematical handbooks, one quickly sees that there are only two general vector derivatives of the vector field $\mathbf{u}$ that fit the bill, and those are $\nabla^2 \mathbf{u}$ and $\nabla(\nabla \cdot \mathbf{u})$.* The force caused by the displacement field $\mathbf{u}$ must therefore be some linear combination of these two second-order vector derivatives, which is exactly what Equation (6.23) implies.

Equation (6.23) looks a lot like a wave equation such as Equation (6.10) except for the $\nabla(\nabla \cdot \mathbf{u})$ term, which complicates things. This term arises because solids can support two types of waves, commonly referred to as longitudinal (compressive waves) and transverse (shear waves). The longitudinal waves are the same as common sound waves that propagate in air. Air as a fluid cannot support shear waves. In a longitudinal wave, the supporting material is alternately compressed and stretched, which gives rise to alternating density changes. Changes in density imply that the divergence of the displacement field $(\nabla \cdot \mathbf{u})$ cannot be zero, thus giving rise to the complicating term in Equation (6.23). With the transverse waves, however, the displacement of the material is perpendicular to the direction of propagation of the wave and does not involve any change in density; thus, $\nabla \cdot \mathbf{u} = 0$.

On recognizing these facts, the pioneers in the wave propagation field came up with the idea of dividing the displacement field into two parts, one with zero divergence and the other with non-zero divergence. In fact, there is a mathematical theorem that states that this can be done in the following way:

$$\mathbf{u} = \mathbf{u}_l + \mathbf{u}_t \tag{6.24}$$

where

$$\nabla \cdot \mathbf{u}_t = 0; \quad \nabla \times \mathbf{u}_t = 0$$

Thus, the transverse shear waves are described by a vector field $\mathbf{u}_t$ with zero divergence and thus involve no density changes. The longitudinal waves, on the other hand, have a nonzero divergence but are irrotational, having zero curl ($\nabla \times \mathbf{u}_l = 0$).

---

* The expression $\nabla \times \nabla \times \mathbf{u}$ does not qualify because of the general vector identity $\nabla \times \nabla \times \mathbf{u} = \nabla^2 \mathbf{u} - \nabla(\nabla \cdot \mathbf{u})$.

The advantage of this particular partition of the displacement vector becomes apparent when Equation (6.24) is substituted into Equation (6.23):

$$\mu\nabla^2\left(\mathbf{u_t}+\mathbf{u_l}\right)+\left(\mu+\lambda\right)\nabla\left(\nabla\cdot\left(\mathbf{u_l}+\mathbf{u_l}\right)\right)=\rho\frac{d^2\left(\mathbf{u_t}+\mathbf{u_l}\right)}{dt^2} \quad (6.25)$$

Taking the divergence of both sides of Equation (6.25) and noting that $\nabla\cdot\mathbf{u_t}=0$ and that the divergence operator $\nabla\cdot$ commutes with all of the operators in Equation (6.25), we obtain

$$\nabla\cdot\left[\left(2\mu+\lambda\right)\nabla^2\left(\mathbf{u_l}\right)-\rho\frac{d^2\left(\mathbf{u_l}\right)}{dt^2}\right]=0 \quad (6.26)$$

Because $\nabla\cdot\mathbf{u_l}\neq0$, the only way for Equation (6.26) to hold is if the two terms within the brackets are equal and subtract to cancel each other out, which immediately implies the following relation:

$$\left(2\mu+\lambda\right)\nabla^2(\mathbf{u_l})=\rho\frac{d^2(\mathbf{u_l})}{dt^2} \quad (6.27)$$

Equation (6.27) is easily recognized as the wave equation for a signal propagating at the velocity $v_l=((2\mu+\lambda)/\rho)^{1/2}$, which, because $\nabla\cdot\mathbf{u_l}\neq0$, we also know to be a compressive longitudinal wave. Now, we can go back to Equation (6.25) and instead of taking the divergence of both sides, we take the curl of both sides. This will eliminate the longitudinal displacement and by performing the same kind of manipulation as was done above, we obtain a second wave equation:

$$\mu\nabla^2(\mathbf{u_t})=\rho\frac{d^2(\mathbf{u_t})}{dt^2} \quad (6.28)$$

Equation (6.28) is the wave equation for a transverse shear wave propagating at velocity $v_t=(\mu/\rho)^{1/2}$. What we have learned to this point is that the general equations of continuum theory summarized in Table 3.1 predict the existence of sound waves in elastic solids, which come in two varieties. One type induces density changes in the material and involves displacements in the direction of propagation. The second type involves shear-type deformations perpendicular to the direction of propagation and does not involve density changes. These two types of basic waves propagate at velocities given by the following formulae:

$$\text{a) } v_l=\sqrt{\frac{\lambda+2\mu}{\rho}} \quad \text{Longitudinal waves}$$

$$(6.29)$$

$$\text{b) } v_t=\sqrt{\frac{\mu}{\rho}} \quad \text{Transverse waves}$$

Equations (6.29) are the basic formulae for the propagation of sound waves in a homogeneous elastic solid. As stated, the velocity of propagation should depend only on the elastic constants and the density, but for real materials there is also a small dependence on the material texture and the level of stress. It is the dependence on the stress that makes acoustic measurements useful for estimating the stress in a solid. The following one-dimensional analysis illustrates how this comes about. From Equation (6.29), we know that the velocity of a transverse wave depends on the shear modulus μ. The shear modulus in turn is related to the stress by the constitutive relation, which in this case is simply Hooke's law $\sigma = \mu\varepsilon$, which was used in the derivation of Equation (6.28). Use of the simple form of Hooke's law implies that the velocity of a shear wave will depend only on the shear modulus. However, for real materials Hooke's simple formula is only a first approximation, and in general there will be higher-order effects that also affect the stress in the solid. In particular, a more accurate constitutive relation would be

$$\sigma = \mu\varepsilon + \mu^a\varepsilon^2 \tag{6.30}$$

The term in ε^2 involves anharmonic effects, as signified by the coefficient μ^a. If we denote $\sigma_0 = \mu\varepsilon$ as the purely linear elastic stress, then Equation (6.30) can be rewritten as

$$\sigma = \varepsilon(\mu + \frac{\mu^a}{\mu}\sigma_0) \tag{6.31}$$

According to Equation (6.31), we can look at the net effect of considering anharmonic contributions as giving rise to an effective shear modulus depending on the stress in the material, which is given by

$$\mu^e = \mu + \frac{\mu^a}{\mu}\sigma_0 \tag{6.32}$$

From Equation (6.29b), the sound velocity will now be given by

$$v_s = \sqrt{\frac{\mu + \frac{\mu^a}{\mu}\sigma_0}{\rho}} \tag{6.33}$$

$$= \sqrt{\frac{\mu}{\rho}\left(1 + \frac{\mu^a}{\mu^2}\sigma_0\right)^{1/2}}$$

TABLE 6.2

Acoustoelastic Constants for Various Materials (%/GPa)[a]

Material	K_1	K_2	K_3	K_4	K_5
Polystyrene[51]	−93.7	−4.49	−4.59	−40.8	−11.46
Aluminum alloy[52]	−7.73	1.12	−2.15	−3.96	0.79
Rail steel[53]	−1.21	0.18	−0.12	−0.75	0.02
Pyrex glass[29]	8.26	−0.99	5.85	4.04	−4.72

[a] Adapted from the data of Thompson et al., Note 28.

The quantity $(\mu^a/\mu^2)\,\sigma_0$ in parentheses is small, so Equation (6.33) can be further simplified by expanding the square root to lowest order in $(\mu^a/\mu^2)\,\sigma_0$, giving

$$v_s = v_0\left(1 + 1/2\,\frac{\mu^a}{\mu^2}\sigma_0 + \dots \text{higher order terms}\right)$$

(6.34)

$$v_s = v_0 + K\sigma_0; \quad K = \frac{v_0\mu_a}{2\mu^2}$$

Thus, to a rough approximation the velocity of sound in an elastic solid is shifted by an amount $K\sigma_0$ in the presence of a static stress field σ_0. The coefficient K is known as the acoustoelastic constant. Sample values are given Table 6.2. The clear implication of Equation (6.34) is that we can get a handle on the stress in a material by measuring the velocity of an ultrasonic wave under both stress-free and loaded conditions. Whereas this is indeed true, we do need to be cognizant of a number of complications and limitations associated with the ultrasonic measurement method before deciding whether it should be the method of choice for a particular application.

First, Figure 6.17 illustrates three basic modes of operation for ultrasonic measurements. In all cases, the basic measurement involves determining the velocity of the acoustic wave by measuring the propagation time from a sending unit to a receiver. Figure 6.17a illustrates the standard through-thickness method by which sender and receiver are in the same unit; it is commonly called the pulse echo technique. Figure 6.17b shows a modification of the through-thickness measurement by which the sender and receiver are now in separate units. This approach clearly samples a much larger volume of the part investigated. Finally, Figure 6.17c depicts a case by which sender and receiver are separate units, but special class of waves is employed that propagates only at the surface of the sample. This approach is typically referred to as the surface-skimming method.

In addition to the three basic experimental setups for performing acoustic measurements, there are three different types of acoustic waves that can be propagated. These are illustrated in Figure 6.18. Figure 6.18a illustrates a longitudinal wave for which the oscillatory displacements are parallel to the direction of propagation.

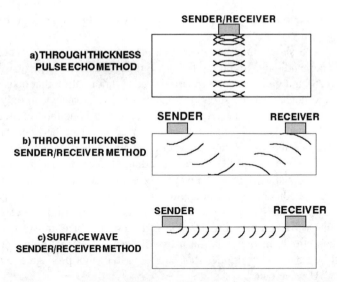

FIGURE 6.17 Three basic configurations of the acoustic wave experiment. (a) A single sender-receiver sends an acoustic pulse through the thickness of the solid, which is reflected back to the receiver. (b) A variation on configuration (a) in which the sender and receiver are separate units placed some distance apart. (c) Separate sender-receiver units using a special wave that propagates only at the surface of the sample, such as a Raleigh wave.

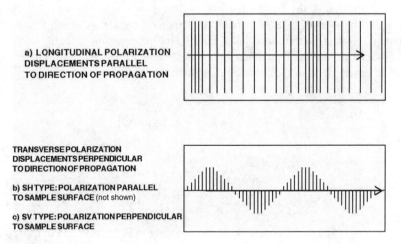

FIGURE 6.18 Three separate types of acoustic waves that can propagate in a solid. (a) Longitudinal polarization with displacements in the direction of propagation. Transverse polarization with displacements perpendicular to direction of polarization. Two types are possible: (b) SH wave with polarization parallel to sample surface; (c) SV wave with polarization perpendicular to sample surface.

Sound waves in air are a common example of this type of wave. This type obeys Equation (6.27) and induces changes in the density of the supporting medium. By common convention, the direction in which the displacements occur is called the polarization direction. The wave in Figure 6.18a would then be said to be longitudinally polarized.

Figure 6.18b shows the case of a transverse wave in which the polarization direction is perpendicular to the direction of propagation. This type of wave obeys Equation (6.28) and gives rise to a shearing type of deformation that does not cause changes in the density of the supporting medium. These waves can only occur in solids that have a nonzero shear modulus. If there is a bounding surface present, such as the free surface of a material object or an interface between layers of a laminate structure, then the transverse waves are classified in two categories. If the polarization direction is parallel to the bounding surface, then one talks about a horizontally polarized wave or an SH wave. The second case is when the polarization direction is perpendicular to the bounding surface, in which case one speaks of a vertically polarized wave or an SV wave. The main reason for using this classification is the fact that waves propagating near the surface of an object must satisfy certain boundary conditions. In particular, at a free surface the normal component of the stress must be zero, which implies that an SV wave cannot propagate at a surface, whereas an SH wave can.

Given that we have three standard setups for ultrasonic experiments and three separate wave types to use, the conclusion is that we can devise at least nine different types of experiments.* For the purposes of this work, we focus only on those experiments that can be applied directly to coatings and laminates and thus support adhesion measurement investigations. However, we first need to look at a more refined derivation of the basic equations of ultrasonic stress measurement. Several authors[26,27] have given a more general and rigorous derivation of the fundamental equations connecting the stress in a solid to the propagation velocity of a sound wave based on the same type of reasoning that lead to Equation (6.34). The essential result is summarized in the following equations[28]:

$$\text{a) } \frac{v_{ii} - v_l^0}{v_l^0} = K_1\sigma_i + K_2\left(\sigma_j + \sigma_k\right)$$

$$\text{b) } \frac{v_{ij} - v_t^0}{v_t^0} = K_3\sigma_i + K_4\sigma_j + K_5\sigma_k$$

(6.35)

where

v_{ij} = Speed of acoustic wave propagating in i direction and polarized in j direction

v_l^0 = Speed of longitudinal wave in stress-free material

v_t^0 = Speed of transverse wave in stress-free material

K_i = ith acoustoelastic constant i = 1, 2, 3, 4, 5

$\sigma_i, \sigma_j, \sigma_k$ = Principal stresses in material

---

* This excludes the possibility of combining different wave types and other possible variations.

Equations (6.35) are the working equations of acoustoelasticity, and the factors K_i are referred to as acoustoelastic coefficients. Table 6.2 gives sample values of these coefficients for various types of materials.

6.2.5.1 Through-Thickness Stress Measurement

The most obvious type of measurement one can think of is the through-thickness setup illustrated in Figure 6.17a. It essentially involves timing the passage of an ultrasonic wave from the top surface to the bottom and back. Using a longitudinally polarized wave and taking the 3 direction as normal to the surface and the 1 and 2 directions in the plane of the surface, then Equation (6.35a) yields the following relation for the expected velocity shift caused by in-plane stresses:

$$\frac{v_{33} - v_l^0}{v_l^0} = K_1 \sigma_3 + K_2 \left(\sigma_1 + \sigma_2 \right) \tag{6.36}$$

The stresses appearing in Equation (6.36) are average stresses through the thickness of the layer. This formula is most useful for the case of a biaxially stressed material where $\sigma_3 = 0$. In this case, measuring the velocity shift ratio and knowing the acoustoelastic coefficient K_2 will yield a direct measure of the sum of the in-plane principal stresses. As an example, assume we have a polymer coating of polystyrene that is biaxially and isotropically stressed to a level of 10 MPa (= 0.01 GPa), then using Equation (6.36) and Table 6.2 we expect a velocity shift of $\Delta v/v_0 = -0.09\%$. This is quite a small but still-measurable shift and implies making time measurements down to the level of nanosecond resolution.

6.2.5.2 Surface Stress Measurement Using Skimming Longitudinal Waves

Figure 6.17c shows an experiment using a transmitter and detector to propagate waves at the surface of a solid. This type of experiment is most useful with respect to adhesion measurements because surface stresses are the ones most likely to contribute to delamination processes. Referring to Table 6.2, we see that for most materials K_1 is the largest acoustoelastic coefficient that, in conjunction with Equation (6.35a), implies that the largest velocity shifts are obtained using longitudinal waves polarized in the direction of the stress field to be measured. Thus, we would like to be able to propagate longitudinal waves at a surface to more accurately measure in-plane stresses. However, a problem arises in that using pure longitudinal waves does not let one satisfy the boundary condition that the component of the stress field normal to the surface, at the surface, needs to be zero. Thus, in theory we cannot propagate a longitudinal wave at the surface.

However, in practice one gets around this problem by propagating a longitudinal wave at a grazing angle to the surface. What this does is to create a weak shear wave that serves to satisfy the surface boundary condition and radiates into the bulk of the solid. Thus, there is energy loss caused by this shear wave but not so much that the longitudinal component cannot propagate for considerable distances. This approach is called the surface-skimming longitudinal wave method. Equation (6.35) can be generalized to the case of a surface-skimming wave propagating at an angle θ regarding the in-plane principal stress direction σ_1, giving

$$\frac{v_l(\theta) - v_l^0}{v_l^0} = \frac{K_1 + K_2}{2}\left(\sigma_1 + \sigma_2\right) + \frac{K_1 - K_2}{2}\left(\sigma_1 - \sigma_2\right)\cos 2\theta \qquad (6.37)$$

Equation (6.37) looks like a very handy tool for measuring both of the in-plane principal stresses by simply making velocity measurements at two different angles, say $\theta = 0$ and $90°$. This is clearly the case when the stress field is isotropic; thus, $\sigma_1 = \sigma_2$. However, if the stress field is not isotropic and the two in-plane principal stresses are not equal, then Equation (6.37) can still be used, but one has to be aware that anisotropies in the material texture can cloud the picture because they can also vary as $\cos 2\theta$ and thus confound our ability to sort out what is caused by stress and what is caused by material texture. The reader interested in exploring this problem further should refer to the work of Thompson et al.[29]

6.2.5.3 Rayleigh Wave Method

It is interesting to note that there is a class of waves that does propagate solely at a surface in the rigorous sense that all the field equations and boundary conditions are satisfied exactly, and thus the wave truly exists at and is guided by the surface of the propagating medium. These waves were discovered theoretically by Raleigh[30] and are thus named after him as Raleigh waves. A detailed treatment of Raleigh wave behavior can be found in the monograph by Viktorov.[31] A relationship in the same form as Equation (6.37) also holds for the case of Raleigh waves; the only difference is that the acoustoelastic coefficients are different. As with the skimming longitudinal waves, there is still the confounding effect of material texture properties, which can interfere with any attempt to use Raleigh waves to determine surface stresses.

6.2.5.4 Surface-Skimming SH Waves

Yet a third approach at using surface waves in addition to skimming longitudinal waves or Raleigh waves is the use of surface-skimming transverse waves or SH waves. It will be recalled that the SH wave is polarized parallel to the material surface. The disadvantage of this approach is that it has less sensitivity to the material stress state than the longitudinal waves. However, when the in-plane principal stresses are not equal, this approach has the distinct advantage in that it is not sensitive to material texture and thus avoids the confounding effect that limits the use of Raleigh waves or longitudinal waves.

6.2.6 PHOTOELASTICITY

When dealing with optically transparent materials, photoelastic measurements offer a powerful means of investigating the intrinsic stress distribution. One possible application might be the study of residual stress in optical coatings on lenses. However, to more fully appreciate the practical details of photoelastic behavior we need some background in electromagnetic theory because all of photoelasticity is firmly based in electromagnetics. This leads us directly to the field equations of electromagnetism or what are commonly known as Maxwell's equations.[32] In Chapter 3, we covered in some detail the field equations of continuum theory, which certainly form one of the great monuments to 19th century science and technology. In Maxwell's equations, however, we must certainly have the apotheosis of 19th century science and indeed a paradigm for all of 20th century science as well. The span of application of these equations in space ranges from the atomic level to the edge of the visible universe and in frequency from the direct current powering the common flashlight to the blindingly rapid oscillations of x-rays. There is quite likely no other theory that has been so complete or successful. The only major improvement was the advent of quantum electrodynamics in the mid-20th century, which made the field equations consistent with the principles of quantum mechanics.* Although named after James Clerk Maxwell, these equations represent the tireless work of many eminent scientists of the 18th and 19th centuries. The most outstanding contributors include the legendary Carl Friedrich Gauss, André Marie Ampère, Michael Faraday, and a host of less-well-known but equally eminent investigators. The equations were essentially deduced by simple experiments on charges on pith balls and currents in wires. Maxwell's main contribution was to gather them all together and add a special term that made them fully consistent with known phenomena, which entitles him to immortality in the pantheon of modern science and technology.

An ab initio derivation of these equations would require a work of encyclopedic proportions, so at this stage there is no recourse other than simply to write down the equations in a usable form. We choose the differential form because it leads most directly to the results of most interest to photoelasticity:

a) $\nabla \cdot \mathbf{E} = \dfrac{\rho}{\varepsilon_0}$ (Gauss's Law)

b) $\nabla \cdot \mathbf{B} = 0$ (No magnetic monopoles)

c) $\nabla \times \mathbf{E} = -\dfrac{\partial B}{\partial t}$ (Faraday's Law)

d) $\nabla \times \mathbf{B} = \mu_0 J + \mu_0 \varepsilon_0 \dfrac{\partial \mathbf{E}}{\partial t}$ (Ampere's Law with Maxwell's term)

(6.38)

---

* The cognizant reader will note that this statement is not quite correct. Quantum electrodynamics (QED) is now contained in the formalism of quantum chromo dynamics (QCD). QCD unites QED with the forces that bind nuclear matter (strong forces) and the electroweak forces, which are responsible for beta decay and other important phenomena. In addition, work is now afoot to include the gravitational force and thus create a theory of everything by the use of string theory. String theory is very much a hypothetical proposition at this time. We need not worry, however, because both it and QCD are light years beyond the scope of this volume.

e) $\nabla \cdot \mathbf{J} = -\dfrac{\partial \rho}{\partial t}$ (Conservation of charge)

f) $\mathbf{F} = q(\mathbf{E} + v \times \mathbf{B})$ (Lorentz force on moving charge)

g) $\dfrac{d\mathbf{p}}{dt} = F; \quad \mathbf{p} = \dfrac{mv}{\sqrt{1 - \dfrac{v^2}{c^2}}}$ (Relativistic version of Newton's Second Law)

where

$\mathbf{E}$ = Electric field (N/C also V/m)
$\mathbf{B}$ = Magnetic field [N sec/(C m) also T]
$\mathbf{J}$ = Current density (C/sec m^2)
$\mathbf{F}$ = Electromagnetic force (N)
$\mathbf{p}$ = Relativistic momentum (kg m/sec)
v = Charged particle velocity (m/sec)
q = Particle charge (C)
m = Particle mass (kg)
c = Velocity of light in vacuo (~3 × 10^8 m/sec)
ρ = Charge density (C/m^3)
ε_0 = 8.854 × 10^{-12} (permittivity free space [farad/m] or C^2/N m^2)
μ_0 = 4 10^{-7} (permeability free space [H/m or N sec^2/C^2])

Maxwell's equations consist of Equations (6.38a)–(6.38d). Equations (6.38e)–(6.38g) have been added for completeness and are required when one has to also account for charged particles moving around. Fortunately for the study of photo-elasticity, only neutral matter needs to be considered when all the charges are firmly nailed down, and Equations (6.38e)–(6.38g) will therefore not be needed. For present purposes, we need only consider light waves traveling through a simple dielectric medium; therefore, the charge density ρ and the current density $\mathbf{J}$ are also zero. For this case, the general equations simplify considerably to the following:

a) $\nabla \cdot \mathbf{E} = 0$

b) $\nabla \cdot \mathbf{B} = 0$

c) $\nabla \times \mathbf{E} = -\dfrac{\partial B}{\partial t}$ (6.39)

d) $\nabla \times \mathbf{B} = \mu_0 \varepsilon_0 \dfrac{\partial E}{\partial t}$

Thus, in empty space the electric and magnetic fields can exist without the presence of any free charges or currents and further satisfy a nicely symmetric set of equations. Equations (6.39) can be simplified further with a little manipulation and judicious use of the identity $\nabla \times \nabla \times \mathbf{A} = \nabla^2 \mathbf{A} - \nabla(\nabla \cdot \mathbf{A})$, which holds for any vector field $\mathbf{A}$. The basic trick is to take the curl ($\nabla \times$) of both sides of Equation (6.39c) and make use of the magic identity and further apply Equations (6.39a) and (6.39d) in the following steps:

$$\nabla \times \nabla \times \mathbf{E} = -\frac{\partial \nabla \times \mathbf{B}}{\partial t} \tag{6.40}$$

or

$$\nabla(\nabla \cdot \mathbf{E}) - \nabla^2 \mathbf{E} = -\mu_0 \varepsilon_0 \frac{\partial^2 \mathbf{E}}{\partial t^2}$$

finally

$$\nabla^2 \mathbf{E} = \mu_0 \varepsilon_0 \frac{\partial^2 \mathbf{E}}{\partial t^2}$$

Thus, the electric field is seen to satisfy a vector form of the wave equation just as we found for the displacement field in the discussion of ultrasonics above:

$$\nabla^2 \mathbf{E} = \mu_0 \varepsilon_0 \frac{\partial^2 \mathbf{E}}{\partial t^2} \tag{6.41}$$

Again, from our work on the displacement field associated with ultrasonic vibrations, we see that the velocity of propagation of the electric field will be $1/(\varepsilon_0 \mu_0)^{1/2}$, which on substituting the values for ε_0 and μ_0 given in Equation (6.38) comes close to 3×10^8 m/sec^2, which is an excellent approximation to the speed of light in vacuum. This is a rather remarkable result considering that the constants ε_0 and μ_0 were originally arrived at by simple experiments on charged pith balls and currents in wires. By adding his extra term $\varepsilon_0 \mu_0 \partial \mathbf{E}/\partial t$ to Ampere's law Equation (6.38d), Maxwell managed to expand the range of electromagnetic theory all the way from simple direct currents to the entire electromagnetic spectrum, including light and x-rays. This was by far the greatest theoretical synthesis of 19th century science. Because the speed of light in vacuum is given identically by the equation $c = 1/(\varepsilon_0 \mu_0)^{1/2}$, we can finally rewrite Equation (6.41) as

$$\nabla^2 \mathbf{E} = \frac{1}{c^2} \frac{\partial^2 \mathbf{E}}{\partial t^2} \tag{6.42}$$

Equation (6.42) tell us that electric fields can propagate through empty space at the speed of light. As noted, this is the foundation of all electromagnetic radiation, from cell phones to late-night television. To obtain a clearer picture of the nature of electromagnetic radiation, we need to find a solution to Equation (6.42). This is most readily achieved in the form of a classic plane wave function as follows:

$$\mathbf{E}(r,t) = \mathbf{E}_0 e^{i(\omega t - k \cdot r)} \tag{6.43}$$

where

> ω = Angular frequency of vibration
> $\mathbf{k}$ = Wave number with a magnitude of $\frac{2\pi}{\lambda}$ (wavelength)
> $\mathbf{r}$ = Position vector locating any particular point on the wavefront
> $\mathbf{E}_0$ = Constant amplitude vector
> i = The imaginary unit $\sqrt{-1}$

To show that Equation (6.43) satisfies Equation (6.42) and Equations (6.39), it is helpful to use the following simple vector identities that apply to the functional form given in Equation (6.43). Thus, given that $\mathbf{E}$ is represented by Equation (6.43), then the following identities hold:

$$\nabla^2 \mathbf{E} = -k^2 \mathbf{E}$$

$$\frac{\partial^2 \mathbf{E}}{\partial t^2} = -\omega^2 \mathbf{E} \tag{6.44}$$

$$\nabla \cdot \mathbf{E} = -i\mathbf{k} \cdot \mathbf{E}$$

$$\nabla \times \mathbf{E} = -i\mathbf{k} \times \mathbf{E}$$

With these identities, it is easy to show that Equation (6.43) satisfies Equation (6.42) only if the following condition holds:

$$\left(\frac{\omega}{c}\right)^2 - k^2 = 0 \tag{6.45}$$

or

$$\omega = ck.$$

We have now shown that plane wave solutions of the form Equation (6.43) clearly satisfy the wave equation for the electric field, Equation (6.42). However, we are not done. The remainder of Equations (6.40) must still be investigated. Again, applying the identities in Equations (6.44) to Equations (6.39), we find the following:

$$\nabla \cdot \mathbf{E} = -i\mathbf{k} \cdot \mathbf{E} = 0$$

$$\nabla \cdot \mathbf{B} = -i\mathbf{k} \cdot \mathbf{B} = 0 \tag{6.46}$$

$$-i\omega \mathbf{B} = -i\mathbf{k} \times \mathbf{E}$$

or

$$k \cdot \mathbf{E} = \mathbf{k} \cdot \mathbf{B} = 0$$

and

$$\mathbf{B} = \frac{1}{\omega} \mathbf{k} \times \mathbf{E}$$

Equations (6.46) tell us that both the electric and magnetic fields are perpendicular to the direction of propagation $\mathbf{k}$; furthermore, the magnetic field is given by a cross product of $\mathbf{k}$ with $\mathbf{E}$. Thus, $\mathbf{B}$ is also perpendicular to $\mathbf{E}$. The vectors $\mathbf{k}$, $\mathbf{E}$, and $\mathbf{B}$ form an orthogonal triad that travels through space in the direction of $\mathbf{k}$ as depicted in Figure 6.19.

The analysis thus far provides a fairly complete description of the propagation of electromagnetic radiation through empty space. Although Equation (6.43) represents the simplest plane wave solution, more complex situations can be represented by an appropriate superposition of plane waves using essentially the theory of Fourier transforms. However, the main question at this point is: What happens when the plane wave enters a region of space occupied by some mass of dielectric material? Be aware that the general question of the interaction of electromagnetic radiation with matter is exceedingly complex. This is particularly the case when there are a multitude of free charges floating around in which case one can have a complex plasma existing where charged particles and electromagnetic fields interact in the most elaborate and convoluted manner imaginable.

Luckily, for the purpose of understanding photoelasticity we do not have to approach anywhere near this level of complexity. In this case. we are typically interested in a simple dielectric solid, such as an amorphous glass or other insulator, in which all charges are strongly bound in atomic or molecular structures, and the entire medium is electrically neutral with no net free charges existing anywhere within the medium. What happens is that the electric field induces a local polarization of the bound charges, and this polarization acts in a manner that tends to decrease the local intensity of the external applied field. Figure 6.20 illustrates the basic effect

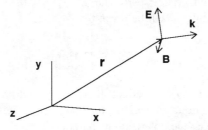

FIGURE 6.19 The electromagnetic field associated with a plane wave at some arbitrary point in space specified by the vector $\mathbf{r}$. The electric field $\mathbf{E}$, the magnetic field $\mathbf{B}$, and the propagation vector $\mathbf{k}$ form an orthogonal triad in space that propagates with the speed of light in the direction of $\mathbf{k}$.

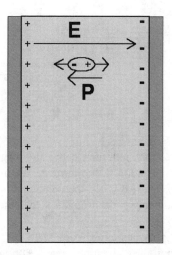

FIGURE 6.20 Schematic of an electric field within a dielectric medium. The net effect of the applied field is to induce a polarization **P** within the medium that acts in the opposite direction to the applied field.

for a dielectric material confined between the plates of a simple capacitor. The applied electric field **E** is seen to induce a certain amount of polarization of the electrically neutral molecules in the medium. This gives rise to a polarization field **P** in a direction opposite to that of **E**. Because we are dealing with relatively small fields, the amount of charge separation is small; in a first-order approximation, the polarization field can be taken as linearly proportional to the applied electric field thus:

$$\mathbf{P} = \varepsilon_0 \chi_e \mathbf{E} \tag{6.47}$$

The proportionality constant χ_e is called the dielectric susceptibility and is taken as a material property of the insulator. Also, if the polarization field is constant, then it is easy to see that there will be no net buildup of charge in the dielectric. However, if the polarization field varies from point to point, then a net polarization charge can be built up within the material. In particular, if $\nabla \cdot \mathbf{P} \neq 0$, then the polarization charge induced within the insulator will be given by

$$\rho_{pol} = -\nabla \cdot \mathbf{P} \tag{6.48}$$

The question now is how this effect shows up in the Maxwell equations. In this case, it is Gauss's law, Equation (6.38a), that is affected. This law is now written as follows to accommodate the polarization charge:

$$\nabla \cdot \mathbf{E} = \frac{\rho_{free} + \rho_{pol}}{\varepsilon_0} \tag{6.49}$$

However, in view of Equations (6.47) and (6.48), the above equation can be written:

$$\nabla \cdot \left(\mathbf{E} + \frac{P}{\varepsilon_0} \right) = \frac{\rho_{free}}{\varepsilon_0} \tag{6.50}$$

or

$$\nabla \cdot (1 + \chi_e) \mathbf{E} = \frac{\rho_{free}}{\varepsilon_0}$$

or

$$\nabla \cdot \kappa_e \mathbf{E} = \frac{\rho_{free}}{\varepsilon_0}$$

or since κ_e is constant

$$\nabla \cdot \mathbf{E} = \frac{\rho_{free}}{\kappa_e \varepsilon_0}$$

Thus, we see that Gauss's law survives more or less in tact with the small revision that the permittivity of free space is replaced by $\kappa_e \varepsilon_0$ where $\kappa_e = 1 + \chi_e$ is a material property commonly called the dielectric constant. Equations (6.38b) and (6.38c) remain unchanged in the presence of any material medium because they are essentially kinematic relationships governing the behavior of the electric and magnetic fields and hold under all circumstances regardless of the presence or absence of free charges and currents. Equation (6.38d) is modified in a manner similar to Gauss's law. In essence, the applied magnetic field B induces local bound currents that give rise to a magnetization $\mathbf{M}$ and a magnetic susceptibility χ_m. The net result is that Ampere's law is modified to the following:

$$\nabla \times \frac{\mathbf{B}}{\mu_0 \kappa_m} = \mathbf{J} + \kappa_e \varepsilon_0 \frac{\partial \mathbf{E}}{\partial t} \tag{6.51}$$

In the absence of free charges, the field equations for electromagnetic wave propagation in a simple dielectric medium become

a) $\nabla \cdot \mathbf{E} = 0$

b) $\nabla \cdot \mathbf{B} = 0$

c) $\nabla \times \mathbf{E} = -\dfrac{\partial B}{\partial t}$ $\qquad (6.52)$

d) $\nabla \times \mathbf{B} = \kappa_e \kappa_m \mu_0 \varepsilon_0 \dfrac{\partial E}{\partial t}$

Thus, knowing that $c^2 = 1/(\mu_0\varepsilon_0)$, the only change is that the speed of propagation of the electromagnetic field is modified to the following:

$$c' = \frac{c}{\sqrt{\kappa_e\kappa_m}} = \frac{c}{n} \tag{6.53}$$

$$n = \sqrt{\kappa_e\kappa_m} = \text{The index of refraction}$$

The index of refraction n is larger than 1 for all simple dielectric media and exactly equal to 1 in vacuum. Thus, the net effect of a dielectric medium is to slow the speed of light.

All of this electromagnetic lore boils down to the following simple facts, which are of direct relevance to the photoelastic behavior of simple dielectric materials:

1. Referring to Figure 6.19, light can be seen to propagate as a plane wave either in vacuum or in a simple dielectric media according to the plane wave equation:

$$\mathbf{E}(r,t) = \mathbf{E}_0 e^{i(\omega t - k \cdot r)} \tag{6.54}$$

2. The vibration frequency and wave number are related according to

$$\omega = ck \tag{6.55}$$

where c = speed of light and $k = \frac{2\pi}{\lambda}$ where λ is the wavelength.

3. In a dielectric medium, the speed of light slows according to the following formula:

$$c' = \frac{c}{n} \tag{6.56}$$

where n is the index of refraction of the medium.

In nature, we find some materials that are naturally birefringent in that the index of refraction is different in different directions. This has interesting consequences for a polarized* light beam traveling through the material in that the direction of the electric field vector will be rotated as the light wave travels through the medium. The fundamental basis of all photoelastic measurements is the fact that a stress field

---

* The term *polarized* as used here refers to the direction of the electric field vector **E**. For present purposes, we always refer to plane-polarized waves for which **E** has a definite fixed direction. More complex polarization states can exist, such as circular and elliptical polarizations, which are beyond the scope of the present treatment.

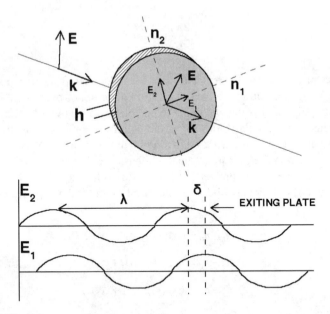

FIGURE 6.21 Effect of a birefringent medium on an electromagnetic wave. The initial electric field **E** is broken into two components, E_1 and E_2, that travel at different speeds through the medium and emerge separated in phase by an amount δ.

can induce temporary birefringence in a material that would otherwise be optically isotropic.

We now discuss how using a simple polariscope allows determination of the stress state of a material based on this principle, but first we need to understand what happens in a simple slab of birefringent material as a beam of polarized light passes through it. Figure 6.21 gives a graphic illustration of the process. A beam of plane-polarized light enters a slab of birefringent material of uniform thickness h from the back side. The light beam is characterized by the electric field vector **E** and the wave vector **k** (magnitude $\frac{2\pi}{\lambda}$) as discussed above. The material is optically characterized by two special directions labeled 1 and 2 for which the index of refraction along Direction 1 is n_1, and that along Direction 2 is n_2. The electric field vector can be resolved into components E_1 and E_2 along Directions 1 and 2. Assuming n_2 is greater than n_1, the field component in the 1 direction will travel at a faster rate c_1 than the component in the 2 direction, which travels at the rate c_2 in accordance with Equation (6.56). Thus, the component E_1 exits the slab ahead of the component E_2 and has a somewhat greater magnitude at that point, which causes the field vector **E** to rotate slightly in the direction of the 1 axis.

The magnitude of the phase shift δ between the two field components is easily calculated from the difference in travel times for each component to traverse the thickness of the slab as $\Delta t = (h/c_2) - (h/c_1) = h((n_2/c) - (n_1/c)) = (h/c)(n_2 - n_1)$. The difference in travel distance is the time difference multiplied by the speed of light, which is $h(n_2 - n_1)$. If we divide by the wavelength of the radiation λ and multiply by 2, we obtain the phase lag Δ in angular radian units:

$$\Delta = \frac{2\pi h}{\lambda}(n_2 - n_1) \tag{6.57}$$

Now, because we know that residual stress can induce birefringence in a material, and that birefringence can rotate the polarization direction of plane-polarized light, then we can easily suspect that we ought to be able to somehow get at the stress by observing the shift in polarization of plane-polarized light passing through the specimen. Again, the seminal work on this problem was done by Maxwell, who, as discussed above, was the first to organize the basic field equations of electromagnetism into a unified and comprehensive theory. Maxwell's theory of the photoelastic effect in a three-dimensional solid is summarized by the following equations[33]:

$$\text{a) } n_1 - n_0 = g_1\sigma_1 + g_2(\sigma_2 + \sigma_3)$$

$$\text{b) } n_2 - n_0 = g_1\sigma_2 + g_2(\sigma_1 + \sigma_3) \tag{6.58}$$

$$\text{c) } n_3 - n_0 = g_1\sigma_3 + g_2(\sigma_1 + \sigma_2)$$

where

n_1, n_2, n_3 = Index of refraction in three orthogonal directions
$\sigma_1, \sigma_2, \sigma_3$ = Principal stresses in solid
g_1, g_2 = Stress optic coefficients
n_0 = Index of refraction of unstressed solid

In principle, Equations (6.58) should allow us to measure the principal stresses in a solid by measuring the appropriate indices of refraction as compared to the index of refraction of the unstressed material.

However, this turns out to be a rather laborious procedure and is rarely carried out. Instead, the standard approach of photoelasticity is to recast Equations (6.58) by cleverly taking the following differences:

$$\text{a) } n_2 - n_1 = g\left(\sigma_1 - \sigma_2\right)$$

$$\text{b) } n_3 - n_2 = g\left(\sigma_2 - \sigma_3\right) \tag{6.59}$$

$$\text{a) } n_1 - n_3 = g\left(\sigma_3 - \sigma_1\right)$$

where $g = g_2 - g_1$ (in Pa^{-1}). In a photoelastic measurement, the index of refraction is measured by observing the retardation of a plane light wave passing through the material. Thus, if we consider a simple slab of material, we can use Equation (6.57) to convert the differences in index of refraction to phase differences induced in the passing light wave as follows:

$$\text{a) } \Delta_{12} = \frac{2\pi h}{\lambda}(n_2 - n_1) = \frac{2\pi hg}{\lambda}(\sigma_1 - \sigma_2)$$

$$\text{b) } \Delta_{23} = \frac{2\pi h}{\lambda}(n_2 - n_3) = \frac{2\pi hg}{\lambda}(\sigma_2 - \sigma_3) \qquad (6.60)$$

$$\text{c) } \Delta_{31} = \frac{2\pi h}{\lambda}(n_3 - n_1) = \frac{2\pi hg}{\lambda}(\sigma_3 - \sigma_1)$$

In principle, one can use Equations (6.60) to ferret out the full three-dimensional stress state of a transparent solid. In practice, this is a rather nontrivial task. However, for the much simpler and quite practical case of plane stress in two dimensions, the problem is a good deal more tractable. Figure 6.22 illustrates the setup of a simple polariscope that can be used to investigate the stress state in a plane slab of material. A vertically polarized beam is created by passing light from a nonpolarized light source through a vertical polarizer. The vertical polarizer is made of a special material that absorbs all of the horizontally polarized component of the beam. Rotating a vertical polarizer 90° turns it into a horizontal polarizer. Thus, the electric field of the light entering the sample slab from the back can be represented by the following formula:

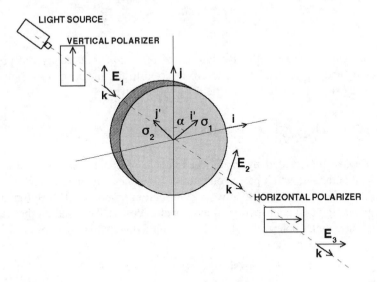

FIGURE 6.22 Schematic of a simple polariscope for measuring the birefringence of a plane slab of material. A beam of vertically polarized light is passed through the sample. If no birefringence is present the light remains vertically polarized, and no light will pass through the horizontal polarizer. If birefringence is present, then the initial field E_1 will be rotated and a horizontal component E_3 will be created that will pass through the horizontal polarizer. The horizontal polarizer can be rotated by some angle such that the E_3 is extinguished, thus revealing the polarization angle α.

$$\mathbf{E}_1 = \mathbf{j}\, E_0 \cos\left(\omega t\right) \tag{6.61}$$

where E_0 is an amplitude factor, and $\mathbf{j}$ is a unit vector in the vertical direction. For the photoelastic experiment, the phase of the incoming light is not important, so it can be left out. The material slab is in a state of biaxial stress, as shown in the figure, with the principal stresses σ_1 and σ_2 acting in the $\mathbf{i}'$ and $\mathbf{j}'$ directions, respectively, where $\mathbf{i}'$ and $\mathbf{j}'$ are unit vectors rotated with respect to $\mathbf{i}$ and $\mathbf{j}$ as shown in the diagram. The incident field vector makes an angle α with the $\mathbf{i}'$ direction and thus has components in the $\mathbf{i}'$ and $\mathbf{j}'$ directions, which can be written as follows:

$$\mathbf{E}_1 = E_0 \cos(\omega t)[\mathbf{i}' \cos\alpha + \mathbf{j}' \sin\alpha] \tag{6.62}$$

Now, because of the birefringence induced by the stresses σ_1 and σ_2, the two components of the field will get out of phase and on emerging from the right side of the plate will have developed phase differences Δ_1 and Δ_2, respectively, which from Equation (6.57) will be given by

$$\Delta_1 = \frac{2\pi h}{\lambda}(n_1 - 1)$$
$$\Delta_2 = \frac{2\pi h}{\lambda}(n_2 - 1) \tag{6.63}$$

where n_1 and n_2 are the effective indices of refraction induced by the principal stresses σ_1 and σ_2, respectively, and are referred to the index of refraction of air, which is close to 1. The emerging field $\mathbf{E}_2$ will be given by

$$\mathbf{E}_2 = E_0 \left[\mathbf{i}' \cos\alpha \cos(\omega t - \Delta_1) + \mathbf{j}' \sin\alpha \cos(\omega t - \Delta_2) \right] \tag{6.64}$$

The wave represented by $\mathbf{E}_2$ proceeds unchanged until it passes through the horizontal polarizer, which passes only those components parallel to the unit vector $\mathbf{i}$. Thus, the $\mathbf{i}'$ component of $\mathbf{E}_2$ gives a contribution proportional to $\sin(\alpha)$, and the $\mathbf{j}'$ component gives a contribution proportional to $\cos(\alpha)$. The field $\mathbf{E}_3$ that emerges from the horizontal polarizer will be in the $\mathbf{i}$ direction and is given by

$$\mathbf{E}_3 = \mathbf{i} E_0 \cos\alpha \sin\alpha \left[\cos(\omega t - \Delta_1) - \cos(\omega t - \Delta_2) \right] \tag{6.65}$$

Equation (6.65) takes a more useful form if we apply the following elementary identities from trigonometry:

$$\sin 2\alpha = 2 \cos\alpha \sin\alpha$$
$$\cos A - \cos B = 2 \sin\frac{B - A}{2} \sin\frac{A + B}{2} \tag{6.66}$$

Applying Equations (6.66) to Equation (6.65) yields

$$\mathbf{E}_3 = i E_0 \sin(2\alpha) \left[\sin \frac{\Delta_1 - \Delta_2}{2} \sin\left(\omega t - \frac{\Delta_1 + \Delta_2}{2} \right) \right] \tag{6.67}$$

The field vector $\mathbf{E}_3$ is what emerges from the horizontal polarizer and passes on to whatever detector is used. However, nearly all detectors, whether the human eye or some sort of photocell, are sensitive only to the intensity of the light falling on them. The intensity in turn is proportional to the square of the electric field vector integrated over one period of oscillation, which in angular units is given by

$$I = \int_0^{2\pi} \mathbf{E}_3^2(\omega t)\, d\omega t = \left[E_0^2 \sin^2(2\alpha) \sin^2 \frac{\Delta_1 - \Delta_2}{2} \right] \int_0^{2\pi} \left[\sin\left(\theta - \frac{\Delta_1 + \Delta_2}{2} \right) \right] d\theta \tag{6.68}$$

$$I = \pi E_0^2 \sin^2(2\alpha) \sin^2 \frac{\Delta_1 - \Delta_2}{2}$$

Equation (6.68) holds for every beam of light passing through the slab of dielectric shown in Figure 6.22 and thus can be used as a formula to map the direction of the principal stresses and the magnitude of $\Delta_1 - \Delta_2$ (or $\Delta_2 - \Delta_1$ because $\sin^2$ is an even function) and thus from Equation (6.60a) the magnitude of $\sigma_1 - \sigma_2$. Dark bands will occur wherever I is zero, which from the properties of the sin function happens whenever the following conditions are met:

a) $\alpha = \dfrac{n\pi}{2} \quad n = 1, 2, 3, \ldots$

$$\tag{6.69}$$

b) $\Delta = \Delta_2 - \Delta_1 = \dfrac{2\pi h}{\lambda}(n_2 - n_1) = \dfrac{2\pi h g}{\lambda}(\sigma_1 - \sigma_2) 2n\pi \quad n = 1, 2, 3, \ldots$

Thus, the direction of the principal stresses can be inferred from dark bands that occur because Equation (6.69a) is satisfied, and the magnitude of the principal stress difference can be deduced from bands that arise from Equation (6.69b). The interference bands that arise from Equation (6.69a) are called the isoclinic fringe pattern, and those arising from Equation (6.69b) are called the isochromatic fringe pattern. Both the isoclinic and isochromatic patterns tend to be present when looking at a typical photoelastic interference diagram. Special techniques are required to distinguish between the two patterns. This tends to be a highly specialized endeavor and is beyond the scope of the present treatment. Those interested are referred to the volume by Dally and Riley[34] for more complete details. However, one can easily detect the stress concentrations in a transparent solid with a simple light source and a pair of polarizer plates. Figure 6.23 shows the superposition of the isoclinic and isochromatic lines in a simple plastic strip used as a separator in a small parts bin. This is

FIGURE 6.23 Interference fringes caused by the photoelastic effect in a plastic strip used as a separator in a small parts bin. This is an injection-molded part, and the birefringence is quite likely caused by residual flow stresses induced during the injection-molding process.

a mass-produced, injection-molded part, and the fringe pattern is indicative of the residual flow stresses incurred during the injection-molding process.

6.2.7 STRAIN RELIEF METHODS

As noted, the stress state of an object is generally revealed, if at all, by the effect it has on the shape of the object or, in less-benign terms, the havoc it wreaks in the form of fractures, delaminations, and excessive warping. However, for an object in a subcritical stress state, there will quite likely be no readily observable phenomenon that would reveal its internal stress condition. Consider, for example, a plane coating on a relatively massive and rigid substrate. The coating might be highly stressed, but its attachment to the substrate keeps it from deforming in any discernible way. However, considering the above remarks, there is one way to make the stress state reveal itself: Make a small hole in the surface and observe the resulting local deformation of the surface as the material relaxes to accommodate the new boundary condition introduced by the presence of the hole. Mathar[35] was apparently the first to employ this method to measure the stress in plates by using extensometers to gauge the strain relief induced by the hole. Since then, the method has been refined considerably, and a standard procedure using strain gauges has been developed.[36]

Figure 6.24 gives a schematic diagram of a typical experimental setup. As shown in the figure, the first step is to mount a triangular array of strain gauges onto the surface at the location where the residual stress is to be measured. A hole is then drilled at the center of the strain gauge rosette, and the resulting relaxation strains are measured with standard test meters. Having measured the strains, the associated stress components can be computed for the case of a through hole by relatively simple formulae. When the hole does not penetrate through the part, one can use specially prepared calibration tables. In more detailed work, use is made of detailed finite element stress calculations to deduce the stress field associated with the measured strains. By and large, the main difficulties with this method do not arise from the calculations required but from a number of experimental concerns. The following are the main points to be noted when carrying out this type of measurement:

1. *Surface preparation*: The strain gauges have to adhere perfectly to the sample surface, so proper cleaning is essential. The use of abrasive papers needs to be carried out very carefully to avoid inducing further residual

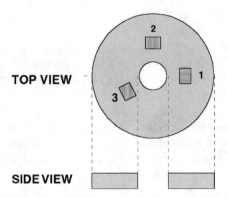

FIGURE 6.24 Placement of strain gauges used to measure the strain relief in a stressed plate induced by drilling a small hole.

stress into the sample surface. Electrochemical cleaning methods are typically preferable.

2. *Hole drilling method*: Quite likely, the most critical step in this procedure is the method used to create the strain-relieving hole. The main concern is to avoid inducing further residual stress into the surface as a consequence of the drilling procedure. The following is a list of usable drilling procedures, noting the pros and cons of each method:

a. Standard drilling with drill press or end mill: Using a good sharp bit at high speed will typically give good results. This approach has problems with hard materials such as stainless steel, for which excessive residual stress is likely to be induced.

b. Abrasive jet system: A high-velocity stream of very fine abrasive particles is used to bore the required hole. This is an excellent approach for drilling very hard materials and leaves very little residual stress. Problems arise in controlling the depth of the hole; furthermore, this method does not leave a flat-bottom hole, which complicates the subsequent data analysis.

c. High-speed turbo drill: A very high-speed turbine is used to develop angular velocities of up to 400,000 rpm. A modified tungsten-carbide dental bit is used, which gives a good flat-bottom hole. Because of the very high drilling speed, little residual stress is left in the part. Problems with this method arise because the drill has very low torque and is easily stopped on very hard surfaces. Frequent bit breakage can also be a problem.

d. Laser drilling: In certain applications, use of a high-power laser may be the best way to prepare the stress relief hole. This method can be very effective even on the hardest surfaces. Care must be taken to choose the correct laser wavelength for the material bored and to control the laser power to avoid overheating. Calibration procedures must be used to obtain the correct hole depth and diameter.

3. *Presence of stress gradients*: Special precautions must be taken if the stress in the part varies rapidly with depth of penetration. In this case, an incremental approach is required in which the depth of the hole is increased in small steps. The calculations become increasingly complex in this case and require advanced analytical methods to fit the observed strain measurements.

In addition to these concerns, one also has to pay attention to a host of further technicalities, such as the appropriate choice of strain gauge, calibration methods, effects of plasticity, and so on. A detailed treatment of all these problems can be found in the work of Schajer et al.[37]

An interesting variation of the strain relief approach that can be used for very small geometries involves the use of laser drilling to create the required hole coupled with optical interferometry to measure the ensuing displacements. Figure 6.25 illustrates a sample application that is relevant to the creation of via holes in multilevel microelectronic structures. A detailed treatment of the deformation and stresses generated near a hole in a coating on a rigid substrate was treated by Lacombe[38] using finite element methods. Note from the figure that the surface of the coating pulls back from the center of the hole and downward simultaneously. This surface deformation is a definitive signature of the residual stress level in the coating that caused it and can be measured by a number of optical methods. The most useful are laser holography, moire interferometry, and laser speckle interferometry. Regardless of which method is used, the final result is a direct determination of the displacement field of the coating surface in a neighborhood of the hole. This displacement data can then be used as a boundary condition for subsequent numerical analysis of the local stress field using either finite element or boundary element methods. Note that this approach differs significantly from the strain gauge method

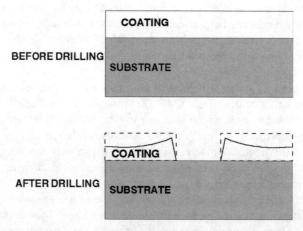

FIGURE 6.25 Local deformation of a coating after drilling a small hole by laser ablation or any other method. Optical interferometry can be used to map the local surface deformation, and these data can be used in conjunction with a finite element analysis of the deformation to determine the neighboring stress field.

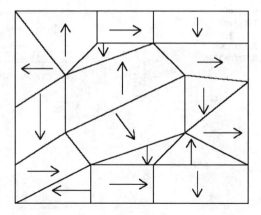

FIGURE 6.26 Schematic of a ferromagnetic material showing the detailed domain structure. Each domain can be considered a small magnet.

in that it is the displacement field that is measured and not the strain field. This approach also offers the advantage of giving much more detailed and accurate information concerning the local stress field.

6.2.8 MAGNETICS

When dealing with magnetic materials such as iron and nickel, one can get a handle on the state of stress via the phenomenon of magnetostriction. To understand this behavior, one first needs to understand that a ferromagnetic material such as iron has an intricate microstructure made up of a large number of magnetic domains, each of which can be thought of as a small magnet, as shown in Figure 6.26. The details of the domain structure of a ferromagnet are essentially a frontier problem in solid-state physics that require a detailed quantum mechanical treatment. For present purposes, however, all that is required is to understand the basic phenomenology associated with domain formation and interactions. The magnetostrictive effect involves the dimensional changes that occur when an external field is applied to the material thus setting up a localized strain and a concomitant stress in the material. Conversely, a stress field will induce localized strains, which in turn alter the magnetic domain structure. It is essentially this coupling between the stress field and the magnetization that is the basis of magnetic stress measurements. In essence, the magnetic domains can be used as microscopic strain gauges for monitoring and measuring the stress state of the material. Thus, the magnetostrictive effect has the potential for use as a nondestructive, portable, and rapid stress measurement tool. Two approaches to using this effect for stress measurements have been reported in the literature[39]: the Barkhausen noise method and the magnetostriction method. We give a brief review of each method next.

6.2.8.1 Barkhausen Noise

In 1919, Barkhausen[40] published on an obscure magnetic effect by which an oscillating magnetic field applied to a ferromagnetic material gave rise to a curious noise

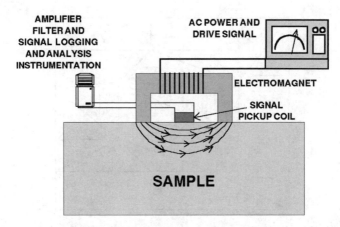

FIGURE 6.27 Simple schematic of the Barkhausen noise method for determining the stress in a ferromagnetic material. An electromagnet is used to agitate the magnetic domains near the surface of the sample. A small coil is used to monitor the subsequent relaxation of the magnetic domains. The local stress in the sample can be inferred from the power spectrum of the noise signal, which is monitored by the pickup coil.

signal that could be monitored by a small pickup coil. Figure 6.27 gives an idealized schematic of the experimental setup. As the applied field varies, it causes the magnetic domains to shift and reorient; this causes the local magnetic field to vary in a complicated manner, which can be detected by an appropriately located pickup coil. The signal detected by the pickup coil looks to all intents and purposes like some kind of noise signal. However, the power spectrum of this noise signal is correlated to the stress/strain in the material and can be measured by monitoring a quantity called the magnetoelastic parameter (MP), which is in essence a measure of the root mean square amplitude of the signal. The key to the Barkhausen method is the correlation between the MP and the local strain in the material, which is shown schematically in Figure 6.28. Curves such as the one shown in the figure are generated by taking a carefully prepared test piece and mounting it in a load frame in which it can be precisely stretched and compressed while the MP is measured using an apparatus similar to that in Figure 6.27. These data then serve as a calibration curve that can be used to determine the strain in a particular object made of metal prepared in a manner identical to the calibration piece. Thus, the advantages and disadvantages of the Barkhausen noise method of determining residual stress in magnetic materials can be summarized as follows:

Advantages
- Rapid nondestructive measurement method
- Can be carried out with convenient portable equipment
- Can be applied to complex shapes under field conditions

Disadvantages
- Requires carefully determined calibration data
- Sensitive to microstructural variations as well as residual strain

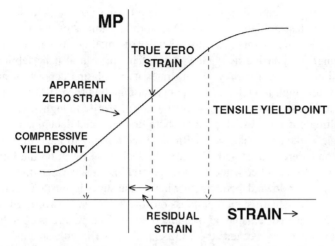

FIGURE 6.28 Dependence of the magnetoelastic parameter (MP) on the local strain level in a ferromagnetic material. Because MP can be computed from the data gathered in the Barkhausen noise experiment shown in Figure 6.27, it forms the basis of the Barkhausen stress/strain measurement procedure.

- Special precautions required if biaxial stress present
- Applicable to ferromagnetic materials only

A detailed description of the Barkhausen noise method was given by Tiitto[39], including an interesting discussion of the use of the method to determine fatigue failure.

6.2.8.2 Magnetostriction Approach

The magnetostriction approach is analogous to the Barkhausen method, but instead of measuring the noise spectrum of shifting magnetic domains, one measures the local magnetization directly. Unfortunately, this approach cannot be recommended because of the numerous difficulties involved; it is mentioned only for the sake of completeness. In the words of Tiitto[39]: "The method is susceptible to various factors, which hinder its progress and applications. Systematic studies are indispensable to overcome the difficulty. The work calls for much effort and perseverence." In essence, all of the caveats mentioned concerning the Barkhausen noise method regarding microstructural technicalities such as grain size, texture, surface condition, and the like are multiplied considerably when trying to measure the local magnetization directly. There is, however, a possible application of this method to thin magnetic films such as those produced in microelectronic structures. Because of the care of the manufacturing process, all of the microstructural properties of these films are well controlled, which alleviates much of the problem with creating calibration samples. The interested reader should refer to Tiitto's article for further details.

6.2.9 RAMAN SPECTROSCOPY

The last major stress measurement method we mention is the use of Raman spectroscopy to evaluate the stress in diamond coatings. Even though the method applies essentially only to diamond, it is included here because of the great technical and commercial interest in diamond films. One normally thinks of diamond as a gemstone used in jewelry or often as a fine powder of small crystals used for grinding and polishing. As amazing as it may seem. diamond can also be fabricated as a polycrystalline film on a variety of substrates using plasma deposition methods.[41] The exceedingly high modulus of single-crystal diamond (near 1000 GPa) plus the fact that it is deposited at very high temperatures gives rise to high-stress conditions in the resulting coatings, which promote both cracking and delamination behavior.

The first workers to make use of Raman spectroscopy to measure the stress in CVD diamond were Knight and White.[42] The technique relies on the fact that the stress in the diamond lattice will cause the spectral line at a wavelength of 1332 cm^{-1} to shift either up or down depending on whether the stress is tensile or compressive, respectively. Under hydrostatic compression, the 1332 cm^{-1} wavelength will shift by about 2.58 cm^{-1} per gigapascal of applied stress. The picture is complicated for CVD diamond coatings because the 1332 cm^{-1} line is triply degenerate (i.e., three distinct vibration modes appear at the same wavelength), and the biaxial stress in the coating causes the line to split into its components.

Gray et al.[43] attacked this problem and came up with a correction formula that accounts not only for the biaxial stress state, but also for domain size and temperature effects. Using this approach, they were able to measure the stress in a 70-μm CVD-deposited film on a molybdenum substrate and obtained results that were within 8% of the value obtained using substrate-bending methods. Ager[44] used the Raman method to measure the stress in both CVD diamond films and amorphous carbon coatings. Finally, Ralchenko et al.[45] used the Raman method to measure the stress in CVD diamond coated onto 11 different substrates. Their results for the stress and frequency shift are reproduced in Table 6.3 as a useful reference for anyone interested in diamond coatings.

TABLE 6.3
Residual Stress in CVD Diamond Coatings on Various Substrates[a]

Substrate Material	Thermal Expansion Coefficient of Substrate (10^{-6} K^{-1})	Measured Stress on Cooling from 950 C (690 C for Silicon) (GPa)
Vitreous silica (SiO$_2$)	0.5	2.3
Silicon	2.5	−1.0
Silicon carbide	2.8	−1.6
Tungsten-cobalt composite	5.2	−3.8
Molybdenum	5.3	−4.1
Steel (R18)	11.2	−7.4
Steel (60S2)	12	−5.5
Nickel	13	−6.9
Iron-nickel alloy	4.5	−10.6
Copper	16.7	−2.9

[a] Adapted from data of Ralchenko et al.[48]

6.2.10 Miscellaneous Methods

We finally mention a handful of stress measurement methods that might come in handy but are more or less peripheral to our main concerns because they mainly pertain to applied loading conditions and are not very useful for measuring the intrinsic stress in an object.

6.2.10.1 Stress Pattern Analysis by Thermal Emission

The stress pattern analysis by thermal emission (SPATE) method is an interesting approach to stress measurement based on a thermodynamic phenomenon first investigated by Kelvin[46] and later refined by Biot.[47–49] The basic relationship on which this method relies can be written as follows:

$$\Delta T = \frac{-\alpha}{\rho C_P} T \Delta\sigma \qquad (6.70)$$

where

α = Thermal expansion coefficient (K^{-1})
ρ = Mass density (kg/m^3)
C_P = Specific heat at constant pressure ($J/(k_g$ degrees kelvin))
T = Temperature (K)
$\Delta\sigma$ = Change in sum of principal stresses ($\sigma = \sigma_1 + \sigma_2 + \sigma_3$) (Pa)
ΔT = Change in local temperature caused by change in stress (K)

Equation (6.70) applies to homogeneous isotropic materials subject to elastic loads under adiabatic conditions. In essence, if you apply a stress to an object and then very quickly measure its temperature, you will find that the temperature has increased by some increment given by Equation (6.70). Thus, if you could find some clever way of rapidly tracking the temperature of the object, you could get a handle on the stress induced by some applied load. Surprisingly, one can do just that with modern infrared camera equipment, which provides a way of detecting very small temperature changes in the range of 10^{-3} K using advanced infrared imaging. Thus, in a typical SPATE experiment one applies a cyclic load to the part under investigation while simultaneously monitoring the temperature rise. Some of the advantages of the SPATE approach include

1. Can give qualitative full-field stress pattern rapidly and with minimal surface preparation
2. Complex geometries evaluated as easily as flat surfaces
3. High spatial resolution attainable down to 0.5 mm
4. Good stress sensitivity compared to other full-field methods
5. Noncontact method that can be applied to surfaces at elevated temperatures

A number of the disadvantages of this approach are as follows:

1. Expensive equipment required, mainly the infrared camera and associated electronics and computer control equipment and software
2. Cannot measure residual stress, only applied stress
3. Requires precise material property evaluation for calibration purposes
4. Estimates only surface stresses
5. Clear line-of-sight view of surface required

6.2.10.2 Photoelastic Coating Technique

Returning to the discussion of the photoelastic method, imagine what would happen if one applied a photoelastic coating to an object and then applied an external load. Reflecting light off the part would have to pass through the full thickness of the coating twice, and because of the stress-induced birefringence, an interference pattern would appear when using polarized light that would be indicative of the surface stress field. Thus, the use of a photoelastic coating can be a quick and inexpensive way of estimating the stress buildup in a part subjected to arbitrary load conditions. Advantages of this approach include

1. Inexpensive as it requires minimal equipment
2. Applicable to parts of arbitrary shape under wide range of loading conditions
3. Can be applied to structures under actual use conditions

The limitations of this method include

1. Significant level of skill required in applying the coating
2. Not applicable to relatively thin sheets or to low-modulus materials because the coating may appreciably affect the stiffness of the sheet
3. Subject to errors arising from through-thickness strain in the coating and ambient temperature fluctuations when working under field conditions

6.2.10.3 Brittle Lacquer Method

A final "quick-and-dirty" method for estimating the stress in an object involves the use of a brittle lacquer coating with known fracture properties. This approach is similar to the photoelastic coating method in many respects except that even less detailed information on the applied stress field is obtained. As a load is applied to the coated part, the coating reaches its fracture threshold, and a pattern of cracks appears on the surface. From this crack pattern, one can see right away the direction of the maximum principal stress at the surface because it is perpendicular to the crack lines. Further, knowing the critical stress level of the coating allows estimation of the magnitude of the maximum principal stress. This method has roughly the same advantages and disadvantages as the photoelastic approach mentioned above.

6.3 SUMMARY

Because all adhesion failure is in one way or another caused by a stress field, either residual or applied, an entire chapter is devoted to the problem of stress measurement. Residual stress fields in particular present an insidious problem because they are effectively invisible until they have worked their mischief, leaving the engineer with only a corpus delicti to examine. Thus, the ability to measure, and thereby anticipate, the level of stress in a structure is critical to the engineering design process not only for purposes of avoiding delamination failure, but also for problems of cracking and excessive warping. The cantilevered beam method is covered in some detail because it is one of the most straightforward and reliable methods available for measuring residual stress in thin films and coatings. By this method, one essentially measures the deflection of the beam by optical, electrical, or x-ray techniques and then employs Stoney's equation or one of its variants to determine the residual stress.

Although nearly every method for measuring residual stress involves the measurement of a displacement of some sort, the suspended membrane approach is unique in that it fundamentally involves the measurement of a frequency instead of a displacement. An additional advantage of this method is that it does not require knowledge of the elastic constants of the material investigated, relying only on knowledge of the density. This has great advantages in terms of improving the accuracy of the measurement because in many cases this is limited by knowledge of the elastic constants.

A variety of other methods are also discussed because they may have distinct advantages in certain applications. These include the following:

- The photoelastic method, which may be used to capture the full-field stress distribution in transparent samples
- Use of ultrasonic waves for relatively thick samples
- Magnetostriction measurements for ferromagnetic materials
- X-ray measurements for crystalline materials
- Use of photoelastic or brittle coatings

This array of measurement techniques offers the practicing engineer a versatile set of tools for investigating stress-related phenomena and further form the foundation for implementing the powerful fracture mechanics analysis methods discussed in Chapter 4 for dealing with coating delamination and other stress-related problems.

NOTES

1. G.G. Stoney, *Proceedings Royal Society (London)*, A32, 172 (1909).
2. A. Brenner and S. Senderoff, *Journal Res. Nat. Bur. Stand.*, 42, 105 (1949).
3. N.N. Davidenkov, *Soviet Phys.—Solid State*, 2, 2595 (1961).
4. S. Timoshenko, *Mechanical Engineering*, 45, 259 (1923).
5. A. Brenner and S. Senderoff, *Journal Res. Natl. Bur. Stand.*, 42, 105 (1949).

6. "An Apparatus for Measuring Stress in Thin Films," E. Klockholm, *The Review of Scientific Instruments*, 40, 1054 (1969).

7. "Intrinsic Stress in Evaporated Metal Films," E. Klokholm and B.S. Berry, *Journal of the Electrochemical Society: Solid State Science*, 115, 823 (1968).

8. On completing a postdoctoral assignment dealing with polymer solution thermodynamics, I took up a position in the microelectronics industry fully expecting to work on problems related to photoresist technology. Much to my surprise, my manager set me to work on problems related to stress measurement because at the time the development laboratory was experiencing grave problems with delamination and cracking in the modules being fabricated. At the time, this took me by surprise because I was under the impression that the only people concerned with stresses were engineers who build tall buildings, bridges, or aircraft. The devices we were making were hardly the size of a fingernail and yet were destroying themselves very effectively. The data in Table (6.1) were a real eye opener that immediately pointed out the source of the problem.

9. See, for instance: H.P. Murbach and H. Wilman, *Proceedings Phys. Soc.,* B66, 905 (1953), and M.A. Novice, *British Journal of Applied Physics*, 13, 561 (1962).

10. P.H. Wojciechowski, *Journal of Vacuum Science and Technology*, A6, 1924 (1988).

11. B.C. Bell and D.A. Glocker, *Journal of Vacuum Science and Technology*, A9, 2437 (1991).

12. B.C. Bell and D.A. Glocker, *Journal of Vacuum Science and Technology*, A10, 1442 (1992).

13. A.E. Ennos, *Applied Optics*, 5, 51 (1966).

14. Details on bridge circuits can be found in most standard texts dealing with electrical measurements, such as *Introduction to Electronics for Students of Physics and Engineering Science*, D.M. Hunten, Ed. (Holt, Reinhart, & Winston, New York, 1964), p. 42.

15. J.D. Wilcock and D.S. Campbell, *Thin Solid Films*, 3, 3 (1969).

16. *The Theory of Sound*, Lord Raleigh (Dover reprint, 1949).

17. The Bessel functions are similar to the much more common sines and cosines with which nearly every student of mathematics is familiar. They form a complete family of orthonormal functions that can be used as a basis for expanding any other function much as the sines and cosines are used in Fourier series. The sines and cosines are so common that the manufacturers of scientific calculators have created special keys to evaluate them. Much the same could have been done (and quite frankly should have been done) for the Bessel functions, but alas I am unaware they ever achieved the requisite level of popularity to command a special key on any calculator. The student is thus forced to use either a table of special functions or a software package to evaluate $J_n(x)$. Fortunately a number of excellent tables are available. One of the best is the *Handbook of Mathematical Functions*, M. Abramowitz and I. Stegun, National Bureau of Standards Applied Mathematics Series 55 (reprint, Dover Publications, New York, 1965). This is an outstanding reference volume that was created during the Depression years of the 1930s by unemployed mathematicians working for the U.S. government as part of the unemployment relief effort. It is an outstanding value and used to be available for free from the U.S. Government Printing Office.

18. There is nothing really mysterious about the numbers λ_{nm}. If one plots $J_n(x)$ for some value of n, one sees that it oscillates about the x axis and decays in magnitude with increasing x much as would be observed for a decaying sine function. The points at which $J_n(x)$ crosses the x axis are called the zeros and are numbered in order of appearance. Thus, λ_{n1} is the first zero of $J_n(x)$. For any given value of n, there are an

infinite number of λ_{nm}. For most practical purposes, only the first half dozen or so values are required and may be found in tables given in reference works such as that of Abramowitz and Stegun mentioned in Note 17.

19. Y.-C. Ku, L.-P. Ng, R. Carpenter, K. Lu, and H.I. Smith, *Journal of Vacuum Science and Technology*, B9, 3297 (1991).

20. Professor Richard Farris and his co-workers have done an extensive amount of work using the laser holographic technique to determine the stress in thin organic films by the suspended membrane method. A short list of relevant references includes "Stress Analysis of Thin Polyimide Films Using Holographic Interferometry," R.J. Farris and M.A. Maden, *Journal of Experimental Mechanics*, June, 178 (1991); "Vibrational Technique for Stress Measurement in Films. Part I: Ideal Membrane Behavior," R.J. Farris, M.A. Maden, A. Jagota, and S. Mazur, *Journal of the American Ceramic Society*, 77, 625 (1994); "Vibrational Technique for Stress Measurement in Films: Part II. Extensions and Complicating Effects," R.J. Farris, Q.K. Tong, M.A. Maden, and A. Jagota, *Journal of the American Ceramic Society*, 77, 636 (1994).

21. A number of references are available concerning the holographic method of imaging. A very nice guide and overview written for the practical enthusiast and hobbyist is *Homemade Holograms: The Guide to Inexpensive and Do-It Yourself Holography*, J. Lovine (TAB Books, Division of McGraw Hill, New York, 1990). At a more technical level, but still quite practical, the reader can refer to *The Complete Book of Holograms: How They Work and How to Make Them*, J.E. Kasper and S.A. Feller (John Wiley & Sons, New York, 1987). Finally, at a deeper level complete with Fourier analysis and other details, one can refer to *Principles of Holography*, by H.M. Smith (John Wiley & Sons, New York, 1975).

22. "Stress Measurements in Thin Films Deposited on Single Crystal Substrates Through X-ray Topography Techniques," E.W. Hearn, in *Advances in X-ray Analysis*, Vol. 20, H.F. McMurdie, C.S. Barrett, J.B. Newkirk, and C.O. Ruud (Plenum Publishing, New York, 1977), p. 273.

23. "X-ray Diffraction Method," M. François, N.M. Sprauel, C.F. Déhan, M.R. James, F. Convert, J. Lu, J.L. Lebrun, and R.W. Hendricks, in *Handbook of Measurement of Residual Stresses,* J. Lu, Ed. (Society for Experimental Mechanics, published by Fairmount Press, Liburn, GA), Chap. 5.

24. Bragg's law is covered in every text on crystallography and most introductory general physics books. A highly recommended general physics text is *Physics, Parts 1 and 2*, D. Halliday and R. Resnick, Eds. (John Wiley & Sons, New York, 1978); see in particular Chapter 47 on gratings and spectra.

25. *The Feynman Lectures on Physics*, Vol. 2, R.P. Feynman, R.B. Leighton, and M. Sands (Addison-Wesley, Reading, MA, 1964), Chapter 39, p. 39-7.

26. *Finite Deformation of an Elastic Solid*, T.D. Murnaghan (John Wiley & Sons, New York, 1951).

27. Second-Order Elastic Deformation of Solids, D.S. Hughes and J.L. Kelly, *Physics Review*, 92, 1145 (1953).

28. A more comprehensive treatment of these equations can be found in "Ultrasonic Methods," R.B. Thompson, W.-Y. Lu, and A.V. Clark, Jr., in *Handbook of Measurement of Residual Stresses*, J. Lu, (Society for Experimental Mechanics, published by Fairmont Press, Lilburn, GA, 1996), Chap. 7, p. 149.

29. "Angular Dependence of Ultrasonic Wave Propagation in a Stressed Orthorhombic Continuum: Theory and Application to the Measurement of Stress and Texture," R.B. Thompson, S.S. Lee, and J.F. Smith, *Journal of the Acoustic Society of America*, 80, 921 (1986).

30. "On Waves Propagated Along the Plane Surfaces of an Elastic Solid," Lord Rayleigh, *Proceedings of the London Mathematical Society*, 17, 4 (1885).

31. For a detailed account of the mathematical development, refer to *Rayleigh and Lamb Waves, Physical Theory and Applications*, I.A. Viktorov (Plenum Press, New York, 1967), Chap. 1.

32. The seminal work in this field is *A Treatise on Electricity and Magnetism*, J.C. Maxwell (Oxford, UK, 1873; also reprinted by Dover in two volumes of the third edition (New York, 1954). Because it is no more likely that anyone would refer to Maxwell's original work to study electromagnetics than they would refer to Newton's *Principia* to study mechanics, we give a few more modern references that are decidedly more accessible. A comprehensive and modern approach to electromagnetic theory is given in *Electromagnetic Fields and Interactions*, R. Becker, (F. Sauter, Ed., Blaisdell Publishing Company, New York, 1964). Although less comprehensive than Becker and Sauter's works, the reference of choice for the practicing scientist has to be *The Feynman Lectures on Physics*, Vol. 2, R.P. Feynman, R.B. Leighton, and M. Sands, (Addison-Wesley, Reading, MA, 1964). Finally, at a more elementary level and definitely the most accessible of all the works, is *Physics, Parts 1 and 2*, combined 3rd ed., D. Halliday and R. Resnick (John Wiley & Sons, New York, 1978); the treatment of electromagnetic theory starts with Chapter 26.

33. "On the Equilibrium of Elastic Solids," J.C. Maxwell, *Transactions of the Royal Society of Edinburgh*, 20, part 1, 87 (1853). Maxwell was indeed one of the titans of 19th century science. In addition to his seminal work on electromagnetism and the photoelastic effect, which are of direct interest here, he also did pioneering work in statistical mechanics and the kinetic theory of gases with the development of the now-famous Maxwell-Boltzmann statistics. On top of this, we see his name occurring in fundamental thermodynamics with the so-called Maxwell relations; in electrical circuit theory with the Maxwell bridge for measuring capacitance and inductance; in image technology with the pioneering demonstration of three-color additive synthesis using three black-and-white negatives; and, presumably in his spare time, his fundamental contribution to viscoelasticity theory with his well-known Maxwell model for the stress relaxation behavior of a viscoelastic solid. Maxwell died in 1879, long before the Nobel Prize in Physics came into existence. Had he lived into the 20th century, he would have garnered at least four of these prizes. Truly it can be said that the 19th century was a time when giants walked the Earth.

34. *Experimental Stress Analysis*, 3rd ed., J.W. Dally and W.F. Riley (McGraw-Hill, New York, 1991).

35. "Determination of Initial Stresses by Measuring the Deformation Around Drilled Holes," J. Mathar, *Transactions of the ASME*, 56, 249 (1934).

36. *Determining Residual Stresses by the Hole-Drilling Strain Gage Method*," ASTM Standard E837-92 (American Society for Testing and Materials, 1992).

37. "Hole Drilling and Ring Core Methods," G.S. Schajer, M.T. Flaman, G. Roy, and J. Lu, in *Handbook of Measurement of Residual Stresses*, J. Lu, Ed. (Society for Experimental Mechanics, published by Fairmont Press, Liburn, GA, 1996). Chap. 2.

38. "Stresses in Thin Polymeric Films: Relevance to Adhesion and Fracture," R.H. Lacombe, in *Surface and Colloid Science in Computer Technology*, K.L. Mittal, Ed. (Plenum Press, New York, 1987).

39. "Magnetic Methods," S. Tiitto, in *Handbook of Measurement of Residual Stresses*, J. Lu, Ed. (Society for Experimental Mechanics, published by Fairmont Press, Liburn, GA, 1996), Chap. 8.

40. "Rauschen der Ferromagnetischen Materialen," H. Barkhausen, *Physics Zeitschrift*, 20, 401 (1919).

41. The basic method of deposition is referred to as plasma-assisted chemical vapor deposition (CVD). In this method, a gas of hydrogen containing a small percentage of methane (1 to 3% roughly) is passed over a substrate such as a silicon single crystal that has been seeded and is also held at a high temperature, between 700 and 1000 C. The gas is ignited near the surface of the substrate by an applied radio-frequency (or sometimes microwave-frequency) electromagnetic field of sufficient power density to excite the gas into a plasma. The plasma field rips apart the methane molecules to form carbon radicals, which manage to deposit themselves onto the seed nuclei on the substrate and then proceed to grow into a relatively uniform film. The role of the hydrogen seems to be that of a carrier gas that provides a reducing atmosphere that prevents oxidation and removes itself cleanly from the growing diamond crystal. Much of this work was pioneered by Russian scientists during the Cold War era of the 1950s and 1960s. Political tensions between East and West at the time greatly hindered publication of this work, and it remains largely obscured. The first apparent publication in the Western literature was by S. Matsumoto, Y. Sato, M. Tsutsumi, and N. Setaka, *Journal of Materal Science*, 17, 3106 (1982). Interest in diamond-coating technology exploded in the 1990s when many companies, both large and small, tried to exploit the unique properties of diamond for everything from semiconductor chips to biological implants. A fairly complete overview of diamond-coating activity during that period is documented in the following: *Applications of Diamond Films and Related Materials: Third International Conference*, NIST Special Publication 885, A. Feldman, Y. Tzeng, W.A. Yarbrough, M. Yoshikawa, and M. Murakawa, Eds. (U.S. Government Printing Office, Washington, DC, 1995).

42. D.S. Knight and W.B. White, *Journal of Material Research*, 2, 385 (1989).

43. "Measurement of Stress in CVD Diamond Films," K.J. Gray, J.M. Olson, and H. Windishmann, in *Mechanical Behavior of Diamond and Other Forms of Carbon*, MRS Materials Research Society Symposium Proceedings, Vol. 383, M.D. Drory, D.B. Bogy, M.S. Donley, and J.E. Field (MRS Press, Pittsburgh, PA, 1995), p. 135.

44. "Residual Stress in Diamond and Amorphous Carbon Films," Joel W. Ager III, in *Mechanical Behavior of Diamond and Other Forms of Carbon*, MRS Materials Research Society Symposium Proceedings, Vol. 383, M.D. Drory, D.B. Bogy, M.S. Donley, and J.E. Field (MRS Press, Pittsburgh, PA, 1995), p. 143.

45. "Stress in Thin Diamond Films on Various Materials Measured by Microraman Spectroscopy," V.G. Ralchenko, E.D. Obraztsova, K.G. Korotushenko, A.A. Smolin, S.M. Pimenov, and V.G. Pereverzev, in *Mechanical Behavior of Diamond and Other Forms of Carbon*, MRS Materials Research Society Symposium Proceedings, Vol. 383, M.D. Drory, D.B. Bogy, M.S. Donley, and J.E. Field (MRS Press, Pittsburgh, PA, 1995), p. 153.

46. "On the Thermoelastic, Thermomagnetic and Pyro-electric Properties of Matter," W. Thompson (Lord Kelvin), *Phil. Mag.*, 5, 4 (1878).

47. "Plasticity and Consolidation in a Porous Anisotropic Solid," M.A. Biot, *Journal of Applied Physics*, 26, 182 (1955).

48. "Irreversible Thermodynamics With Application to Viscoelasticity," M.A. Biot, *Physics Review*, 97, 1463 (1955).

49. "Thermoelasticity and Irreversible Thermodynamics," M.A. Biot, *Journal of Applied Physics*, 27, 240 (1956).

50. The units of the acoustoelastic coefficient is inverse gigapascals. The percent sign indicates that using these values in Equation (6.35) and expressing the stresses in units of gigapascal will give the ratio of the velocity shift to the zero-stress velocity as a percentage. To obtain the actual value of the ratio, one must divide by 100. Actual data have been adapted from Note 27.
51. Source is Note 26.
52. Source is R.F.S. Hearmon, in *Landolt-Bornstein: Numerical Data and Functional Relationships in Science and Technology*, Vol. 3/1 (Springer, Berlin, 1966), p. 12.
53. "Measurement of Acoustoelastic and Third-Order Elastic Constants for Rail Steel," D.M. Eagle and D.E. Bray, *Journal of the Acoustics Society of America*, 60, 741 (1976).

7 Case Studies from the Field

The trouble with things like these is that many of the real situations which are apt to arise are so complicated that they cannot be fully represented by one mathematical model. With structures there are often several alternative possible modes of failure. Naturally the structure breaks in whichever of these ways turns out to be the weakest — which is too often the one which nobody had happened to think of, let alone do sums about.

J. E. Gordon: *Structures or Why Things Don't Fall Down*

This chapter attempts to flesh out the overall picture of adhesion measurement by presenting a few "real world" examples of how particular measurement techniques can be employed, how the results are analyzed, and what the consequences are in terms of impact on manufacturing processes. As Prof. Gordon points out in the quotation, these are complicated situations in which the failure mode cannot be anticipated in advance, much less accounted for by rigorous calculations. Nevertheless, when appropriate measurement methods are at hand, one can proceed to systematically do experiments that eventually reveal the cause of the problem and further provide a quantitative basis for applying corrective measures.

The first example deals with the problem of how very low levels of contamination can have very large effects on the adhesion strength of a coating. Because adhesion is basically a surface phenomenon, it does not take much in the way of an impurity level relative to bulk concentrations to upset the surface interactions that provide the binding force that establishes the adhesion of the coating to the substrate. This is clearly a situation that can give rise to wholly unanticipated behavior.

The second example deals with another subtle case in which failure to cure a coating properly led to surprisingly high levels of stress that brought about massive delamination problems in a manufacturing process. This problem was not only unanticipated, but also turned out to be an adhesion problem of the second type, which requires careful attention to controlling the stress level in the coating. In contrast to the above-mentioned contamination problem, which could be attacked by semiquantitative methods, this case required fully quantitative fracture mechanics methods to overcome the problem.

A third and final example deals with the technology of adhering pins to a ceramic substrate to make an electrical connection to an epoxy board. Here, we find a surprisingly counterintuitive situation in which the tensile load applied to the pin amounted to a relatively minor part of the overall driving force that led to pin failure. In this case, detailed fracture mechanics calculations revealed that it was the stress in the bonding solder or braze material that dominated the driving force for failure. This particular example also leads us to the subject of the closing section of this chapter: stability maps. In essence, a stability map is a sort of guide for the design engineer for avoiding thermal-mechanical failure problems. The stability map plots out a region in a multidimensional space (which includes product dimensions, material properties, manufacturing specifications, etc.) that enjoys a property known as unconditional stability regarding certain stress-driven failure phenomena. In other words, the stability map gives a guide to the design and process engineers that will allow them to devise and manufacture a part that will not fail under any conditions of typical use, excluding deliberate abuse or sabotage, of course.

7.1 A STUDY IN ADHESION SENSITIVITY
TO CONTAMINATION

The work discussed in this section was published in the technical literature in two articles by Buchwalter and Lacombe.[1,2] This section presents a much less formal account of this episode in adhesion science by giving more attention to the day-to-day vagaries and complications of doing adhesion work than would ever be allowed in the formal technical literature. The focus is slanted toward giving the reader the flavor of what goes on in the real world as opposed to presenting a dry technical tract *fait accompli*.

It all began with an attempt to perform a systematic study (using the peel test and the methods discussed in Chapter 5, Section 5.1) of the effect of the mechanical properties of polyimide films on their adhesion to silicon substrates. A systematic study was carried out along the following lines:

1. The surface properties of plasma-cleaned silicon wafers were character-ized using x-ray photoelectron spectroscopy (XPS).[3]
2. After cleaning and characterization, the silicon wafers were coated with a polyimide film.[4]
3. The polyimide film supported on a silicon wafer was then cut into strips for peel testing as shown in Figure 5.1.
4. A number of strips were carefully removed for tensile testing to establish the elastic properties of the material and the yield stress and strain.
5. Other strips were peeled off the silicon substrates in the manner discussed in detail in Chapter 5 to establish the apparent peel strength of the coatings as deposited on the well-characterized substrates.
6. It was anticipated that, using the above data, some sort of elastic-plastic analysis might be performed that would establish the "true" surface frac-ture energy of the coatings.
7. The entire study, if successful, would establish a baseline value for the adhesion of polyimide to clean silicon wafers.

The entire project as outlined above seemed to be relatively straightforward aside from performing the elastic-plastic peel analysis; the bulk of the sample preparation and testing was carried out by technical staff. However, what promised to be a routine study quickly turned into a mystery as the peel test data became available. On some days, the data came in at a nominal 200 J/m^2; on other days, the measured peel strength was more like 20 J/m^2. Thus, there was an apparent order of magnitude drop in the peel strength of the polyimide strips prepared under identical conditions and peeled off the same substrate. It was clear that something rather unusual was going on, and the first suspicion was that the test equipment was malfunctioning. An investigation was launched that did not uncover any particular misbehavior of the equipment but did reveal that the tests that were run up to that time were done under prevailing ambient conditions at the time any particular test was run. This observation immediately triggered the suspicion that possibly day-to-day variations in humidity might be the cause of the strange peel test data.

The test equipment was housed in an enclosure that could be sealed and flushed with dry nitrogen so that the peel testing could be performed under carefully controlled humidity conditions. To test this hypothesis, a sample was mounted for testing on a fairly humid day (roughly 50% relative humidity [RH]) with the test apparatus and sample fully exposed to the ambient air. The strip chart recorder quickly came to a steady peel value near 20 J/m^2 and remained stable for several minutes, indicating no apparent problem with the test apparatus. The door on the enclosure was then sealed shut, and a stream of dry nitrogen was passed through, eventually bringing the relative humidity in the chamber down to roughly 15% RH. However, immediately after the humidity in the chamber began to fall, the pen on the strip chart recorder started to climb to higher and higher peel force values as the humidity dropped. As the humidity in the chamber settled to its lowest value, the apparent peel strength settled to a stable value somewhat above 200 J/m^2. Continuing the experiment, the nitrogen flow was shut off, and the enclosure was opened to the ambient conditions. Immediately, the pen on the strip chart started to descend and in a matter of a few minutes was back to its original value as the humidity in the enclosure came back up to ambient conditions. A typical recorder trace of peel load versus time is shown in Figure 7.1.

These results piqued a much-heightened interest in what was happening with these samples. The authors were well aware that moisture can have a strong effect on adhesion, but the results showed an order of magnitude difference between high and low humidity, and the response was very rapid. At this point, it was decided that all of the tests would have to be run again on a fresh set of samples to confirm what was happening. Thus, an entirely new batch of samples was prepared according to the outlined procedure with the apparently insignificant difference that the samples were plasma cleaned with a brand new plasma cleaner that the laboratory had just received. However, this seemingly inconsequential change turned out to have rather significant repercussions, as revealed by the peel test data shown in Figure 7.2. Bear in mind that the only difference between the samples used in Figure 7.1 and in Figure 7.2 is the change in plasma cleaners; nevertheless, the difference in the peel load data and the sensitivity to moisture could hardly be more dramatic. Under dry conditions, the second set of samples showed more than double the peel strength of the first. What is even more surprising is that the apparent moisture sensitivity also nearly disappeared. So, instead of confirming the validity of the first set of results,

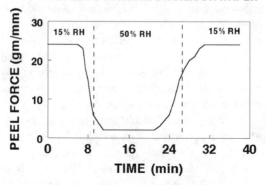

FIGURE 7.1 Tracing of the strip chart recorder output of the peel load versus time/distance of a 90° peel test of a 1-mm wide strip of polyimide off a clean silicon wafer.

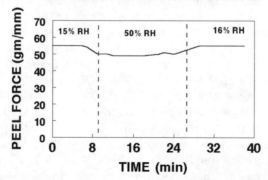

FIGURE 7.2 Strip chart recorder output for a 90° peel test of polyimide off silicon wafer. All conditions are identical to those in Figure 7.1 except that the silicon wafer was cleaned in a different plasma reactor.

the second batch of samples only served to further deepen the mystery of what was happening.

It was fortunate, however, that there were XPS data for the cleaned silicon substrates for both sets of samples. It was clearly time to have a look at the surface chemistry of the as-prepared substrates to see if that could shed any light on the conundrum we were facing. Figure 7.3 and Figure 7.4 show the XPS spectra of the as-prepared silicon substrates for the two sets of samples. At first glance, Figure 7.3 reveals the XPS spectrum of a nominally clean silicon wafer. The main central line is that of electrons ejected from an oxygen 1s orbital that arises from the native oxide layer that exists on nearly every single crystal silicon surface. The very top

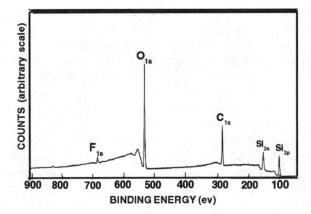

FIGURE 7.3 XPS spectrum of a plasma-cleaned silicon wafer. The oxygen and silicon peaks are exactly as expected from the native oxide layer on the wafer. Residual organic contamination is indicated by the presence of the carbon 1s peak. A nearly invisible peak to the left of the main oxygen peak is caused by a small amount of fluorine contamination.

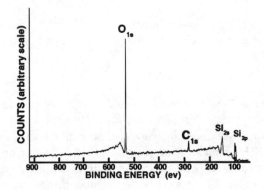

FIGURE 7.4 XPS spectrum of a plasma-cleaned silicon wafer identical to that in Figure 7.3 except that a new and previously unused plasma reactor was used. Note the absence of any fluorine contamination.

surface of the crystal is in fact a layer of about 1000 Å SiO_2. This layer is formed very rapidly as soon as any clean silicon surface is exposed to oxygen. In the semiconductor industry, this oxide layer serves as a convenient insulator to support wiring layers on a silicon wafer that has been patterned with semiconductor elements such as diodes and transistors. In fact, this oxide layer is one of the key properties of silicon that has given it the predominant position it has held in the semiconductor industry for the past 50 years. The next peak to the right of the oxygen line is the carbon 1s peak, which quite naturally arises from the 1s electrons that are ejected from residual carbon contamination still left on the as-cleaned silicon substrate. One finds that it is nearly impossible to remove all carbon contamination from a surface even under the severe scouring imposed by plasma cleaning. Finally, the last two peaks to the right of the carbon are the silicon 2s and silicon 2p peaks, which arise from the respective orbitals in the silicon residing on the surface. At this point, one might ask why the XPS spectrum of a sample that is nearly 100% silicon show peaks that tend to give the impression the sample is primarily made of oxygen and carbon. The answer to this question is simply the fact that the XPS experiment is sensitive mainly to what is on the surface, which in this case is silicon dioxide along with some carbon contamination.

The XPS data for the second set of samples shown in Figure 7.4 show essentially the same surface chemistry as in Figure 7.3 aside from the smaller carbon peak, which indicates less contamination on the surface, apparently because a new plasma cleaner was used in this case. However, careful comparison of the spectra in the two figures does reveal one small fly in the ointment. The spectrum for the first set of samples reveals a tiny peak to the left of the major oxygen peak; this is caused by a small amount of fluorine contamination. The spectrum for the second sample set reveals no such peak.

The presence of the fluorine contamination peak on the surface of only the first set of samples clearly pointed the finger at fluorine as the cause of the extraordinary moisture sensitivity of those samples. A number of questions then immediately arose:

1. What was the source of the fluorine?
2. How were the moisture and the fluorine interacting to disrupt the coating adhesion?
3. How was moisture getting to the interface?

The first question was answered by talking with the process engineers who routinely carry out plasma cleaning and etching. It turns out that the old plasma cleaner had multiple uses, including substrate cleaning with argon/oxygen gas and etching with CF_4. The problem arose with the use of CF_4 for etching because all of the racks and other hardware in the cleaner are made from aluminum, which basically absorbed small amounts of fluorine from the CF_4 plasma much like a sponge absorbs water. Once the fluorine got into the aluminum hardware, there was essentially no way of removing it short of throwing out all the hardware and anything else within the chamber that contained aluminum. Thus, when the first set of samples was cleaned, the fluorine in the aluminum hardware leached out into the oxygen/argon plasma that was supposedly cleaning the silicon substrate but now was also depositing a small contamination level of fluorine. Subsequent experiments on yet a third set of samples that were cleaned in a fluorine-contaminated reactor showed a return of the high moisture sensitivity shown in Figure 7.1, which clearly confirmed the fact that low-level fluorine contamination was a key factor in the moisture sensitivity puzzle.

A straightforward diffusion calculation settled the question of how the moisture was getting to the interface. The polyimide films used for the peel test samples were 17 μm thick. Moisture is known to penetrate through polyimide by Fickian diffusion; thus, the moisture concentration in the polyimide coating was governed by the classical Fickian diffusion equation:

$$\frac{\partial c}{\partial t} = D \frac{\partial^2 c}{\partial x^2} \tag{7.1}$$

where

c = Moisture concentration (arbitrary units)
x = Distance penetrated into coating from top surface (m)
t = Elapsed time from initial exposure to ambient moisture (sec)
D = Diffusion coefficient (m²/sec)

For the problem of moisture diffusing through the polyimide coating down to the silicon substrate, we need to apply the following boundary and initial conditions to Equation (7.1):

At $t = 0$, $c(x) = c_0$ at $x = 0$

$c(x) = 0$ for all $x > 0$

For all $t > 0$, $C(x) = c_0$ at $x = 0$

$\partial c/\partial x = 0$ at $x = 1$ (silicon substrate barrier)

Equation (7.1) together with these boundary conditions form a well-defined boundary value problem that can be readily solved by standard methods, giving the following result[5]:

$$\frac{c}{c_0} = 1 - \frac{4}{\pi} \sum_{n=0}^{\infty} \frac{(-1)^n}{2n+1} \exp\left[-(2n+1)^2 \frac{t}{t_0}\right] \cos\left[(2n+1)\pi \frac{\mu}{2l}\right]$$

$$t_0 = \frac{4l^2}{\pi^2 D} \tag{7.2}$$

$$\mu = x - l$$

Setting $x = 1$ in Equation (7.2) allows us to estimate the buildup in moisture concentration at the polyimide silicon interface. In addition, if we look only at the first two terms in the infinite sum, Equation (7.2) simplifies to the following:

$$\frac{c}{c_0} = 1 - \frac{4}{\pi}\left[\exp\left(-\frac{t}{t_0}\right) - \frac{1}{3}\exp\left(-9\frac{t}{t_0}\right) + \ldots\right] \tag{7.3}$$

Equation (7.3) clearly shows that at long times the moisture buildup at the polyimide/silicon interface is governed by the first term with time constant t_0 given in Equation (7.2). The time constant t_0 can be easily estimated from the known diffusion constant for moisture in polyimide, which is in the neighborhood of 7×10^{-13} m^2/sec and the known coating thickness of 17×10^{-6} m. Using these values, we estimate t_0 as about 167 sec, which comes to something between 2 and 3 min. Referring to Figure 7.1, we see that this is roughly the response time for moisture to attack the polyimide/silicon interface. Thus, it is clear that the moisture is penetrating to the interface through the coating thickness by straightforward Fickian diffusion.

The final question regarding just how the fluorine and moisture conspire to severely weaken the adhesion of the polyimide coating to the silicon substrate is somewhat more enigmatic than the first two questions. However, it was clear that moisture diffusing to the polyimide/silicon interface was somehow interacting with the few atomic percent of fluorine contamination present in a way that adversely affected the coating adhesion. One possible mechanism that is consistent with the available data is the following: During the plasma cleaning process, fluorine molecules that have leached out of the aluminum hardware in the plasma cleaner make their way to the silicon surface, where they form Si-F bonds. This in itself is apparently enough to weaken the silicon dioxide glass structure of the native oxide interlayer to cause the coating adhesion, even under dry condition, to be less that half of what is observed for an uncontaminated surface. Now, as water molecules start diffusing to the interface they can abstract the fluorine from the silicon to form SiOH and hydrofluoric acid (HF) according to the following simple reaction:

$$SiF + H_2O \rightarrow SiOH + HF \tag{7.4}$$

This reaction is rather sinister in that it forms HF, which is known to attack glass vigorously and is in fact a well-known etchant for glass. The HF is now free to diffuse away and destroy the glass structure by attacking the Si–O–Si bonds via the following reaction:

$$Si-O-Si + HF \rightarrow Si-OH + Si-F \tag{7.5}$$

The net result of this reaction is to weaken the glass structure by breaking an Si–O–Si bond and leaving over a Si–F group, which can then wait for another water molecule to come along and repeat the process. By this mechanism, each fluorine molecule is used over and over in a series of reactions that progressively eat away at the native oxide glass structure that is key to maintaining the adhesion strength of the polyimide coating. This clearly also explains why only a very small amount of fluorine is required for this process to proceed.

In closing, it should be pointed out that the original objective of the investigation, which was to see if some kind of elastic-plastic analysis of polyimide peeling could be established as was done for copper in Chapter 5, was not achieved. The simple fact is that polyimide is not a straightforward elastic-plastic material like copper. When stretching polyimide, only half of the energy of deformation goes into viscous flow, which is essentially nonrecoverable. The remainder of the work done on the polyimide goes into internal energy of bond stretching, strained entanglements, and

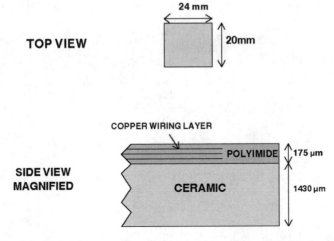

FIGURE 7.5 Schematic of a ceramic substrate coated with a polyimide layer containing copper thin film wiring.

other nonequilibrium configurations. This internal energy of deformation can be at least partially recovered by heating the material to close to its glass transition, which will cause the chains to relax to their former equilibrium state. Peeling a material such as polyimide requires a full viscoelastic analysis, which was beyond the scope of the project.

7.2 CASE OF THE IMPROPERLY CURED FILM

The investigation of the fluorine contamination problem was typical of a wide class of adhesion problems that can be successfully attached using semiquantitative adhesion measurement methods by which one simply needs to rank the adhesion strength of one coating system compared to another. In this case, peel testing clearly exposed the effect of low levels of fluorine contamination on the adhesion of polyimide to silicon substrates. The case to be studied in this section, however, raises the stakes a notch higher in that a full fracture mechanics analysis was required to sort out what was happening. As with the previous study, the problem was wholly unanticipated. In addition, this case was significantly more serious because it threatened to derail an important development program that would have to be shelved unless the delamination problem was solved.

It all started when ceramic substrates* coated with a fairly thick polyimide layer started to exhibit catastrophic delamination of the polyimide coating. Figure 7.5 gives a schematic representation of the structure that was causing the problem. The device shown is essentially a small square of ceramic that serves as a support base

---

* Be aware that the term *ceramic* covers a wide range of material compositions, ranging from porcelains used to fabricate fine china to alumina-based formulations typically used in insulator applications in the electronics industry. In this section, we refer to a particular formulation tailored for its dielectric properties and ease of processing. For present purposes, we need only be concerned with certain elastic and fracture properties of the ceramic material in question.

TABLE 7.1

Comparison of Decohesion Behavior of Three Hypothetical Polyimide Films and Plane Copper

Coating	Modulus (GPa)	Poisson Ratio	Assumed Residual Stress (MPa)	Critical Thickness to Decohere Ceramic $(K_1 = 1.4 \text{ MPa}\sqrt{m})$	Critical Thickness to Decohere Ceramic $(K_1 = 1.0 \text{ MPa}\sqrt{m})$
Low-stress polyimide	10	0.4	7	8.65 mm	4.4 mm
Medium-stress polyimide	6.6	0.4	25	450 μm	230 μm
High-stress polyimide	3.0	0.4	45	65 μm	35 μm
Copper	100	0.33	140	210 μm	105 μm

for a top layer of polyimide/copper thin film wiring. Delamination of the polyimide always started at the edge and proceeded to the interior of the module, eventually undermining the copper/polyimide wiring layers and thus destroying the entire module. Also, closer examination of the delaminated layer indicated that, regardless of where the delamination started, the crack front eventually veered into the ceramic material and then proceeded to propagate at a fixed depth slightly below the polyimide/ceramic interface.

Figure 7.6 shows a micrograph of a failed substrate with a decohesion* crack apparently starting at the polyimide/ceramic interface and subsequently diving into the ceramic and propagating all the way to the copper wiring layers. This is precisely the type of crack behavior that was discussed in Chapter 4, Section 4.2.2, and has been studied theoretically by Suo and Hutchinson[6] and experimentally by Hu et al.[7] The problem can be analyzed using the simple formulae developed in Chapter 4, Section 4.2.2. Table 7.1 gives the results of such a calculation applied to the structure in Figure 7.5 for just the polyimide coating on the ceramic substrate, ignoring the embedded copper wiring. Because of the thickness of the polyimide layer shown in Figure 7.5, the polyimide chosen for this application was the low-stress material listed in Table 7.1. A quick glance at the table, however, indicates that there should be no way that the polyimide coating could cause decohesion of the underlying ceramic substrate because the table lists a worst-case critical thickness of 4.4 mm and the actual coating thickness was some 30 times thinner. It was clear at this stage that we had an unusual situation, and that some careful analytical work would be

---

* A note on usage is in order. The failure mode in Figure 7.6 is clearly cohesive failure of the ceramic. However, at first glance the failure path is so close to the polyimide/ceramic interface that it appears to be a delamination, which for practical purposes is what it is. The technically correct term for the failure in Figure 7.6 is decohesion; however, because the net effect is to remove or delaminate the polyimide from the ceramic, we use the term delamination interchangeably with the term decohesion when referring to the overall failure process.

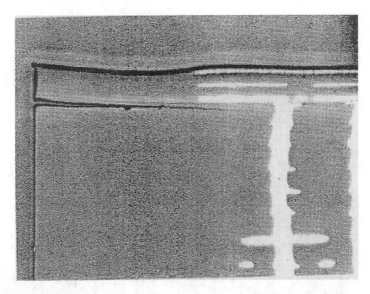

FIGURE 7.6 Micrograph showing the delamination/decohesion of a polyimide layer from a ceramic substrate.

required. In particular, it would be important to determine the level of stress in both the polyimide and copper materials. This particular problem is just one concrete instance for which knowing the level of stress in the coating materials is critical to the ability to evaluate the stability of a structure with regard to delamination and fracture processes. The reader might want to review some of the material in Chapter 6, which deals with this question. In particular, for this problem the x-ray method was used to evaluate the stress in the copper layers, and substrate bending was used to determine the stress in the polyimide.

Although this case was clearly a fracture problem because in all instances the delamination front dove into the ceramic and then propagated parallel to the interface, it is for all practical purposes an adhesion problem also. Imagine if the adhesion specialists involved in solving the problem had gone to management and told them that they could not be of any help because in fact the problem was one of fracture in the ceramic substrate. I do not think the management would have been impressed. Thus, to tackle this problem a number of experiments were put in place to evaluate the stress level in the various material layers; in addition, a modeling effort was initiated to investigate why the original "back-of-the-envelope" calculations done using the analytical methods of Suo and Hutchinson[6] did not seem to predict the behavior of this system.

At first, it was believed that the analytical calculations may not have been adequate because they could not take into account the effect of the copper wiring layers and had to assume a long crack under steady-state conditions. Therefore, a more realistic model was created using the finite element method to handle the increased mathematical complexity. A sample cross section of the model is shown in Figure 7.7. The finite element method is in effect a numerical analysis tool for

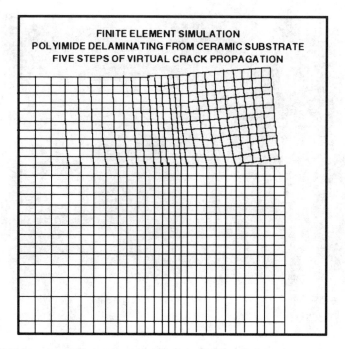

FIGURE 7.7 Sample of a finite element mesh for modeling the delamination of a polyimide coating off a rigid ceramic substrate. Shown is the result of a five-step virtual crack propagation calculation. This is a quite general and versatile procedure for estimating the driving force for delamination/decohesion of the coating from the substrate caused by high intrinsic stress levels in the coating.

solving the field equations of elasticity developed in Chapter 3 and exhibited in all their glory in Table 3.1.

The basic approach is to break the structure in question into a large number of small rectangular elements for which the analysis is fairly simple and can be solved by analytical methods. Some fairly sophisticated matrix methods are then employed to assemble all the small elements into a more complex structure that closely approximates the actual device modeled. Appropriate boundary conditions are then applied, and the resulting set of coupled equations is solved for all the displacements, stresses, and strains. With these data in hand, one can easily compute the total strain energy locked up in the system for any given set of loads and boundary conditions.

The model in Figure 7.7 actually goes a step further and performs what is known as a virtual crack propagation analysis. This method performs a series of step-by-step complete model calculations by which a virtual crack is propagated in the model by the node release method. Thus, starting with the full intact model, all stresses and displacements are computed, and the total strain energy in the system is then computed from these results and saved. The boundary conditions are then changed by relaxing the condition that certain nodes that hold the model together be coupled. This node decoupling is performed where one suspects a crack will form. A further complete analysis of this slightly modified model is run again, all stresses and

EDGE DELMINATION OF POLYIMIDE
(LOW THERMAL EXPANSION MATERIAL)

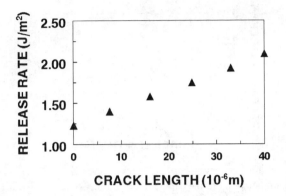

FIGURE 7.8 Plot of strain energy release rate versus crack length for the virtual delamination shown in Figure 7.7.

displacements and the total strain energy are calculated, and the results are saved for further use. This whole procedure is repeated as many times as necessary, developing a virtual crack while keeping track of how the strain energy in the structure is changing as the crack propagates. Thus, the model shown in Figure 7.7 represents the last of five complete analyses that simulate the propagation of an interface crack between the polyimide coating and the underlying ceramic for a five-step virtual crack.

Now, knowing the strain energy in the structure at each step, we can easily compute the strain energy as a function of crack length and then from these data compute the derivative, which is the strain energy release rate (i.e., rate of change of total strain energy with crack length).* Figure 7.8 plots the strain energy release rate as a function of crack length for the virtual crack shown in Figure 7.7. There are essentially two things to notice about this plot. The first is that the strain energy release rate is relatively low, less than 2.2 J/m^2 for a delamination up to 40 μm long. Second, the release rate is also steadily increasing. Because the driving force is low, one does not expect the delamination crack to go anywhere; however, because the driving force increases with crack length, this type of delamination could become dangerous if the crack front were to grow too long.

As the data stands, however, it cannot explain the catastrophic decohesion shown in Figure 7.6. In addition, Figure 7.6 shows that the actual failure mode is decohesion in the underlying ceramic as opposed to clean delamination between the polyimide and the ceramic. However, this case can also be easily investigated by the model in Figure 7.7 simply by lowering the virtual crack into the ceramic. In fact, a number of virtual crack simulations can be run at different depths within the ceramic. The

---

* The strain energy concept is discussed in more detail in Chapter 4, Section 1.1.1. For the model discussed here, it essentially represents the driving force available to drive a crack through the structure by virtue of expending built-up strain energy that was locked in the structure as a result of the manufacturing process.

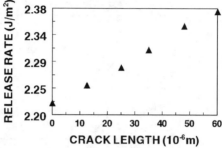

FIGURE 7.9 Plot of strain energy release rate versus crack length for a decohesion crack. All details of the computation are the same as for Figure 7.8 except that the crack front is now 50 μm below the polyimde/ceramic interface.

result of such a set of simulations suggested that the maximum driving force occurs at a depth of close to 50 μm into the ceramic.

Figure 7.9 shows the driving force versus crack length for a decohesion crack 50 μm below the polyimide/ceramic interface. Again, the driving fore is still low at roughly 2.38 J/m² for a crack 60 μm long. The fracture toughness for this ceramic material would require a driving force of close to 15 J/m² to create a crack such as that shown in Figure 7.6, thus indicating that the assumptions of the model were not in accord with what was seen in Figure 7.6.

At this stage of the investigation, a series of stress measurements on the polyimide pointed to what was going wrong. Remember that it was the low-stress polyimide material that was assumed for all of the above analyses. All of the process engineers swore that this was the case, so their word was taken. However, stress measurements done on failed parts indicated that the actual stress in the polyimide was closer to the 45 MPa that characterizes the high-stress material. If we use this value for the stress level, then the plot of the driving force versus crack length comes out as shown in Figure 7.10. One look at the data in this figure immediately explains why the ceramic was cracking. For the same crack path as illustrated in Figure 7.9, the fracture energy is now some six times larger, bringing it up to the critical fracture energy for the ceramic. The crack associated with the data in Figure 7.10 is still in the transient stage in that the driving force keeps increasing as the crack grows. Such cracks eventually reach a steady state at which the driving force levels off at some fixed value. Figure 7.11 shows the data for the fully developed crack that evolved out of that shown in Figure 7.10. The driving force is now well above the critical strength of the ceramic, and the crack has propagated well into the copper wiring region of the module.

Figure 7.12 illustrates a model simulation for the fully developed crack associated with the data in Figure 7.11. The displacements are magnified for visual purposes. The main feature to note is the difference in residual stress in the detached ligament as opposed to the still-adhered film. This is signified by the dark color (low

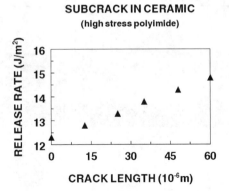

FIGURE 7.10 Strain energy release rate versus crack length for the same virtual crack as plotted in Figure 7.9 except that the stress level in the polyimide is consistent with a standard polyimide material as opposed to the low-stress variety.

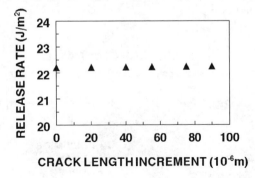

FIGURE 7.11 Data from Figure 7.10 extended into the steady-state regime in which the driving force is essentially constant. The steady-state value of 22.2 J/m^2 will readily drive a crack through the ceramic substrate material at acoustic velocities.

stress) of the detached film as compared to the light color (high stress) of the adhered film region. As the delamination/decohesion progresses, strain energy from the high-stress region is literally dumped into driving the crack, leaving the remaining detached film relatively stress free. Thus, one can imagine the high residual stress region as a reservoir of locked up strain energy waiting to be set loose through whatever failure mode is kinematically most accessible.

At this point, it is instructive to look at the contributions of the various material layers to the overall driving force for crack propagation. One of the advantages of the virtual crack propagation model approach is that it is quite easy to perform this analysis, and the results for the delamination shown in Figure 7.12 are illustrated in Figure 7.13. Note that the main contribution to the driving force comes from the polyimide material, with a small contribution made by the ceramic substrate and

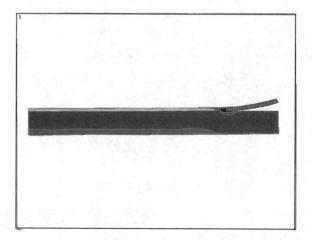

FIGURE 7.12 Sample finite element model of the displacements incurred by the delamination/decohesion driving force plotted in Figure 7.11. Displacements are magnified for ease of visualization. The dark color of the detached ligament indicates a low stress level, thus signifying the strain energy has been drained out of this region. The light color of the attached coating indicates a much higher level of stress and thus strain energy available to drive the decohesion process further.

SUBCRACK IN CERAMIC
(high stress polyimide, steady state propagation)

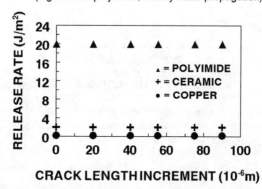

FIGURE 7.13 Relative contribution of the various materials that make up the laminate structure in Figure 7.5. Note that the contribution to the overall driving force for decohesion is in the order polyimide > ceramic > copper. This is precisely the reverse order of the stress level in the respective material layers, which underlines the need to be very careful in relying on the stress level alone to understand fracture problems.

virtually nothing contributed by the copper wiring layers. This is somewhat counterintuitive because the highest stress is in the copper, followed by the ceramic, followed by the polyimide, which is the precise opposite of the order of contribution to the actual driving force for delamination.

This points out the need to be careful when thinking about the stress in a structure as opposed to the driving force that can potentially tear the structure apart. The two are quite separate, albeit related, entities. The driving force is a fracture mechanics concept and is derived by considering the total stress in the structure and how it changes as a crack is propagated. The stress in the structure is a continuum mechanics concept and essentially reflects the strain energy density in the structure at any given point. The polyimide material dominates the fracture process because it makes up the greatest volume of the thin film structure as opposed to the copper. Thus, although the strain energy density of the polyimide is much lower than the copper, there is so much more polyimide present compared to copper that the overall strain energy contribution of the polyimide dominates. Also, this is more than an academic consideration in real-life situations because, during some of the early engineering meetings held to address this problem, a number of individuals thought that the copper was the real villain because of its comparatively high stress level and suggested that the major thrust of the task force should be to address the copper problem. It was not until the fracture mechanics simulations clearly demonstrated that it was the polyimide that was causing the problem and not the copper that the task force effort was put on the correct track.

Finally, the attentive reader may be wondering at this point how was it that the stress level in the polyimide became so high when everyone thought that it should have been low. It is here that the vagaries of real-world engineering intercede and make matters both complicated and interesting. It so happens that the low-stress polyimide is of a special type that contains very stiff chains that tend to align in the plane of the coating when properly cured, thus giving the material a surprisingly low thermal expansion coefficient, which leads to low curing strains and thus the low-stress state. However, the key to the problem lies in the caveat "properly cured." After application by spin coating, the material must be dried for approximately 1 h at a relatively low temperature, near 80°C, to drive off the solvent and thus allow the stiff chains to relax into their in-plane alignment. After the drying step, the temperature is then ramped up to the final cure temperature near 400°C at which point complete imidization occurs, and everything is frozen in place forever. Now, it so happened that on the production line the drying furnace was very busy a large part of the time, but some nearby furnaces tended to be available. When this occurred, the technician in charge of curing the part would make use of the neighboring oven instead of the proper one that was in use, thus saving time and maintaining the ever-pressing need to keep up the production rate. The net result was what you might expect. The critical drying step was completely skipped, and the parts were heated at a high cure temperature, thus creating a high-stress polyimide instead of the intended low-stress material. Straight out of the oven, the difference could hardly be detected, and the parts went merrily on their way to eventual self-destruction.

7.3 CASE OF THE STRESSED PIN

The example covered in this section deals with an application of the pull test for which again the methods of fracture mechanics turned out to be key in making the correct engineering design decisions. Although the application in question belongs to the rather prosaic domain of pin-to-board interconnect technology, it still contains a number of unanticipated surprises, and the analysis required to uncover those surprises is anything but prosaic. Many microelectronic devices consist of a hierarchy of wiring and interconnect schemes. At the very lowest level, individual transistors on a silicon chip are connected to one another by thin film wiring schemes that can involve several layers or wiring separated by an insulator with the individual layers joined by "via columns". To get the signals generated by the chip to other chips in the same device or to the outside world, the chip is wired to a larger substrate using either solder balls (flip chip technology) or wire bonding. The substrate in question may support one or more separate chips.

For the problem under consideration here, the substrate actually supported 100 or more chips. These were multichip modules made from ceramic and used in high-end mainframe machines. The problem at this point becomes how to connect the ceramic module to the next wiring level, which in this case was a large epoxy board. One common solution is the use of pins soldered to the ceramic substrate, which can then be plugged into the appropriate receptacle on the epoxy board. Now, to appreciate the magnitude of the problem one has to understand that for these large multichip modules the number of pins involved is on the order of 1000 per module. In addition, each of the pins has to perform its appointed task flawlessly because one critical pin failure could turn a valuable multichip module into a near-valueless paperweight. The main concern was the fact that, during its service life, the module is likely to be plugged and unplugged from its board many times for a variety of reasons having to do with everything from maintenance to module upgrades. Each time the module is plugged or unplugged, there is a chance that one or more of the pins can get dinged. Now, this does not necessarily have to be a fatal event. If the pin is only bent, then it can be straightened. If the shaft breaks, then the remaining stub can be removed and a new pin soldered into place. However, one very disturbing failure mode can occur in which the entire pin comes off, taking a piece of the ceramic substrate with it and leaving behind a small crater. This is a nonrepairable failure, and the module involved is thereby instantly turned into a ceramic tile.

With this background in mind, one can understand that a large amount of testing was carried out using the pull test to check on the mechanical integrity of the pins that were soldered to the ceramic modules. Pins were pulled off the ceramic substrate using a suitably modified tensile test frame. The load on the pin was ramped up until it failed. The test was considered successful if the pin either cleanly delaminated from the substrate or failed somewhere in the shank. Any tear out of the underlying ceramic was considered a serious failure and required further investigation into the apparent cause. The main cause of this fatal failure mode is investigated in what follows.

Figure 7.14 gives a schematic representation of a typical pin configuration. This figure shows a greatly magnified view of a typical pin that in practice might be roughly 0.5 cm long by 0.5 mm in diameter. The pin in the diagram has been swaged to give a flared out bottom, which facilitates soldering to the substrate. On the substrate, there is a pad typically made of vapor-deposited nickel, which makes electrical connection with the metal line buried in the substrate and serves as a

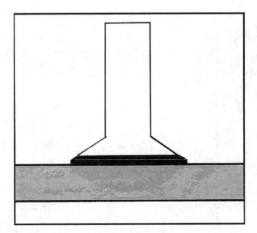

FIGURE 7.14 Schematic of a common pin-to-substrate configuration. Note the flared out bottom and the existence of a solder layer on top of a solder pad.

diffusion barrier and to contain the solder used to hold the pin in place. An appropriate solder or braze material sits on top of the nickel pad and envelops the bottom of the pin, making a strong electrical and mechanical connection.

As mentioned, the main concern with the pin soldering scheme was the ceramic tear out problem. This implied that the stresses in the ceramic close to the pin were becoming too high. In this case, the only way to get a handle on this problem is with an appropriate stress model. Simple experience suggested that the source of the problem lay at the edge of the pin just under the nickel pad where the stress field can get quite high because of the nickel. Further, it was known that the ceramic substrate was riddled with micrometer-size flaws and cracks that could be induced by various grinding and polishing procedures incurred during the manufacturing process.

With these thoughts in mind a finite element model was developed; a few sections are shown in Figure 7.15. Compared to the analysis performed in Section 7.2, this model is two levels more difficult and belongs to the realm of the stress analysis specialist. Panel 1 on the left is a true scale model magnified 10 times showing the ceramic substrate and the nickel pad with the pin and solder left off for clarity. Note that even at this magnification the nickel pad is virtually invisible compared to the thickness of the ceramic substrate. One of the first problems to be overcome with this type of model is that the stress field of interest exists in a region some 10,000 times smaller than the size of the device modeled. To obtain accurate results in the high-stress region, a very fine mesh of elements is required on the order of 0.01 μm in size.

Now, to mesh the entire structure shown in panel 1 with elements of this size would require some 10^{11} elements, which would easily overwhelm even the largest computers. Fortunately, this fine resolution is required only in the region close to the high-stress region, which allows one to employ a graded element mesh as shown in the figure. Note the large triangular elements at the bottom of the model, which

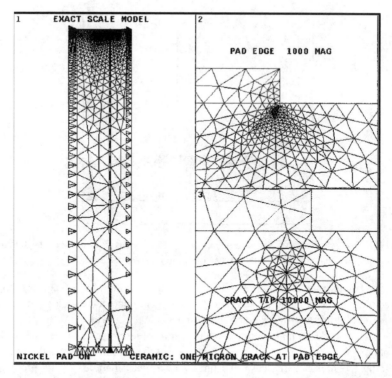

FIGURE 7.15 Finite element model of a pin on a ceramic substrate with the pin and solder layer removed for enhanced visual clarity. Note the tremendous range of scales required for this model, ranging from millimeter size at full scale to nearly 10,000 times smaller at the scale required to model subcrack propagation under the nickel pad.

grow smaller as you go toward the top of the figure where the nickel pad resides. As can be seen, the mesh size becomes so small that it fades into a solid continuum near the pad because of the limited resolution of the figure.

Panel 2 in the upper right corner of Figure 7.15 shows the region near the edge of the nickel pad magnified 1000 times. At this resolution, the nickel pad is now clearly visible. A 1-μm long crack has been inserted into the model starting from the bottom edge of the nickel pad and protruding into the ceramic substrate at roughly a 45° angle to the ceramic surface. Note that the graded mesh is still becoming smaller as it approaches the crack tip and again fades into a blacked out continuum because of lack of resolution in the figure. Panel 3 in the figure raises the magnification to 10,000, and at this resolution we can finally make out even the smallest elements that define the crack tip. The tip of the crack is at the center of a rosette of wedge-shape elements that look like a circular pie cut into 12 pieces. These elements are of a special type designed to accurately model the singular stress field we know to exist at the crack tip itself. With this carefully graded mesh and using the special crack tip elements, the model can be used to estimate the stress field driving the 1-μm crack starting from the pad edge and diving into the ceramic.

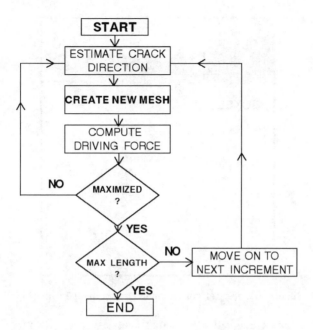

FIGURE 7.16 Simplified flow diagram of the steps required to generate a virtual crack in the finite element model shown in Figure 7.15.

But, this is only the start because we do not know *a priori* what angle the crack will take. A first approximation can be made by computing the direction of the maximum principle stress vectors near the edge of the nickel pad and starting the crack off perpendicular to this direction. The approach taken with this model, however, improves on this method somewhat by setting the crack at four different directions and from the resulting strain energy data computing the direction that gives the maximum driving force (strain energy release rate) for crack propagation. This requires that the entire model with the fully graded mesh be generated de novo for each iteration. Fortunately, because of the modern miracle of automatic mesh generation,* all this work could be done by a computer, making the whole process of propagating the crack entirely automatic.

Figure 7.16 shows a simplified flow chart of the steps required to generate a virtual crack in the model that will follow the direction dictated by the maximum driving force principle. At each step of the crack propagation process, a complete stress analysis is performed to evaluate the driving force for four different crack directions. From these data, the direction of maximum driving force can be evaluated, and a fifth analysis is performed for this direction. The whole process is then repeated, which lengthens the crack another increment. Figure 7.17 shows the result of a 50-step simulation. This diagram required 250 complete stress analyses to complete, and it shows the shape of the crack that should be generated by a ceramic tear out event. Electron micrographs of actual tear outs confirm this result.

---

* All of this work was done using the *ANSYS®* programming system. This is truly an industrial-strength stress analysis code of great power and flexibility.

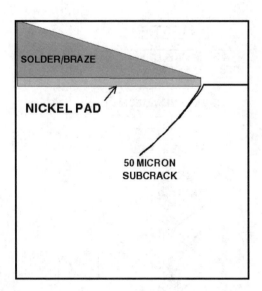

FIGURE 7.17 Simulated crack path for the finite element model depicted in Figure 7.15.

The virtual crack generation modeling analysis gave us a powerful tool for investigating the ceramic tear out problem. One of the key questions to be answered concerned the source of the problem: the stress in the nickel pad, the stress in the solder, or the mechanical load on the pin? As a first step, a simulation was run to look at the effect of the stress in the nickel pad. The stress in nickel coatings can become very high and in this case was believed to be near 700 MPa, which is relatively moderate for nickel but is an otherwise extremely high stress level. Perhaps the stress in the nickel pad could be the source of the tear out problem. The results of a virtual crack propagation simulation with only the nickel pad in place is shown in Figure 7.18. The main feature of this plot is the fact that although the crack driving

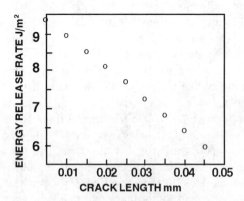

FIGURE 7.18 Driving force for crack propagation caused by residual stress in the nickel pad alone computed from the finite element model in Figure 7.15. The main feature of this plot is the fact that the driving force is decreasing as the crack grows longer, indicating that this crack will always stop and become arrested.

force starts out at a fairly vigorous 9 J/m^2, it then proceeds to decrease as the crack elongates. This clearly demonstrated that the stress in the nickel pad alone was insufficient to propagate a tear out crack because the driving force propelling any initial crack keeps decreasing as the crack gets longer and will eventually reach a point at which it is insufficient to drive the crack any further. Thus, the strain energy locked in the nickel pad was insufficient to cause the ceramic tear out problem even though the stress in the nickel is very high.

Having convinced ourselves that the high-stress nickel layer alone is insufficient to cause a ceramic tear out, we then proceeded to run a complete model using a full pin structure with pin on top of solder on top of a nickel pad. Figure 7.19 shows the plot of driving force versus crack length for a 50-μm long crack similar to the one shown in Figure 7.17. A single glance at the driving force data in this figure clearly shows that the cause of the ceramic tear out problem has been found. Not only is the driving force for crack propagation high, but also it is increasing with crack length, which makes for an unstable condition. As an example of what can happen, consider a pin that has a crack similar to the one depicted in Figure 7.17. Under dry conditions, the strength of the ceramic is about 15 J/m^2, which is just above the maximum driving force of 12 J/m^2 shown in Figure 7.19. This means that the crack would be arrested under these conditions, and the pin structure would appear to be stable. However, it is well known that the ceramic material in question is moisture sensitive and should the module containing the pin in question be moved into a humid environment, moisture can condense on the ceramic surface and eventually diffuse along the preexisting crack to the crack tip, where it can proceed to attack the ceramic and reduce the effective cohesive strength to 12 J/m^2 or less. At this point, the crack becomes unstable and will start to lengthen in short increments. Assume, for instance, the crack advances another micrometer in the direction

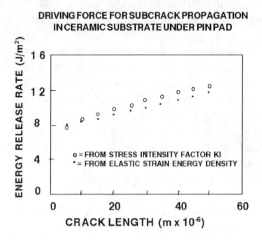

FIGURE 7.19 Driving force for crack propagation for the full pin-on-substrate model including tensile load on the pin. Because the driving force is both high and increasing with crack length, this plot indicates the existence of a very dangerous situation that will lead to ceramic tear out. *Note:* Results of two entirely different computational methods are plotted to ensure the accuracy and consistency of the calculation.

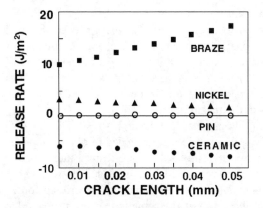

FIGURE 7.20 Decomposition of the driving force for crack propagation shown in Figure 7.19 into the individual contributions of each of the material layers present. Note that it is the solder braze material that makes the dominant contribution, followed by the nickel layer. The ceramic makes a negative contribution, indicating that it tends to blunt the progress of the crack. Finally, there is virtually no contribution from the pin, indicating somewhat counterintuitively that the tensile load on the pin is only a minor contributor to the ceramic tear out process.

of maximum driving force. It now encounters fresh dry ceramic material that is stronger than the old material weakened by moisture attack. The crack will again be arrested; however, because the crack is now longer, the driving force at the new crack tip is increased, making the situation even more unstable than previously. In addition, the moisture in the environment will continue on its course and start attacking the freshly created ceramic surface near the new crack tip. It is clear that this process will continue as long as the humid condition prevails, and the crack will progress increment by increment until a point is reached at which the driving force becomes larger than that needed to propagate a crack in the dry ceramic, at which point there is nothing to stop further crack growth, and the crack will take off at acoustic velocity and create a ceramic tear out crater.

At this point, it is instructive to resolve the total driving force into the components contributed by each of the different materials in the model. This can easily be done, and the result is shown in Figure 7.20. This plot breaks out the five components of the total load that is driving the tear out crack. The contributions include the tensile load on the pin, the residual stress in the solder/braze material, the stress in the nickel pad, and finally the stress in the ceramic substrate. Curiously, the tensile load on the pin contributes next to nothing toward the driving force for crack propagation even though the 12-lb load imposed for this simulation is close to the ultimate load for tensile failure of the pin. The stress in the nickel pad is clearly only a minor player in the overall crack propagation process, and the ceramic material actually imposes a significant resisting force. The clear culprit driving the tear out crack is the solder/braze material, which even though the stress in this material is four times lower than that in the nickel pad, accounts for 90% or more of the total driving force. This is yet another instance when one simply cannot go by the level of stress alone to uncover the source of trouble. A full fracture mechanics analysis is required to obtain an accurate picture of what is going on.

7.4 STABILITY MAPS

The examples related in the previous sections give rise to the obvious question: What is the best way to engineer around stress-related delamination and decohesion phenomena? Although the examples given above relate to totally different structures and process conditions, they share a number of points in common. In particular, we note the following:

1. The source of the failure mode tends to arise from an unexpected and nonobvious source.
2. The absolute level of stress in a particular material is not necessarily an accurate guide to its contribution to the overall failure process.
3. There is always a distribution of flaws in any real structure that are susceptible to environmental corrosion and other insidious phenomena. Conditions can converge on one particular flaw and drive it to critical size, which then initiates catastrophic failure.
4. In a stable structure, the driving force for crack propagation decreases with increasing crack length. The opposite is true for an unstable structure.
5. A complete and accurate fracture mechanics analysis is required to determine the state of stability of any given structure.

These points bring up the notion of creating a stability map, which is essentially a plot in a multidimensional space of the driving force for flaw propagation versus flaw size for a number of different parameter values that directly affect the driving force. Using the pin problem as an example, the engineer is compelled to design the pin attach process subject to a number of constraints, ranging from availability and compatibility of different material sets to economic cost considerations imposed by the marketplace. An array of myriad technical questions must be answered, ranging from how thick the nickel pad should be to what is the best material to use for pin fabrication. However, regardless of all other considerations, the pin structure has to be mechanically stable, and the analysis shown in Figure 7.20 clearly shows that attention has to be paid to the solder/braze material used to attach the pin to the nickel substrate.

One obvious question that immediately comes to mind is what is the effect of varying the stress in the solder on the driving force for crack propagation. This question is clearly answered by the elementary stability map shown in Figure 7.21. What is given is the driving force for crack propagation versus crack length at several different stress levels in the solder/braze material. In a full-blown stability map study, this could be one section of a multidimensional diagram in which other parameters such as solder volume, pad thickness, pin dimensions, and the like are held constant. What the figure shows is that the stress in the solder has a very strong effect on the crack driving force, and that indeed there is a stress level of approximately 12 kpsi (84 MPa) below which a state of unconditional stability exists. In effect, if the stress in the solder is kept below 12 kpsi, then the driving force for crack propagation will decrease with increasing crack size, and thus whatever flaws are present in the ceramic will eventually arrest and remain stable. The stress levels in Figure 7.21 are

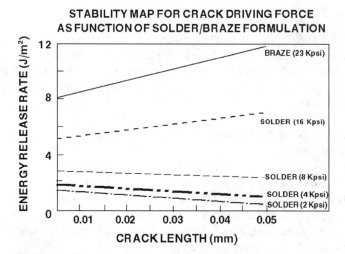

FIGURE 7.21 Sample stability map for the pin structure in Figure 7.14. Lines of constant solder stress are plotted on the plane of crack driving force versus crack length. The map divides naturally into two regimens, one unstable region in which the driving force increases with crack length and a stable portion in which the driving force decreases with crack length. The border between the two domains occurs at a solder stress level of approximately 12 kpsi (84 MPa).

controlled by varying the solder/braze formulation and the process conditions used to fabricate the final joint. Thus, the data in this figure are an invaluable guide to the engineer in selecting the correct solder material and process conditions to employ in forming the final pin joint.

7.5 SUMMARY

In this chapter, three examples of adhesion/decohesion failure phenomena taken from actual practice are presented to give an overall perspective regarding how all the supporting material developed in previous chapters comes together to solve real-world problems. There are a number of common threads that link all of the above examples; they can be summarized as follows:

1. Delamination/decohesion failure processes occur because of varied and largely unanticipated causes that require the systematic application of the physical principles governing material behavior coupled with appropriate measurement methods to decipher the underlying causes.
2. Failure in a material structure inevitably occurs by whatever mode offers the least resistance, whether it is delamination or decohesion, so the practicing engineer has to be prepared to address the full range of thermal-mechanical/physicochemical phenomena that could be operating.
3. Excessive stress in a structure is often the source of most failure behavior. Caution must be exercised, however, in that it is not always the most

highly stressed material that is the culprit. A careful fracture mechanics analysis is required to determine which material or set of conditions is contributing the most to the crack driving force.

4. The use of stability maps can be a very effective tool to help the design engineer make the appropriate material/process choices that will avoid anticipated failure problems.

In addition to the above list of admonitions, it should be added that quite likely the most important step one can take in countering delamination/decohesion problems is to have a set of measurement tools in place before getting heavily involved in resource-consuming manufacturing processes. If failure problems occur during manufacturing, then the first thing one must be able to do is measure what is occurring. If excessive stress is the problem, then one has to have a way of knowing what it is. Without appropriate measurement methods, you will have no way of knowing where in the game you are much less be able to discern which direction to go in to get on top of the situation.

Notes

1. "Adhesion of PNDA-ODA Polyimide to Silicon Oxide Surface: Sensitivity to Fluorine Contamination," L.P Buchwalter and R.H. Lacombe, *Journal of Adhesion Science and Technology,* 2–6, 463 (1988).
2. "Adhesion of Polyimide to Fluorine-Contaminated SiO_2 Surface. Effect of Amino-propyltriethoxysilane on the Adhesion," L.P. Buchwalter and R.H. Lacombe, *Journal of Adhesion Science and Technology,* 5–6, 449 (1991).
3. Although typically abbreviated as XPS, there is another school that follows the suggestion of the founder of the technique and calls the method ESCA for electron spectroscopy for chemical analysis. Regardless of what you call it, the method amounts to bombarding the surface in question with x-rays, which effectively kick electrons out of their atomic orbits into free space. The energy spectrum of the ejected electrons is then analyzed using sophisticated electromagnetic lenses and electronic detection equipment. Typically, a number of peaks are detected that uniquely characterize the atomic species on the surface. Because the escape depth of electrons from below the sample surface is no more than about 100 Å, XPS is considered a surface analysis tool. Those interested in a more detailed description of this technique can consult the following volume edited by D. Briggs and M.P. Seah, *Practical Surface Analysis, Second Edition, Volume 1: Auger and X-ray Photoelectron Spectroscopy* (John Wiley and Sons, New York, 1990).
4. The polyimide used was pyromellitic dianhydride-oxydianiline (PMDA-ODA). This is perhaps one of the most common polyimides available. One can obtain preformed sheets from the DuPont company under the Kapton® trademark. PMDA-ODA is also available as an uncured viscous liquid, which is called the amic-acid form. The amic acid can be readily spin coated onto a wafer and then dried and heated to as high as 400°C, which then induces the imidization reaction that forms the final PMDA-ODA coating. After imidization, the material becomes insoluble, infuseable, and largely chemically unreactive. The polyimide materials generally are attractive as insulators in the microelectronic industry because of their thermal and chemical stability and favorable dielectric properties.

5. The interested reader can refer to the following standard reference for dealing with this type of problem: *The Mathematics of Diffusion*, J. Crank (Oxford University Press, Oxford, UK, 1964).
6. "Steady State Cracking in a Brittle Substrate Beneath Adherent Films," Z. Suo and J. Hutchinson, *International Journal of Solids Structures*, 25, 1337 (1989).
7. "The Decohesion of Thin Films From Brittle Substrates," M.S. Hu, M.D. Thouless, and A.G. Evans, *Acta Metallica,* 36, 1301 (1988).

Appendix A
Vectors and Vector Calculus

I understand what an equation means if I have a way of figuring out the characteristics of its solution without actually solving it.

P.A.M. Dirac

A.1 INTRODUCTION

To make this volume reasonably self-contained, it is convenient to include some sort of introductory overview to vectors and vector calculus, especially because Chapters 3 and 4, which deal with continuum theory and fracture mechanics, draw heavily on this mathematical machinery. Ideally, we could all meet Dirac's criteria for understanding mathematical formulae by simply looking at them and by some divine intuition figuring out what they mean. This type of intuition is developed by the hard-core practitioners of the art who work with this type of material nearly daily. The rest of us have to struggle to obtain a firmer grasp of what is presented. The aim of this appendix is to assist both the beginner who has been exposed only to elementary algebra and calculus and the more experienced practitioner whose skills have lapsed from neglect. Thus, a quick overview of vectors and vector calculus is presented here along with a fairly extensive table of useful formulae and identities that will help one follow the mathematical developments in Chapters 3 and 4. In essence, it is hoped that the reader who is trying to follow one of the more involved derivations should be able to consult this appendix and find a satisfying answer.

All of the material in this section is available in textbooks, some of which go back as far as the 19th century. This overview relies on several relatively modern 20th century texts. A useful and entertaining introduction to vectors and vector calculus can be found in the second volume of the *Feynman Lectures*.[1] An informative and informal text devoted to vector calculus has been written by Schey.[2] Useful tables of formulae can be found in the handbook by Anderson.[3] A detailed treatment of tensor calculus with useful tables of formulae can be found in the extensive chapter by Gurtin[4] and the text by Gould.[5] Finally, a fairly comprehensive overview of vector calculus as applied to continuum theory can be found in the Schaum Outline by Mase.[6]

We start from the beginning by giving an essentially intuitive definition of what a vector is. However, it helps first to define a related but lower-order quantity known as a scalar. For our purposes, a *scalar* is any quantity that requires only one number in its specification. Thus, the temperature of a fluid at some particular point is a scalar quantity because it requires only one number. At a slightly higher level of abstraction, we can talk about a scalar field, which is essentially a whole collection of scalars. A concrete example would be the collection of all temperatures as a function of position required to specify precisely the temperature distribution in some large container of fluid.

Please note that we use the term *field* in a rather loose manner. To the mathematician, the term field is more than just a collection of numbers. A mathematical field is a collection of objects that behave in a very special way and in essence obey what are know as the field axioms. Our scalar temperature field belongs to this class of mathematical objects, and we fortunately do not have to worry too much about the mathematical details to obtain a workable level of understanding. It is only when you have to prove something rigorously that you have to know the precise definition of a field; the purpose of this appendix is more to provide a heuristic as opposed to a rigorous understanding of the mathematics with which we are dealing. I believe it was Sir Arthur Eddington who quipped, "Proof is an idol before which the mathematicians torture themselves" or words roughly to that effect. In this work, we really have not the luxury of understanding vector calculus at a rigorous mathematical level, so we will have to be satisfied with a level of understanding that is intuitive and plausible but be assured that the results are rigorous and have been worked out in great detail elsewhere.

Getting back to our container of liquid, we know that to understand the temperature distribution we need the temperature at each point, which amounts to a scalar field. Now, let us assume that the container is stirred somehow, that the entire liquid is circulating about, and that we want to know the velocity of the fluid at every point. The velocity is a vector quantity because it requires that three numbers be specified at each point, which essentially amounts to knowing the speed at which the liquid is traveling and the direction. Thus, the velocity distribution requires what we call a vector field for its complete description. It is clear that the velocity field is a good deal more complicated than the temperature field. However, it gets worse the more detailed the description that you require. We know that to set up a circulation in the liquid there has to be some sort of paddle rotating about that sets up a shear stress in the liquid, which causes it to circulate. What if we need to know the shear stress distribution in our liquid? The shear stress is a tensor quantity requiring in general nine numbers to specify in three dimensions. Thus, a tensor field is called for to specify our shear stress distribution. Clearly, the level of mathematical complexity is climbing at an alarming rate, and we have yet to see the worst of it.

If instead of a container of liquid we had a block of some crystalline solid and we were interested in its elastic properties, in general this would require specifying the elasticity tensor, which is a tensor quantity of rank 4 requiring a maximum of 81 numbers. Now, the reader should not panic because, for the present purposes, the discussion will go no further than second-rank tensors and symmetric ones at that, which require no more than 6 numbers to specify. Did you ever wonder why there are not many people working in the field of general relativity of gravitational fields? In general relativity all the vectors are four dimensional, as are the tensors, and quantities like the elasticity tensor are considered child's play. We will not even mention how many numbers are involved in working out the relativistic dynamics of a spiral galaxy.

A.2   ELEMENTARY DEFINITIONS OF SCALARS, VECTORS, AND TENSORS

A.2.1 SIMPLE SCALARS

With the introductory discussion as a background, we can now set about defining scalars, vectors, tensors, and the mathematical fields comprised of them. Because they are single numbers, the scalars present little problem, and the most important scalar field is the real line comprising all numbers from negative infinity to positive infinity, compactly designated as $(-\infty, \infty)$ and often signified by the designation R1. Although students of abstract algebra fully realize the real line is much more complex than it seems, this discussion avoids all these subtleties and simply uses the real numbers as a convenient way of designating simple physical quantities such as temperature, density, linear distance, and the like. By a scalar field, we understand a large collection of numbers, one associated with each point in a three-dimensional space, that could describe things such as a temperature distribution, a density distribution, or any other physical quantity specifiable by a single number.

A.2.2 VECTORS

Next up the scale are the vectors, which require at least two numbers to specify. The position vector in two dimensions is certainly the most intuitive and well-known vector quantity that points from the origin to a point in the plane with coordinates x and y. There are three common ways of specifying this vector, each of which has certain advantages and disadvantages. We use the bold symbol $\mathbf{r}$ to represent this vector. Given the components x and y that comprise $\mathbf{r}$, the three common notations in use are

- Matrix notation: $\mathbf{r} = (x, y)$ (row notation) or $\begin{pmatrix} x \\ y \end{pmatrix}$ (column notation)
- Vector (monadic) notation: $\mathbf{r} = x\mathbf{i} + y\mathbf{j}$ ($\mathbf{i}, \mathbf{j}$ = unit vectors in the x and y directions, respectively)
- Indicial notation: $\mathbf{r} = r_i$, i = 1, 2 where $r_1 = x$ and $r_2 = y$

These definitions for a two-dimensional vector have obvious extensions to three or more dimensions simply by tacking on more components. Each of the notations has some distinct advantages depending on the problem at hand. Matrix notation has the advantage of the ability to employ all the machinery of matrix manipulation developed in the extensive lore of linear algebra.[7] On the other hand, the vector notation is desirable because it has the direct visual appeal of arrows pointing in various directions plus the fact that many of the most useful formulae of vector calculus have compact and easily manipulable representations in vector form.

Finally, the indicial notation is a favorite of theorists who have to deal with tensor quantities of rank 2 and higher. In this work, we adopt the expeditious if somewhat peripatetic approach of using whatever notation is most convenient for the application at hand.

A.2.2.1 Vector Operations

Vectors can be manipulated by the following operations, which are both useful and serve to define precisely what a vector is. The operations are defined in each of the above-mentioned formats: matrix, vector, and indicial. In what follows, we consider the behavior of the three vectors, which in vector format are designated $\mathbf{a}$, $\mathbf{b}$, and $\mathbf{c}$. For simplicity, we take these vectors to be two dimensional; therefore, their designation in matrix format would be $\mathbf{a} = (a_1, a_2)$, $\mathbf{b} = (b_1, b_2)$, and $\mathbf{c} = (c_1, c_2)$. The indicial format is compact, and the vector $\mathbf{a}$ would be written as a_i where $i = 1, 2$ and likewise for the vectors $\mathbf{b}$ and $\mathbf{c}$. One can see why the indicial notation would be preferred by those who have to work in the higher dimensional spaces because the range of the subscript needs only to be specified once, and from that point all vectors require only two characters to represent. The obvious drawback to this notation is that subscripts can have a wide range of meanings that can only be inferred from context. The potential for confusion is obvious, and the reader therefore has to be careful.

Vector Addition
- *Vector notation*: Given vectors $\mathbf{a}$ and $\mathbf{b}$, a third vector $\mathbf{c}$ exists that can be either the sum or difference of $\mathbf{a}$ and $\mathbf{b}$; thus, $\mathbf{c} = \mathbf{a} \pm \mathbf{b}$
- *Matrix notation*: $(c_1, c_2) = (a_1 \pm b_1, a_2 \pm b_2)$
- *Indicial notation*: $c_i = a_i \pm b_i$ where $i = 1, 2$

Scalar Product

If a designates the magnitude of $\mathbf{a}$ and b the magnitude of $\mathbf{b}$, then the scalar product of $\mathbf{a}$ and $\mathbf{b}$ is a scalar quantity designated as $\mathbf{a} \bullet \mathbf{b}$ and given by the formulae

- *Vector notation*: $\mathbf{a} \bullet \mathbf{b} = a\,b \cos \theta$ where θ is the angle between $\mathbf{a}$ and $\mathbf{b}$.
- *Matrix notation*: In matrix notation, this formula has a somewhat more useful representation that does not require knowing the angle θ. At this point, we need to introduce a somewhat different brand of vector called the column vector. A row vector $\mathbf{a}$ is represented by a simple row of elements such as (a_1, a_2) for which there can be any number of entries all separated by commas. Vectors can also be arranged in columns, such as the vector

$$\mathbf{b} = \begin{pmatrix} b_1 \\ b_2 \end{pmatrix}.$$

Again, any number of elements can be included. The scalar product of two vectors is then defined as the following matrix product in which the row vector is always on the left and the column vector is always on the right:

$$\mathbf{a} \cdot \mathbf{b} = \left(a_1, a_2\right)\begin{pmatrix} b_1 \\ b_2 \end{pmatrix} = a_1 b_1 + a_2 b_2$$

- *Indicial notation*: Again, indicial notation is compact, and the scalar product $\mathbf{a} \bullet \mathbf{b}$ is given by $a_i b_i$. Because the index i is repeated, it is assumed to be summed over. Thus, the explicit formula for the scalar product is

$$\sum_{i=1}^{2} a_i b_i \ ,$$

which makes it precisely equivalent to the matrix definition. The repeated index rule, although allowing for a compact representation, again opens the door for confusion, which has the effect of limiting this notation to those applications dealing with higher spaces, such as are encountered in general relativity theory.

Absolute Magnitude of a Vector

Using the definition of the scalar product, we can now succinctly define the magnitude of a vector as follows:

- *Vector notation*: magnitude of $\mathbf{a} = |\mathbf{a}| = (\mathbf{a} \cdot \mathbf{a})^{1/2}$
- *Matrix notation*: $|\mathbf{a}| = (a_1^2 + a_2^2)^{1/2}$
- *Indicial notation*: $|\mathbf{a}| = (a_i a_i)^{1/2}$

Vector Product

- *Vector notation*: Given vectors $\mathbf{a}$ and $\mathbf{b}$, a third vector $\mathbf{c}$ exists that is the vector product of $\mathbf{a}$ and $\mathbf{b}$ given by $\mathbf{c} = \mathbf{a} \times \mathbf{b}$. The vector product is properly defined only in three dimensions or higher, so we have to make our vectors three dimensional by adding one more component. With this done, the vector product of $\mathbf{a}$ and $\mathbf{b}$ is $\mathbf{c} = (a_2 b_3 - a_3 b_2)\mathbf{i} - (a_1 b_3 - a_3 b_1)\mathbf{j} + (a_1 b_2 - a_2 b_1)\mathbf{k}$. The vector $\mathbf{c}$ turns out to be perpendicular to both $\mathbf{a}$ and $\mathbf{b}$, although this is not obvious from the definition just given. In fact, not much is obvious about the vector product at first sight, and it is definitely one of the trickier concepts in vector algebra. Its main use is in electromagnetic theory when dealing with magnetic fields and in fluid dynamics when dealing with the concept of vorticity, so there is really no way to get around it. Those familiar with determinants, to be discussed later, will find the following formula to be a handy way of remembering the vector product:

$$\mathbf{c} = \mathbf{a} \times \mathbf{b} = \det \begin{bmatrix} \mathbf{i} & \mathbf{j} & \mathbf{k} \\ a_1 & a_2 & a_3 \\ b_1 & b_2 & b_3 \end{bmatrix} \qquad\qquad \text{(A.1)}$$

- *Matrix notation*: In matrix notation, the components of the vector product of **a** and **b** are $\mathbf{c} = (a_2 b_3 - a_3 b_2, -(a_1 b_3 - a_3 b_1), a_1 b_2 - a_2 b_1)$.
- *Indicial notation*: Things start to get fairly arabesque when using indicial notation. One first has to start by defining the following third-rank tensor, commonly called the Levi-Cevita tensor density:

$\varepsilon_{ijk} = 1$ if i, j and k are a cyclic permutation of the indices 1, 2 and 3

$\qquad = -1$ if i, j and k are a noncyclic permutation of the indices 1, 2 and 3

$\qquad = 0$ if any two indices are equal

- With this definition in place, the vector product **c** of **a** and **b** has a compact form as follows: $c_i = \varepsilon_{ijk} a_j b_k$. It is hard to believe, but it works. The skeptical reader should work out the details remembering that the repeated indices j and k are summed over, and that the components of ε_{ijk} are nonzero only when i, j, and k are all different. Thus, $\varepsilon_{123} = 1$ because 123 is a cyclic permutation as are 312 and 231, where the sequence is simply shifted one step forward, with the last index going to the front. The sequence 132 is not cyclic because it involves the positional swapping of just two indices, and this sequence cannot be obtained by any cyclic shift of the index sequence 123. Thus, $\varepsilon_{132} = -1$. Any sequence with a repeated index number such as 113 gives a zero value. Try it; you will be amazed that it works.

A.2.2.2 Tensor Operations

The above list covers essentially all the vector operations one needs to know to do vector algebra. In continuum theory, one also has to deal with higher rank quantities known as tensors because a quantity such as a shear stress requires that two directions in space be specified at any point. The first direction is that of the applied force, and the second direction is the normal to the surface on which the force is acting. In vector notation, this problem is best approached by starting with an orthonormal set of unit vectors **i**, **j**, **k** that form a triad looking like a badly bent pitchfork. These vectors are uniquely defined by the following set of properties:

$$\mathbf{i} \cdot \mathbf{j} = \mathbf{j} \cdot \mathbf{k} = \mathbf{k} \cdot \mathbf{i} = 0$$

$$\mathbf{i} \cdot \mathbf{i} = \mathbf{j} \cdot \mathbf{j} = \mathbf{k} \cdot \mathbf{k} = 1$$

$$\text{(A.2)}$$

The fact that the scalar product between two different unit vectors is zero makes them all orthogonal, thus giving rise to the ortho prefix; the fact that the scalar product of any of the vectors with itself is 1 means that they are normalized and thus the normal suffix in the designation orthonormal. By juxtaposing any two of the monadic quantities **i, j, k**, we can form a dyadic quantity such as **ij**. The quantity **ij** points in both the i and the j direction simultaneously and can thus be considered a tensor type of quantity. Suppose we take the standard vectors $\mathbf{a} = a_1\mathbf{i} + a_2\mathbf{j} + a_3\mathbf{k}$ and $\mathbf{b} = b_1\mathbf{i} + b_2\mathbf{j} + b_3\mathbf{k}$ and form the dyadic quantity **ab**. We would get a general tensor quantity $\underline{\mathbf{A}}$ of the form

$$\underline{\mathbf{A}} = \left(a_1\,\mathbf{i} + a_2\,\mathbf{j} + a_3\,k\right)\left(b_1\,\mathbf{i} + b_2\,\mathbf{j} + b_3\,\mathbf{k}\right) =$$

$$a_1b_1\,\mathbf{ii} + a_1b_2\,\mathbf{ij} + a_1b_3\,\mathbf{ik}$$

$$a_2b_1\,\mathbf{ji} + a_2b_2\,\mathbf{jj} + a_2b_3\,\mathbf{jk} \qquad\qquad (A.3)$$

$$a_3b_1\,\mathbf{ki} + a_3b_2\,\mathbf{kj} + a_3b_3\,\mathbf{kk}$$

The above example shows that a general tensor quantity $\underline{\mathbf{A}}$ can be formed in a straightforward manner by the composition of two vectors.* However, we could simply write down a general tensor quantity $\underline{\mathbf{A}}$ without resorting to vector composition as follows:

$$a_{11}\,\mathbf{ii} + a_{12}\,\mathbf{ij} + a_{13}\,\mathbf{ik}$$

$$\underline{\mathbf{A}} = a_{21}\,\mathbf{ji} + a_{22}\,\mathbf{jj} + a_{23}\,\mathbf{jk} \qquad\qquad (A.4)$$

$$a_{31}\,\mathbf{ki} + a_{32}\,\mathbf{kj} + a_{33}\,\mathbf{kk}$$

The dyadic form of the tensor $\underline{\mathbf{A}}$ has a number of nice properties. First, it clearly exhibits the multidirectional nature of the tensor; second, using the rules in Equation (A.2) it is intuitively obvious how one would go about taking a scalar product of a tensor with a vector. One drawback of this notation is that one is forever writing down the dyadic compositions **ij**, **ik**, and so on, which can become cumbersome. Thus, we come to the matrix notation, which dispenses with the dyadics and writes down only the coefficients that are the essence of the tensor quantity. In matrix notation, $\underline{\mathbf{A}}$ becomes

$$\underline{\mathbf{A}} = \begin{pmatrix} a_{11} & a_{12} & a_{13} \\ a_{21} & a_{22} & a_{23} \\ a_{31} & a_{32} & a_{33} \end{pmatrix} \qquad\qquad (A.5)$$

Although the matrix notation clearly streamlines the representation of a tensor quantity by exhibiting only the essential components required for an unambiguous

---

* For simplicity, we compose two vectors by simple juxtaposition as in the simple tensor quantity **ii**. Other texts use the $\otimes$ symbol to indicate that two vectors are composed into a tensor; thus, the simple tensor **ii** would be written $\mathbf{i} \otimes \mathbf{i}$.

definition, it is still somewhat cumbersome, especially if one wants to represent quantities of rank higher than 2. This leads us to the ultimately compact indicial notation, which defines $\underline{\mathbf{A}}$ by simply declaring what its components are in the following fashion:

$$\left(\underline{\mathbf{A}}\right)_{ij} = a_{ij} \quad i, j = 1, 2, 3, \ldots N \tag{A.6}$$

Equation (A.6) states that the tensor $\underline{\mathbf{A}}$ is a two-dimensional array of coefficients a_{ij} where the indices i and j run from 1 to some integer N. This is very compact notation, indeed, compared to Equation (A.5) because it represents an array of arbitrary order in less space than Equation (A.5) requires to represent a simple array of order 3. Also, it is clear that Equation (A.6) can easily be generalized by simply adding more subscripts and thus representing quantities of arbitrary rank as well as arbitrary order.* Thus, it is clear why indicial notation is preferred by those who work in a field such as general relativity, which involves the routine use of higher rank quantities such as the Ricci tensor or the Christoffel symbols. The practical engineer, however, studiously avoids using any quantity of rank higher than 2 or order greater than 3 if possible.

A.2.2.3 Special Tensors and Operations

There are a number of special operations that are used over and over in manipulating tensor arrays. The most useful is the transpose of an array typically signified by the notation $\underline{\mathbf{A}}^T$ by which the off-diagonal elements are swapped with one another thus:

Transpose of an array $\underline{\mathbf{A}}$:
　　If an array $\underline{\mathbf{A}}$ has elements a_{ij}, then its transpose $\underline{\mathbf{A}}^T$ has elements a_{ji} by which corresponding off-diagonal elements are simply swapped with one another. Thus, $a_{ij} \to a_{ji}$ and $a_{ji} \to a_{ij}$ for all i and j.

Symmetric arrays:
　　An array is said to be symmetric if it is equal to its transpose, that is, $\underline{\mathbf{A}} = \underline{\mathbf{A}}^T$.

The identity matrix:
　　The identity matrix $\underline{\mathbf{I}}$ has the property that multiplying it by any other matrix leaves that matrix unchanged. Thus, $\underline{\mathbf{A}}\underline{\mathbf{I}} = \underline{\mathbf{I}}\underline{\mathbf{A}} = \underline{\mathbf{A}}$. The identity matrix has the following representations:
　　• *Vector notation*: Using the dyadic notation, the identity array can be written $\underline{\mathbf{I}} = \mathbf{ii} + \mathbf{jj} + \mathbf{kk}$.

---

* As used here, the rank of an array designates the number of different indices required, and the order of the array indicates the largest integer value that any index can take on. Note that these definitions are similar to but not entirely congruent with those used in standard linear algebra.

- *Matrix notation*: In three dimensions, the identity can be written

$$\underline{\mathbf{I}} = \begin{pmatrix} 1 & 0 & 0 \\ 0 & 1 & 0 \\ 0 & 0 & 1 \end{pmatrix}.$$

The generalization to higher-order arrays is obvious.
- *Indicial notation*: In indicial notation, one defines what is known as the Kronecker symbol:

$$\delta_{ij} = 1 \quad \text{for } i = j \text{ and } i, j = 1, 2, 3, \dots N$$

$$\delta_{ij} = 0 \quad \text{for } i \neq j$$

Trace of an array:

A useful quantity defined for any square array is the trace that is defined as the sum of all the diagonal elements. Thus, if $\mathbf{A}$ has elements a_{ij}, then we designate the trace as $tr\,\mathbf{A} = a_{11} + a_{22} + \dots a_{NN}$. In indicial notation, this can be represented as $tr\,\mathbf{A} = a_{ii}$ where summation over the repeated index i is assumed. Note that this is one instance for which the indicial notation becomes ambiguous because it may not be clear whether a_{ii} represent the trace of the array or some general diagonal element. As usual, the reader has to be cautious in such instances.

Determinant of an array:

I can recall as a student in high school studying the mysteries of matrix arrays and coming across the utterly enigmatic concept of the determinant of an array. The formal definition seemed completely byzantine as to its use.* It turns out that determinants arise quite naturally in the problem of solving simultaneous sets of linear equations. This is most definitely a practical problem arising in such diverse applications as determining the electrical performance of a complex circuit. In the simplest case, one has two equations in two unknowns as follows:

$$a_{11}x + a_{12}y = c_1$$
$$a_{21}x + a_{22}y = c_2$$

(A.7)

---

* The reader might find amusement in interpreting the following formal definition of the determinant of an array a_{ij} where i, j run from 1 to some integer n taken from a standard text: "The value of the determinant is: $\Sigma\, (-1)^k\, a_{1s1}\, a_{2s2} \dots a_{nsn}$ where the indices $s_1, s_2, \dots s_n$ form a permutation of the integers 1, 2, $\dots n$ and are all distinct. The integer k is equal to the number of inversions in the sequence $s_1, s_2, \dots s_n$, that is, the number of times a large integer precedes a smaller one." Whereas this is a perfectly correct definition, we wish you luck in trying to use it to evaluate determinants.

In the above system, x and y are considered unknown variables, and the coefficients a_{ij} and c_i are known constants. The problem is to determine x and y in terms of the known constants. For the simple case of two equations in two unknowns, the solution process is simple. For example, to solve Equations (A.7) for the variable y, one first divides both sides of the top equation by a_{11} and both sides of the bottom equation by a_{21}, which leaves the variable x with a coefficient of 1 in both equations. The next step is to subtract the bottom equation from the top, which leaves an equation involving y alone, which can be readily solved as follows:

$$y = \frac{c_2 a_{11} - c_1 a_{21}}{a_{22} a_{11} - a_{12} a_{21}} \tag{A.8}$$

A similar solution is easily obtained for the variable x using the same kind of manipulations:

$$y = \frac{c_1 a_{22} - c_2 a_{12}}{a_{22} a_{11} - a_{12} a_{21}} \tag{A.9}$$

Note the repeated appearance of the expression $(a_{22} a_{11} - a_{12} a_{21})$. This type of cross product appears again and again in the manipulation of matrix arrays and is in fact the determinant of the 2×2 array:

$$\det \begin{pmatrix} a_{11} & a_{12} \\ a_{21} & a_{22} \end{pmatrix} = a_{11} a_{22} - a_{12} a_{21} \tag{A.10}$$

It is easy to remember that the determinant of a 2×2 array is simply the difference of the two diagonal products. Thus, the problem of evaluating the determinant of 2×2 arrays is simplicity itself. There is in fact only one more rule to remember to evaluate the determinant of an array of arbitrary size. It turns out that the determinant of any square array can be reduced to the evaluation of the determinant of a collection of 2×2 subarrays using the method of expansion in the minors of the first row.

Take the case of a general 3×3 array. We want to find the determinant. The basic rule is as follows: Focus on the top row of elements and start from the left, with the first element a_{11}. The first term in the result is obtained by multiplying a_{11} by its associated minor determinant, which is the determinant of the array one gets by deleting the first row and first column of the original array. Because this is a 2×2 array, we have no problem finding its determinant. Moving on to the next element a_{12}, we do essentially the same thing, form the determinant of the 2×2 array created by deleting the first row and second column of the original array and multiply it by a_{12}. A small twist enters here in that this product is subtracted from the first term involving a_{11}. The general procedure is to alternately add and subtract as one proceeds through the elements of the first row. There is only one more element to go: a_{13}. Multiply this element by the determinant of the array created by deleting the first row and third column and add the result to the first two terms. The final result is the determinant of the 3×3 array, which is depicted symbolically as follows:

$$\det\begin{pmatrix} a_{11} & a_{12} & a_{13} \\ a_{21} & a_{22} & a_{23} \\ a_{31} & a_{32} & a_{33} \end{pmatrix} = a_{11}\det\begin{pmatrix} a_{22} & a_{23} \\ a_{32} & a_{33} \end{pmatrix} - a_{12}\det\begin{pmatrix} a_{11} & a_{13} \\ a_{31} & a_{33} \end{pmatrix} + a_{13}\det\begin{pmatrix} a_{21} & a_{22} \\ a_{31} & a_{32} \end{pmatrix} \quad (A.11)$$

Equation (A.11) can be readily evaluated because it involves a sum of products of scalars with determinants of 2×2 arrays that we know how to compute from Equation (A.10). The pattern continues in much the same fashion for 4×4 arrays and higher. Focusing on the first row, one starts with the product of a_{11} times the determinant of the minor array formed by deleting the first row and column of the original array. The next term is the negative of a_{12} times the determinant of the minor array formed by deleting the first row and second column of the original array. The procedure continues in this way, with the signs of successive terms alternating between negative and positive, with those involving odd-numbered columns positive and those involving even-numbered columns negative. Therefore, starting with an array of arbitrary order one can whittle it down eventually to a string of products of scalars with 2×2 determinants, which can then be fully evaluated using Equation (A.10).

There is one little catch, however. The number of terms one has to deal with goes up at an alarming rate as the order of the matrix worked on goes up. The number of terms in fact increases as n! (read n factorial), which equals n(n − 1)(n − 2) ... 1. Thus, for the case n = 3, we have 3! = 3 × 2 × 1 = 6, and thus there are six terms in the determinant expansion, as can be readily verified from Equation (A.11), which is not that bad. Going to n = 4, we get 4! = 24 terms, which is a rather dramatic increase over the n = 3 case. For n = 5, we have 5! = 120 terms, and it is clear that already, for rather small-size arrays, things have gotten completely out of hand. It should be pointed out that as a matter of practical computation, one rarely works with determinants or order much greater than 3 or 4. In solving systems of linear equations, for instance, one would use Gaussian elimination with singular value decomposition as the best numerical method.[8] Use of determinants finds more use in purely theoretical work, for which one is more interested in deriving analytical formulae than smoking out numerical results.

A.2.2.4 Tensor/Matrix Multiplication

Vector Notation

From the rules in Equation (A.2), we can clearly work out what the scalar product of two tensors ought to be. For instance, $\mathbf{ij} \cdot \mathbf{jk} = \mathbf{i}(\mathbf{j} \cdot \mathbf{j})\mathbf{k} = \mathbf{ik}$ or $\mathbf{ij} \cdot \mathbf{kj} = \mathbf{i}(\mathbf{j} \cdot \mathbf{k})\mathbf{j} = \mathbf{0}$. Thus, if we have two tensor quantities $\underline{\mathbf{A}}$ and $\underline{\mathbf{B}}$, then they may be represented as follows:

$$
\begin{array}{ll}
a_{11}\,\mathbf{ii} + a_{12}\,\mathbf{ij} + a_{13}\,\mathbf{ik} & b_{11}\,\mathbf{ii} + b_{12}\,\mathbf{ij} + b_{13}\,\mathbf{ik} \\
\underline{\mathbf{A}} = a_{21}\,\mathbf{ji} + a_{22}\,\mathbf{jj} + a_{23}\,\mathbf{jk} \quad & \underline{\mathbf{B}} = b_{21}\,\mathbf{ji} + b_{22}\,\mathbf{jj} + b_{23}\,\mathbf{jk} \\
a_{31}\,\mathbf{ki} + a_{32}\,\mathbf{kj} + a_{33}\,\mathbf{kk} & b_{31}\,\mathbf{ki} + b_{32}\,\mathbf{kj} + b_{33}\,\mathbf{kk}
\end{array}
\quad (A.12)
$$

Taking the scalar product of these two arrays yields a third tensor, $\underline{C} = \underline{A} \cdot \underline{B}$:

$$\underline{C} = \tag{A.13}$$

$$\left(a_{11}b_{11} + a_{12}b_{21} + a_{13}b_{31}\right) \mathbf{ii} + \left(a_{11}b_{12} + a_{12}b_{22} + a_{13}b_{32}\right) \mathbf{ij} + \left(a_{11}b_{13} + a_{12}b_{23} + a_{13}b_{33}\right) \mathbf{ik}$$

$$\left(a_{21}b_{11} + a_{22}b_{21} + a_{23}b_{31}\right) \mathbf{ji} + \left(a_{21}b_{12} + a_{22}b_{22} + a_{23}b_{32}\right) \mathbf{jj} + \left(a_{21}b_{13} + a_{22}b_{23} + a_{23}b_{33}\right) \mathbf{jk}$$

$$\left(a_{31}b_{11} + a_{32}b_{21} + a_{33}b_{31}\right) \mathbf{ki} + \left(a_{31}b_{12} + a_{32}b_{22} + a_{33}b_{32}\right) \mathbf{kj} + \left(a_{31}b_{13} + a_{32}b_{23} + a_{33}b_{33}\right) \mathbf{kk}$$

From Equation (A.8), one can see that matters are already getting out of hand, and these are only second-rank quantities or order 3. Clearly, if we start to crank up the rank, we quickly begin to run out of space on the page. Fortunately, for most purposes rank 2 and order 3 are about as high as we ever have to go. It turns out that the elements of the product tensor $\underline{C}$ are fairly easy to derive if we think about it a little with regard to the multiplication rules in Equation (A.2). Say we want to know what the coefficient of the dyadic $\mathbf{ij}$ ought to be. From the rules, it is clear that this element is going to come from terms in the array $\underline{A}$ that begin with $\mathbf{i}$ and terms from $\underline{B}$ that end with $\mathbf{j}$. Thus, this element is essentially going to be the scalar product of the tensor formed by taking just the first row of $\underline{A}$ and forming the dot product with the tensor generated by taking only the second column of $\underline{B}$ thus:

$$\underline{C}_{ij} = \left(a_{11} \ \mathbf{ii} + a_{12} \ \mathbf{ij} + a_{13} \ \mathbf{ik}\right) \cdot \left(b_{12} \ \mathbf{ij} + b_{22} \ \mathbf{jj} + b_{32} \ \mathbf{kj}\right)$$

$$= \left(a_{12}b_{12} + a_{12}b_{22} + a_{13}b_{32}\right) \mathbf{ij} \tag{A.14}$$

Thus, any element of the tensor product given in Equation (A.13) can be obtained as the scalar product of two simple tensors taken from the rows and columns, respectively, of the two initial tensors.

Matrix Notation

Again, matrix notation works in much the same manner as the vector version with the omission of the dyadic symbols; the product of two matrices can be represented as follows:

$$\underline{A} = \begin{pmatrix} a_{11} & a_{12} & a_{13} \\ a_{21} & a_{22} & a_{23} \\ a_{31} & a_{32} & a_{33} \end{pmatrix} \quad \underline{B} = \begin{pmatrix} b_{11} & b_{12} & b_{13} \\ b_{21} & b_{22} & b_{23} \\ b_{31} & b_{32} & b_{33} \end{pmatrix} \tag{A.15}$$

$$\underline{C} = \underline{A} \cdot \underline{B} \begin{pmatrix} \left(a_{11}b_{11} + a_{12}b_{21} + a_{13}b_{31}\right)\left(a_{11}b_{12} + a_{12}b_{22} + a_{13}b_{32}\right)\left(a_{11}b_{13} + a_{12}b_{23} + a_{13}b_{33}\right) \\ \left(a_{21}b_{11} + a_{22}b_{21} + a_{23}b_{31}\right)\left(a_{21}b_{12} + a_{22}b_{22} + a_{23}b_{32}\right)\left(a_{21}b_{13} + a_{22}b_{23} + a_{23}b_{33}\right) \\ \left(a_{31}b_{11} + a_{32}b_{21} + a_{33}b_{31}\right)\left(a_{31}b_{12} + a_{32}b_{22} + a_{33}b_{32}\right)\left(a_{31}b_{13} + a_{32}b_{23} + a_{33}b_{33}\right) \end{pmatrix}$$

Again each of the elements of the product matrix $\underline{\mathbf{C}}$ can be obtained as the scalar product between the corresponding row of A and column of B. For instance, the matrix element $(\underline{\mathbf{C}})_{12}$ can be obtained as the scalar product of the first row of $\underline{\mathbf{A}}$ with the second column of $\underline{\mathbf{B}}$ treated as row and column vectors, respectively:

$$\left(\underline{\mathbf{C}}\right)_{12} = \left(a_{11}\ a_{12}\ a_{13}\right) \cdot \begin{pmatrix} b_{12} \\ b_{22} \\ b_{32} \end{pmatrix} = \left(a_{11}b_{12} + a_{12}b_{22} + a_{13}b_{32}\right) \qquad (A.16)$$

Indicial Notation

As usual, the indicial notation gives the most compact representation of tensor multiplication. Thus, given the tensor arrays a_{ij} and b_{kl} where the indices run from 1 to some arbitrary integer N, we can succinctly define the tensor product c_{il} by the following formula: $c_{il} = a_{ij}\,b_{jl}$ where summation over the index j from 1 to N is understood. If we take the special case of N = 3, then we can readily see that this little formula will reproduce each of the elements of the matrix $\underline{\mathbf{C}}$ given in Equation (A.15). The indicial notation is clearly the most powerful and compact method for representing and manipulating higher dimensional arrays. The main drawback, as always, is the potential for confusing which indices are summed over and which are not. Another problem arises from the tendency of some authors to get carried away using four or more indices both as subscripts and superscripts appearing both to the left and to the right of the symbol to which they are attached. One can easily get dizzy dealing with such a baroque notation.

A.2.2.5 Product of a Matrix with a Vector

Vector Notation

To take the scalar product of a dyadic array with any vector, one simply follows the multiplication rules given in Equation (A.2). Note, however, that in general one gets different results when multiplying on the left as opposed to the right.

Matrix Notation

Multiplication of a matrix by a vector is simply a special case of Equation (A.15). Thus, given the array $\underline{\mathbf{A}}$ and the vector $\mathbf{b}$, we find that $\underline{\mathbf{A}} \cdot \mathbf{b}$ is another vector $\mathbf{c}$ given by

$$\underline{\mathbf{A}} = \begin{pmatrix} a_{11} & a_{12} & a_{13} \\ a_{21} & a_{22} & a_{23} \\ a_{31} & a_{32} & a_{33} \end{pmatrix} \quad \mathbf{b} = \begin{pmatrix} b_1 \\ b_2 \\ b_3 \end{pmatrix}$$

$$\mathbf{c} = \underline{\mathbf{A}} \cdot \mathbf{b} = \begin{pmatrix} a_{11}b_1 + a_{12}b_2 + a_{13}b_3 \\ a_{21}b_1 + a_{22}b_2 + a_{23}b_3 \\ a_{31}b_1 + a_{32}b_2 + a_{33}b_3 \end{pmatrix}$$

$$(A.17)$$

Indicial Notation

Indicial notation handles this operation quite succinctly by specifying that the ith component of the vector **c** obtained by multiplying the array with elements a_{ij} by the vector **b** with elements b_i is as follows: $c_i = a_{ij}b_j$ with summation over the repeated index j understood.

A.2.2.6 Tensor/Matrix Scalar Product

Having defined the product of two tensor quantities $\underline{A}$ and $\underline{B}$ that gives rise to third tensor $\underline{C}$, one wonders if there is some operation by which one multiplies two tensors to come up with a scalar quantity as one does for the scalar product of two vectors. Sure enough, such an operation exists and goes under the names of inner product and double dot product, depending on your favorite notation. This is a useful notation for continuum theory because it can be used to compactly express the elastic energy density in a stressed solid.

Vector Notation (Double Dot Product/Inner Product)

If we have two vector dyadics such as **ij** and **jk,** we can form the double dot product as **ij:jk**, which evaluates as **i·(j·j)k** = **i** · **k** = 0 using the rules of Equation (A.2). It is clear from this simple example that the double dot product of any of the unit dyadic pairs will evaluate to either 0 or 1. Using dyadic notation, we define the inner product of two arrays $\underline{A}$ and $\underline{B}$ as $[\underline{A}, \underline{B}] = \underline{A}^T{:}\underline{B}$. This quantity can be easily evaluated from Equation (A.12) by substituting $\underline{A}^T$ for $\underline{A}$ and following the product rules in Equation (A.2).

Matrix Notation

The double dot product has a much more compact representation in matrix format as:

$$\underline{A}{:}\underline{B} = tr\left[\left(\underline{A}\right)^T \cdot \underline{B}\right]$$

(A.18)

$$= a_{11}b_{11} + a_{21}b_{21} + a_{31}b_{31} + a_{12}b_{12} + a_{22}b_{22} + a_{32}b_{32} + a_{13}b_{13} + a_{23}b_{23} + a_{33}b_{33}$$

The notation tr signifies the trace of the matrix product or simply the sum of all the diagonal elements as defined above.

Indicial Notation

As is always the case, the indicial notation gives the most compact definition of the double dot product as $\underline{A}{:}\underline{B} = a_{ij}b_{ij}$ with summation over the repeated indices understood.

A.2.2.7 Assorted Special Arrays and Operations

There are a number of special arrays and operations that need to be mentioned before ending this brief overview of vector and tensor algebra. Because we are assuming that all arrays are made up of real numbers, we do not cover such subjects as Hermetian matrices and self-adjoint matrices, which occur when the elements of an array can be complex numbers. Those who go on to study quantum mechanics will definitely meet these creatures, which are beyond the scope of the present work. Also, we deal exclusively with square arrays for which the indices all have the same range. In the study of linear algebra, one routinely comes across nonsquare arrays that have applications in the statistics of fitting curves to experimental data.

In what follows, we give brief descriptions of the concepts of the matrix inverse, matrix transformations, and the eigenvalue problem for real symmetric matrices. The discussion resumes using only the matrix and indicial notations, leaving behind the dyadic vector notation, which is little used and essentially redundant.

Matrix Inverse

Matrix Notation

Subject to certain conditions, any array $\underline{A}$ has a corresponding inverse denoted as $\underline{A}^{-1}$ by which $\underline{A}\cdot\underline{A}^{-1} = \underline{A}^{-1}\cdot\underline{A} = \underline{I}$ where $\underline{I}$ is the identity matrix defined above. The conditions for the inverse array to exist are fairly technical but essentially amount to stating that the array $\underline{A}$ is not singular. In simplest terms, trying to find the inverse of a singular matrix is the equivalent of trying to divide by zero in simple arithmetic. The full details of what constitutes a singular array and under which conditions one can always find the inverse of a matrix are covered in any text on linear algebra.

Indicial Notation

If the matrix $\underline{A}$ has the elements a_{ij}, then the inverse array with elements a_{ij}^{-1*} satisfies the condition $a_{ik}a_{kj}^{-1} = \delta_{ij}$ where summation over the index k is understood.

Matrix Transformations

Matrix Notation

Say we have a matrix S and its inverse S^{-1}. We know that the product of the two matrices yields the identity that, when multiplied with any other array, leaves that array unchanged. However, say we form the product $\underline{A}\cdot\underline{S}$, we know that we will get some different array $\underline{B}$. What will happen if we try to undo the action of multiplying by $\underline{S}$ by multiplying by $\underline{S}^{-1}$ to form the product $\underline{S}^{-1}\cdot\underline{A}\cdot\underline{S}$? Multiplying on the left by $\underline{S}^{-1}$ does not get us back to the original matrix $\underline{A}$ because matrix multiplication is noncommutative in general (i.e., $\underline{A}\cdot\underline{B} \neq \underline{B}\cdot\underline{A}$). However, the new matrix $\underline{A}' = \underline{S}^{-1}\cdot\underline{A}\cdot\underline{S}$ is similar to the original matrix $\underline{A}$ in many ways. In particular, det $\underline{A}' = $ det $\underline{A}$ and $tr\,\underline{A}' = tr\,\underline{A}$ where the notation det symbolizes the determinant of the matrix. Thus, the matrices $\underline{A}$ and $\underline{A}'$ are said to be similar, and the operation $\underline{S}^{-1}\cdot\underline{A}\cdot\underline{S}$ is called a similarity transformation. A very important class of similarity transformations occurs when the matrix $\underline{S}$ belongs to a special class of orthogonal matrices defined by the property that $\underline{S}^T = \underline{S}^{-1}$. An important subset of orthogonal matrices is the rotation matrices $\underline{R}$ in three dimensions. An example is the rotation matrix $\underline{R}_z$ given by

$$\underline{R}_z = \begin{pmatrix} \cos\theta & -\sin\theta & 0 \\ \sin\theta & \cos\theta & 0 \\ 0 & 0 & 1 \end{pmatrix} \tag{A.19}$$

---

* *Note:* a_{ij}^{-1} is not $\dfrac{1}{a_{ij}}$. The superscript -1 is just a label in this case.

The reader can readily verify that $\underline{\mathbf{R}}_z \cdot \underline{\mathbf{R}}_z^T = \underline{\mathbf{I}}$, so $\underline{\mathbf{R}}_z$ is indeed orthogonal. $\underline{\mathbf{R}}_z$ also has the useful property that it will rotate any three-dimensional position vector (x, y, z) about the z axis by an angle θ. From Equation (A.19), one can easily construct companion matrices that will perform similar rotations about the x or y axis.

Indicial Notation

In indicial notation, the above formulae have compact representations. Given an array a_{ij}, it can be used to transform a vector b_j into another vector c_i by the simple formula $c_i = a_{ij} b_j$ where summation over the repeated index j is assumed. Likewise, an array b_{ij} can be transformed into another array d_{lk} by the following operation: $d_{lk} = a_{li} b_{ij} a_{jk}$ where again summation over the repeated indices i and j is assumed. As a matter of information, the indices that are summed over are called dummy indices in that one can use any symbol one likes to represent them because they go away after the summation is carried out, as opposed to the nonsummed indices, which remain and specify a specific matrix element. Note that in the preceding examples the range of the indices was not specified. Thus, the indicial formulae hold if we are talking about two-dimensional arrays or N-dimensional arrays, and the notation is fully compact regardless of the dimension.

We can also talk about entities of higher rank, such as the array b_{ijk}. This array can be transformed into another third-rank array by the formula $c_{onl} = a_{ok} a_{nj} a_{li} b_{ijk}$. It is clear that we do not have to stop at rank 3 because generalization to rank 4 and higher is obvious by inspection. Fortunately for this work, we rarely need to go higher than rank 2, but it is nice to know that the mathematical machinery is there to manipulate higher rank entities if the need arises. Before leaving the subject of matrix transformations, it should be noted that one can use the transformation properties of an array to define vectors, matrices, and higher rank entities. Thus, we can use the transformation $c_i = a_{ij} b_j$ as the basic definition of a vector. Any array that transforms according to this formula is a vector by definition. Likewise, matrices are defined by their transformation formula $d_{lk} = a_{li} b_{ij} a_{jk}$ and so on to all the higher rank quantities of interest.

The Eigenvalue Problem and Spectral Theorem

The reader has by now noticed that manipulating matrix arrays can be quite cumbersome to say the least, especially when dealing with objects of high order or high rank. Even if we stick to rank 2, a 3×3 matrix has 9 items to be dealt with in general. The 4×4 array has 16 elements, and it is clear that the number of elements is increasing as the square of the matrix order. However, there is a special class of rank 2 arrays called the diagonals, which greatly simplify matters. A diagonal array has nonzero elements only on the diagonal as follows:

$$\underline{\mathbf{D}} = \begin{pmatrix} d_{11} & 0 & 0 & 0 & \cdots \\ 0 & d_{22} & 0 & 0 & \cdots \\ 0 & 0 & d_{33} & 0 & \cdots \\ 0 & 0 & 0 & d_{44} & \cdots \\ \cdots & \cdots & \cdots & \cdots & d_{NN} \end{pmatrix} \qquad (A.20)$$

Aside from containing far fewer elements than a general array, the diagonals have a number of properties that make them far easier to manipulate than their more general cousins. In particular, they behave very much like simple scalars regarding elementary mathematical operations such as multiplication and finding the inverse. Thus, if $\underline{\mathbf{D}}_1$ and $\underline{\mathbf{D}}_2$ are distinct diagonal arrays of the same order, then the following are true:

1. $\underline{\mathbf{D}}_1$ and $\underline{\mathbf{D}}_2$ commute under multiplication, that is, $\underline{\mathbf{D}}_1 \cdot \underline{\mathbf{D}}_2 = \underline{\mathbf{D}}_2 \cdot \underline{\mathbf{D}}_1$.
2. If the array $\underline{\mathbf{D}}_1$ has elements $d_{11}, d_{22}, d_{33}, \ldots, d_{NN}$, then the inverse array $\underline{\mathbf{D}}_1^{-1}$ is obtained by simply taking the inverse of the individual elements $d_{11}^{-1}, d_{22}^{-1}, d_{33}^{-1}, \ldots, d_{NN}^{-1}$.* Note that this simple property is not true for general arrays. Even if the array is diagonal except for one off-diagonal element, this simple property is destroyed. Finding the inverse of a general array, assuming that it exists, is a doable but rather elaborate procedure covered in texts on linear algebra and numerical analysis.
3. Finding the value of a diagonal array raised to some power say $\underline{\mathbf{D}}_1^x$ where x is some number, not necessarily an integer, is as simple as raising each element of $\underline{\mathbf{D}}_1$ to the power x. Thus, for instance, one can obtain the square root of $\underline{\mathbf{D}}_1$ simply by taking the square root of each of its elements.

It needs to emphasized again that general matrix arrays do not share the nice properties listed above, but a large subset has the interesting property that they can be diagonalized by a similarity transformation as mentioned above. Thus, if $\underline{\mathbf{A}}$ is some general square array subject to fairly loose restrictions, then there exists an array $\underline{\mathbf{S}}$ such that $\underline{\mathbf{S}}^{-1} \cdot \underline{\mathbf{A}} \cdot \underline{\mathbf{S}} = \underline{\mathbf{D}}$ where the array $\underline{\mathbf{D}}$ is diagonal.** Finding the array $\underline{\mathbf{S}}$ for some given matrix is again a doable but rather involved problem in linear algebra and numerical analysis.[9] For present purposes, we need only know that quite generally we can find the diagonal representation of an array $\underline{\mathbf{A}}$. This can come in handy for carrying out tricky manipulations, such as finding the square root of an array. Given an array $\underline{\mathbf{A}}$, the square root is the array that when multiplied by itself gives $\underline{\mathbf{A}}$ back again. We suspect in general that such an array must exist, but finding such an array is nonobvious, to say the least. However, we know that our array has a diagonal representation given by $\underline{\mathbf{S}}^{-1} \cdot \underline{\mathbf{A}} \cdot \underline{\mathbf{S}} = \underline{\mathbf{D}}$. If we multiply on the right by $\underline{\mathbf{S}}$ and on the left by $\underline{\mathbf{S}}^{-1}$, we get the equivalent formula $\underline{\mathbf{A}} = \underline{\mathbf{S}} \cdot \underline{\mathbf{D}} \cdot \underline{\mathbf{S}}^{-1}$. Now, what if we were to take the square root of $\underline{\mathbf{D}}$, which is trivial as shown above, and form the array $\underline{\mathbf{S}} \cdot \underline{\mathbf{D}}^{1/2} \cdot \underline{\mathbf{S}}^{-1}$? Multiplying this array by itself gives $\underline{\mathbf{S}} \cdot \underline{\mathbf{D}}^{1/2} \cdot \underline{\mathbf{S}}^{-1} \cdot \underline{\mathbf{S}} \cdot \underline{\mathbf{D}}^{1/2} \cdot \underline{\mathbf{S}}^{-1} = \underline{\mathbf{S}} \cdot \underline{\mathbf{D}} \cdot \underline{\mathbf{S}}^{-1} = \underline{\mathbf{A}}$. Thus, the array $\underline{\mathbf{S}} \cdot \underline{\mathbf{D}}^{1/2} \cdot \underline{\mathbf{S}}$ is in essence the square root of $\underline{\mathbf{A}}$. It is clear that the ability to determine the diagonal representation of an array allows one to perform rather complex operations that might otherwise be undoable.

As mentioned, the process of finding the array $\underline{\mathbf{S}}$ that will reduce a given array $\underline{\mathbf{A}}$ to diagonal form via the similarity transformation $\underline{\mathbf{S}}^{-1} \cdot \underline{\mathbf{A}} \cdot \underline{\mathbf{S}}$ is an involved procedure best covered in a course on linear algebra and numerical analysis. However,

---

* *Note:* In this case, d_{11}^{-1} does equal $\dfrac{1}{d_{11}}$ and likewise for remaining elements.

** The mathematically sophisticated will note that for some arrays the best you can do is achieve what is called Jordan canonical form. Further discussion of this topic would lead us far afield.

for the sake of completeness, we outline some of the details of this process. It all starts with what is called the eigenvalue problem, by which one tries to find a vector **v** that satisfies the equation $\underline{\mathbf{A}}\mathbf{v} = \lambda\mathbf{v}$. A little reflection quickly reveals that any column of the array $\underline{\mathbf{S}}$ that diagonalizes $\underline{\mathbf{A}}$ will satisfy the eigenvalue problem, and in fact if $\underline{\mathbf{A}}$ is of order N, then there will be exactly N numbers λ_i* with associated eigenvectors $\mathbf{v}_i$ that solve the eigenvalue problem; the vectors $\mathbf{v}_i$ in fact form the columns of $\underline{\mathbf{S}}$, and the numbers λ_i, called the eigenvalues of $\underline{\mathbf{A}}$, are in fact the elements of the diagonal array $\underline{\mathbf{D}}$ that results from performing the similarity transformation on $\underline{\mathbf{A}}$.

Although powerful numerical methods exist for determining $\underline{\mathbf{S}}$ and the numbers λ_i, we simply outline the classical approach to this problem. The first step is to bring all the terms to the left-hand side and form the equivalent eigenvalue equation $(\underline{\mathbf{A}} - \lambda\underline{\mathbf{I}})\mathbf{v} = 0$. From linear algebra, one knows that this equation can have a nontrivial solution only if $\det(\underline{\mathbf{A}} - \lambda\underline{\mathbf{I}}) = 0$. On expanding the determinant, this equation gives rise to a scalar polynomial equation of the form

$$a_1\lambda^N + a_2\lambda^{N-1} + a_3\lambda^{N-2} + \dots a_N = 0 \qquad (A.21)$$

Equation (A.21) is called the secular equation for the eigenvalue problem and, because it is an Nth-order polynomial in λ, it is clear that in general there will be exactly N values of λ that satisfy it. These are of course the eigenvalues we seek, and they form the elements of the diagonal matrix $\underline{\mathbf{D}}$. Unfortunately, in practice the problem of smoking out the zeros of an Nth-order polynomial is another nontrivial problem best left to a course on numerical analysis. However, for low-order matrices, say four or fewer, the problem can be worked out by hand. Assuming we have the eigenvalues in hand, we can then go back to the relation $(\underline{\mathbf{A}} - \lambda\underline{\mathbf{I}})\mathbf{v} = 0$ to get the corresponding eigenvectors. Plugging in the eigenvalues one at a time, we get a set of linear equations for the components of the corresponding eigenvector. Because the problem is indeterminant, we do not get a unique solution but rather a set of relationships that express each of the first $N - 1$ elements in terms of the coefficients a_i, λ, and an undetermined Nth coefficient. Uniqueness can be obtained by imposing a further condition such as normalization, by which it is stipulated that $(\mathbf{v}\cdot\mathbf{v})^{1/2} = 1$. This, in principle at least, completes the solution of the problem. We conclude this section with Table A.1, which gives a short list of useful vector identities.

A.3   VECTOR CALCULUS

The vector calculus essentially comes down to three basic operations and three fundamental theorems, with the remainder essentially applications and extensions of these concepts. In what follows, the basic operations and theorems are explained,

---

* Not to dwell too much on detail, one should note that the λ_i need not all be distinct. It can happen that several of the λ_i are identical, in which case the matrix is said to be degenerate. This basically means that it can be represented in a lower dimensional space. The issue of degeneracy arises often in physical problems, for which certain symmetries effectively reduce the number of independent dimensions required to represent the problem.

TABLE A.1

Vector Identities[a]

Identity	Comments
$\mathbf{a} \cdot (\mathbf{b} \times \mathbf{c}) = \mathbf{c} \cdot (\mathbf{a} \times \mathbf{b}) = \mathbf{b} \cdot (\mathbf{c} \times \mathbf{a})$	The scalar–vector product can be cyclically permuted
$\mathbf{a} \times (\mathbf{b} \times \mathbf{c}) = (\mathbf{c} \times \mathbf{b}) \times \mathbf{a} = \mathbf{b}(\mathbf{a} \cdot \mathbf{c}) - \mathbf{c}(\mathbf{a} \cdot \mathbf{b})$	Double vector product best remembered as the bac minus cab rule
$\mathbf{a} \times (\mathbf{b} \times \mathbf{c}) + \mathbf{c} \times (\mathbf{a} \times \mathbf{b}) + \mathbf{b} \times (\mathbf{c} \times \mathbf{a}) = 0$	The sum of all the cyclic permutations of the vector double product comes to zero
$(\mathbf{a} \times \mathbf{b}) \cdot (\mathbf{c} \times \mathbf{d}) = (\mathbf{a} \cdot \mathbf{c})(\mathbf{b} \cdot \mathbf{d}) - (\mathbf{a} \cdot \mathbf{d})(\mathbf{b} \cdot \mathbf{c})$	Handy formula for reducing a complex vector/scalar product to a string of simple scalar products
$(\mathbf{a} \times \mathbf{b}) \times (\mathbf{c} \times \mathbf{d}) = ([\mathbf{a} \times \mathbf{b}] \cdot \mathbf{d})\mathbf{c} - ([\mathbf{a} \times \mathbf{b}] \cdot \mathbf{c})\mathbf{d}$	Vector triple product reduced to two simpler vector/scalar products

[a] In the table, $\mathbf{a}$, $\mathbf{b}$, $\mathbf{c}$, and $\mathbf{d}$ are assumed to be vector quantities.

and the remainder are summarized in Table A.2, which has useful formulas and identities.

A.3.1 FUNDAMENTAL OPERATIONS OF VECTOR CALCULUS

There are basically three operators in the vector calculus that appear repeatedly in a number of applications, ranging from electrostatics to fluid dynamics. In elementary calculus, one learns all about taking the derivatives of scalar functions of a single variable commonly represented by that standard chestnut $y = f(x)$ where x is considered the independent variable and y the dependent variable. The function f is some rule or formula that allows you to determine in an unambiguous way the value of y associated with any given value of x in the allowed range. A question that immediately comes up is, What happens if y depends on more than one independent variable? At this point, one jumps into the realm of vector calculus.

A.3.1.1 The Gradient Operator

The first question one might like to ask concerns the three-dimensional equivalent of the simple derivative in one dimension. To be more precise, if we have some function of three independent variables, say $f(x, y, z)$, how do we go about specifying its rate of change as we move about in three dimensions? The first problem we note is that, unlike the simple one-dimension problem in which one can in essence go in only one direction, in three dimensions there are an infinite number of different directions. Thus, the first thing we need to specify is in which direction we want to know the rate of change of our function. With these thoughts in mind, we formulate the problem in the following manner. We have some function f defined in a three-dimensional space in which each point (x, y, z) is represented by a radius vector $\mathbf{r}$ connecting (x, y, z) to the origin of some convenient Cartesian coordinate system. We let $d\mathbf{r}$ be some infinitesimal vector extending from $\mathbf{r}$ in the direction of interest.

The amount df by which the function f changes along the vector $d\mathbf{r}$ is given by the following formula:

$$df = \nabla f \cdot d\mathbf{r} \qquad\qquad (A.22)$$

where

$$\nabla f = \frac{\partial f}{\partial x}\mathbf{i} + \frac{\partial f}{\partial y}\mathbf{j} + \frac{\partial f}{\partial z}\mathbf{k}$$

As defined here, the gradient ∇ is a vector operator that converts a scalar function f into a vector. The scalar product of this vector with some infinitesimal $d\mathbf{r}$ gives the change of f in the direction $d\mathbf{r}$. One can easily show from the definition of the scalar product that the vector ∇f points in the direction of maximum change in the function f. This property is useful in numerical analysis when one is trying to find a local maximum or minimum of a function f. Starting at some point not too far away from the maximum or minimum, one first calculates the gradient of f. This then indicates the direction in which f is changing most rapidly and thus the direction in which one wants to proceed to locate the maximum or minimum that is sought.

Indicial Notation

The gradient of a function also has a compact representation in indicial notation using what is known as the comma convention. Thus, we define $(\nabla f)_i = f_{,i}\ i = 1, 2, 3$. The comma preceding the index i indicates differentiation with respect to the ith variable. Thus, $\partial f/\partial x$ would be represented as $f_{,1}$, and $\partial f/\partial y$ would be $f_{,2}$, and so on.

A.3.1.2 The Divergence Operator

The above discussion answers the question of how we compute the rate of change of a scalar function. The next obvious question is, What happens if we are dealing with a vector function? Say, instead of a scalar field such as the temperature, we are dealing with a vector field like the velocity of a fluid. The velocity of a fluid at any given point in space is specified by giving both a magnitude and a direction, and the question of how a vector field is varying in space starts to become more involved. There are in fact at least three different aspects to this question, with three different answers.

The first question gives a scalar answer, the second a vector answer, and the third a tensor answer. We focus now on the first question. If we have a fluid moving in space, we imagine that near any given point the velocity of the fluid is represented by some collection of arrows pointing in various directions that indicate both the magnitude and the direction of the flow. Say, we choose some point and enclose it with a small surface ΔS that is made up of a large number of infinitesimal patches of area dS. Each patch is further specified by the unit normal vector $\mathbf{n}$, which is perpendicular to the tangent plane at the patch dS and pointing outward. We now ask, What is the net flow into and out of the surface ΔS surrounding our point? Thus, if we take an inventory of all the patches that make up the surface ΔS, we find that the velocity vector at some patches is pointing into the surface ΔS, and some are pointing out; we want to know what is the net flow into or out of the surface. To answer this question, we sum the net flow into or out of the surface at each infinitesimal patch. Because we are only interested in that component of the velocity vector that is pointing either directly into or out of the surface, we first form the scalar product $\mathbf{v} \cdot \mathbf{n}$. Assuming our surface ΔS surrounds a volume ΔV, we get the following definition of the divergence of a vector field $\mathbf{v}$ at a point in space:

$$\nabla \cdot \mathbf{v} = \lim_{\Delta V \to 0} \frac{1}{\Delta V} \int_{\Delta S} \mathbf{v} \cdot \mathbf{n} \, dS \qquad (A.23)$$

where

$$\nabla \cdot \mathbf{v} = \frac{\partial v_x}{\partial x} + \frac{\partial v_y}{\partial y} + \frac{\partial v_z}{\partial z}$$

The similarity of Equation (A.23) to Equation (A.22) suggests that we can define a gradient operator as follows:

$$\nabla = \mathbf{i} \frac{\partial}{\partial x} + \mathbf{j} \frac{\partial}{\partial y} + \mathbf{k} \frac{\partial}{\partial z} \qquad (A.24)$$

Thus, when the ∇ operator operates on a scalar function one obtains a vector field that serves as a sort of three-dimensional derivative. On the other hand, the scalar product of ∇ with a vector yields a scalar field that is representative of the net flow of the vector quantity.

Indicial Notation

The divergence of a vector quantity in indicial notation v_i can be conveniently represented using the comma notation as $v_{i,i}$ where summation over the repeated index i is understood. Again, this is a very compact notation, but it can be hard on the eyes and sometimes ambiguous regarding whether summation over a repeated index is to be carried out. By and large, this notation is best left to those who must work with quantities of high rank or high-order arrays in which standard vector notation would be too cumbersome.

A.3.1.3 The Curl Operator

If we can take the scalar product of the ∇ operator with a vector and wind up with a scalar quantity called the divergence that is indicative of the net flow of a vector field into or out of a point, then surely we can cook up something interesting by taking the vector product of ∇ with a vector, but what would be the point? Yes, indeed, we can take the vector product of ∇ with some vector $\mathbf{v}$ and interestingly enough come up with a quantity that is both interesting and useful, which essentially amounts to the circulation or vorticity of the vector field. However, to understand what is going on we have to take a small detour and define what is known as a line integral.

Consider a point in our vector space lying on a small element of surface ΔS, which in the simplest case could be a disk but does not have to be. The boundary of ΔS is some curve in space that we label as C. When talking about the surface element ΔS, we specify a unit normal vector $\mathbf{n}$ perpendicular to ΔS that indicates its orientation. Likewise, the curve C is specified by a collection of unit vectors that are tangent to C as one goes around the entire circumference. These vectors we label as $\mathbf{t}$.* What we are interested in is the net sum of the vector field $\mathbf{v}$ starting from some arbitrary point on C and continuing all the way back to the starting point. Another way of putting it is as follows: At every point on the curve C, the vector field $\mathbf{v}$ has a certain magnitude and direction. What we want to know is what do we get if we add all such vectors. If the sum comes to some vector $\mathbf{q}$ with a scalar product with $\mathbf{n}$ that is positive, then we have a positive circulation of the vector field $\mathbf{v}$ operating within the surface element ΔS. If the scalar product is a negative number, then the circulation is termed negative. What this means physically is that some kind of vortex is operating within ΔS.

On the contrary, if the sum comes to zero, then there is no vorticity, and the vector field has no net circulation within ΔS. Think of the velocity field of water sucked down a drain hole as opposed the same water flowing through a straight pipe. The water going down the drain clearly has a net circulation, whereas the water going through the pipe does not. To put all this on a firm mathematical footing, we have to do some calculus. First, when adding all the vectors on the boundary curve C, we have to be careful to take only the component of $\mathbf{v}$ that is tangent to C. Thus, at each point we take the scalar product of $\mathbf{v}$ with the unit tangent vector $\mathbf{t}$, $(\mathbf{v}\cdot\mathbf{t})$, at each point to get just that component pointing along the curve and add this quantity all around the circumference of C. Symbolically, we get a formula like this:

$$circ = \oint_C \mathbf{v}\cdot\mathbf{t} \; ds \qquad (A.25)$$

The infinitesimal ds is a small element of the curve C so that Equation (A.25) essentially represents a standard one-dimensional integral of the scalar quantity $\mathbf{v}\cdot\mathbf{t}$ taken over the entire circumference of C. The loop on the integral sign is standard notation that the integration is to be carried out over a complete closed circuit. Now, by analogy with the fundamental definition of the divergence given in Equation (A.23), we can write the following definition of the curl as

---

* The directions of $\mathbf{n}$ and $\mathbf{t}$ follow the "right hand rule". Thus, take your right hand and point the thumb in the direction of $\mathbf{n}$. The fingers of the right hand now curl in the direction of the vector $\mathbf{t}$.

$$\mathbf{n}\cdot\nabla\times\mathbf{v}=\lim_{\Delta S\to 0}\frac{1}{\Delta S}\oint_C \mathbf{v}\cdot\mathbf{n}\;ds \tag{A.26}$$

where

$$\nabla x\mathbf{v}=\mathbf{i}\left(\frac{\partial v_z}{\partial y}-\frac{\partial v_y}{\partial z}\right)+\mathbf{j}\left(\frac{\partial v_x}{\partial z}-\frac{\partial v_z}{\partial x}\right)+\mathbf{k}\left(\frac{\partial v_y}{\partial x}-\frac{\partial v_x}{\partial y}\right)$$

By combining the definition of a vector cross product given in Equation (A.1) with the vector definition of the ∇ operator given in Equation (A.24), we get the following handy formula for $\nabla\times\mathbf{v}$ in the form of a determinant:

$$\nabla\times\mathbf{v}=\det\begin{bmatrix}\mathbf{i} & \mathbf{j} & \mathbf{k}\\[4pt] \dfrac{\partial}{\partial x} & \dfrac{\partial}{\partial y} & \dfrac{\partial}{\partial z}\\[6pt] v_1 & v_2 & v_3\end{bmatrix} \tag{A.27}$$

A.3.1.4 The Gradient of a Vector

Vector Notation

The final vector derivative we want to discuss is the gradient of a vector. Note that this is not going to be a scalar quantity like the divergence or a vector quantity like the curl but rather a higher-order tensor quantity. This operation popped up in Chapter 3, in which the infinitesimal strain tensor was defined in terms of the gradient of the displacement field $\mathbf{u}$. Using dyadic notation with the definition of the ∇ operator as given in Equation (A.22), we can easily conjure up the following definition for $\nabla\mathbf{u}$:

$$\nabla\mathbf{u}=\left(\frac{\partial}{\partial x}\,\mathbf{i}+\frac{\partial}{\partial y}\,\mathbf{j}+\frac{\partial}{\partial z}\,\mathbf{k}\right)(u_1\,\mathbf{i}+u_2\,\mathbf{j}+u_3\,\mathbf{k})$$

$$=\begin{aligned}&\frac{\partial u_1}{\partial x}\,\mathbf{ii}+\frac{\partial u_2}{\partial x}\,\mathbf{ij}+\frac{\partial u_3}{\partial x}\,\mathbf{ik}\\[6pt] &\frac{\partial u_1}{\partial y}\,\mathbf{ji}+\frac{\partial u_2}{\partial y}\,\mathbf{jj}+\frac{\partial u_3}{\partial y}\,\mathbf{jk}\\[6pt] &\frac{\partial u_1}{\partial z}\,\mathbf{ki}+\frac{\partial u_2}{\partial z}\,\mathbf{kj}+\frac{\partial u_3}{\partial z}\,\mathbf{kk}\end{aligned} \tag{A.28}$$

Matrix Notation

Clearly, we can also cast Equation (A.28) into the somewhat more convenient matrix notation. However, it helps to switch from using the standard (x, y, z) notation to represent a point in three space to the more matrix oriented (x_1, x_2, x_3) notation where the subscripts 1, 2, 3 stand in place of the usual x, y, z. Equation (A.28) now takes the obvious form

$$
\nabla \mathbf{u} = \begin{pmatrix} \dfrac{\partial u_1}{\partial x_1} + \dfrac{\partial u_2}{\partial x_1} + \dfrac{\partial u_3}{\partial x_1} \\[3mm] \dfrac{\partial u_1}{\partial x_2} + \dfrac{\partial u_2}{\partial x_2} + \dfrac{\partial u_3}{\partial x_2} \\[3mm] \dfrac{\partial u_1}{\partial x_3} + \dfrac{\partial u_2}{\partial x_3} + \dfrac{\partial u_3}{\partial x_3} \end{pmatrix} \tag{A.29}
$$

Indicial Notation

Using the comma notation discussed above, it is fairly easy to divine that the indicial notation for $\nabla \mathbf{u}$ will be $u_{i,j}$ where $i,j = 1, 2, 3, \ldots, N$. Again, the powerful indicial notation achieves incredible compactness and generality.

A.3.2 BASIC THEOREMS OF VECTOR CALCULUS

One of the fascinations of the vector calculus is the relationship between line integrals, surface integrals, and volume integrals, which involve the basic divergence and curl operations. The main purpose of this appendix is simply to write down the basic theorems for purposes of quick reference. The reader interested in a derivation should consult one of the many references available. The work of Schey[2] is particularly readable and highly recommended for the more applied reader who simply wants a better understanding of what is going on as opposed to establishing some sort of rigorous proof. The two basic theorems are the divergence theorem and Stokes' theorem, which are outlined next.

A.3.2.1 Divergence Theorem

Given some vector field $\mathbf{v}$ in three dimensions and assuming it is reasonably well behaved without sinister singularities or other kinds of devious behavior, we focus attention on a volume V that is enclosed by a surface S. We also assume that we can define a unit normal vector $\mathbf{n}$ everywhere on the surface S. With these quantities in hand, we can write down the divergence theorem as

Divergence Theorem

$$
\int_V \nabla \cdot \mathbf{v} \, dV = \int_S \mathbf{v} \cdot \mathbf{n} \, dS \tag{A.30}
$$

This theorem finds wide application in theoretical physics, ranging from electromagnetic theory to fluid dynamics. The main application is to convert a global integral formulation of the laws of physics into a local differential form.

A.3.2.2 Stokes Theorem

Given a vector field **v** in three dimensions and a surface S bounded by a curve C, we require that a unit normal **n** be well defined everywhere on the surface, and that the bounding curve C have a well-defined tangent **t** everywhere along its length. Assuming no abnormal behavior of any of the aforementioned quantities, we can write Stokes' theorem as

Stokes' Theorem

$$\int_S (\nabla \times \mathbf{v}) \cdot \mathbf{n} \, dS = \oint_C \mathbf{v} \cdot \mathbf{t} \, dl \qquad (A.31)$$

Stokes' theorem essentially provides the same function as the divergence theorem for the lower dimensions of lines and surfaces. Table A.2 presents useful theorems and identities and is provided as a convenient reference for use in short calculations and as an aid to digesting various and sundry theoretical derivations.

TABLE A.2
Useful Identities and Formulae of Vector/Tensor Calculus

Number	Formula/Identity	Comments
1	$\nabla[u(\mathbf{x})w(\mathbf{x})] = w(\mathbf{x})\nabla u(\mathbf{x}) + u(\mathbf{x})\nabla w(\mathbf{x})$	Gradient of product of two scalar functions of **x**. Note that this formula follows directly from the chain rule of standard one-dimensional calculus.
2	$\nabla \cdot (u\mathbf{v}) = u\nabla \cdot \mathbf{v} + \mathbf{v} \cdot \nabla u$	Divergence of a scalar function u times a vector function **v**
3	$\nabla \times (u\mathbf{v}) = u\nabla \times \mathbf{v} + \nabla u \times \mathbf{v}$	Curl of a scalar function u times a vector function **v**
4	$\nabla \cdot (\mathbf{g} \times \mathbf{v}) = \mathbf{v} \cdot \nabla \times \mathbf{g} - \mathbf{g} \cdot (\nabla \times \mathbf{v})$	Divergence of the cross product of two vectors **g** and **v**
5	$\nabla \times (\mathbf{g} \times \mathbf{v}) = \mathbf{g}(\nabla \cdot \mathbf{v}) - \mathbf{v}(\nabla \cdot \mathbf{g}) + (\mathbf{v} \cdot \nabla)\mathbf{g} - (\mathbf{g} \cdot \nabla)\mathbf{v}$	Curl of the cross product of two vectors **g** and **v**
6	$\mathbf{g} \times (\nabla \times \mathbf{v}) = (\nabla \mathbf{v}) \cdot \mathbf{g} - \mathbf{g} \cdot (\nabla \mathbf{v})$	Cross product of vector **g** with curl of **v**. Note that $\nabla \mathbf{v}$ is a second-rank tensor.
7	$\nabla(\mathbf{g} \cdot \mathbf{v}) = \mathbf{g} \times (\nabla \times \mathbf{v}) + \mathbf{v} \times (\nabla \times \mathbf{g}) + (\mathbf{g} \cdot \nabla)\mathbf{v} + (\mathbf{v} \cdot \nabla)\mathbf{g}$	Gradient of scalar product of two vectors.
8	$\nabla^2 w = \nabla \cdot \nabla w$	The divergence of the gradient of a scalar function w is the Laplace operator ∇^2 del squared.

TABLE A.2 (continued)
Useful Identities and Formulae of Vector/Tensor Calculus

Number	Formula/Identity	Comments
9	$\nabla \times (\nabla \times \mathbf{v}) = \nabla(\nabla \cdot \mathbf{v}) - \nabla^2 \mathbf{v}$	Double curl of a vector related to the gradient of the divergence and the Laplacian.
10	$\nabla \times (\nabla w) = 0$	The curl of the gradient of a scalar function is always 0. Foundational formula of potential theory.
11	$\nabla \cdot (\nabla \times \mathbf{v}) = 0$	The divergence of the curl of a vector field is always zero. Defines the vector potential in electromagnetic theory.
12	$(\nabla \cdot \underline{\mathbf{A}})_i = \sum_j \dfrac{\partial A_{ij}}{\partial x_j}$	Definition of the divergence of second-rank tensor $\underline{\mathbf{A}}$. Result is a vector field.
13	$\nabla \cdot (\mathbf{vu}) = (\nabla \cdot \mathbf{v})\mathbf{u} + (\mathbf{u} \cdot \nabla)\mathbf{v}$	Divergence of general tensor formed as composition of two vectors $\mathbf{vu}$.
14	$\nabla \cdot (w\underline{\mathbf{A}}) = (\nabla w) \cdot \underline{\mathbf{A}} + w(\nabla \cdot \underline{\mathbf{A}})$	Divergence of a scalar function w times a tensor $\underline{\mathbf{A}}$.
15	$\mathbf{r} = x\mathbf{i} + y\mathbf{j} + z\mathbf{k} = (x, y, z)$ Magnitude of $\mathbf{r} = \|\mathbf{r}\| = +\left(x^2 + y^2 + z^2\right)^{1/2}$	Definition of the position vector $\mathbf{r}$ in three dimensions. Next five formulae derive from this definition.
15a	$\nabla \cdot \mathbf{r} = 3$	Divergence of $\mathbf{r}$ is the dimension of the defining space.
15b	$\nabla \times \mathbf{r} = 0$	Curl of $\mathbf{r}$ is 0 because it is the gradient of the scalar function $1/2(x^2 + y^2 + z^2)$.
15c	$\nabla\|\mathbf{r}\| = \dfrac{\mathbf{r}}{\|\mathbf{r}\|}$	Gradient of the magnitude of $\mathbf{r}$ is a unit vector in direction of $\mathbf{r}$.
15d	$\nabla \dfrac{1}{\|\mathbf{r}\|} = -\dfrac{\mathbf{r}}{\|\mathbf{r}\|^3}$	Standard formula from potential theory.
15e	$\nabla^2 \dfrac{1}{\|\mathbf{r}\|} = \nabla \cdot \dfrac{\mathbf{r}}{\|\mathbf{r}\|^3} = 4\pi\delta(\mathbf{r})$	Laplacian of $1/\|\mathbf{r}\|$ is proportional to Dirac delta function. Thus, $1/\|\mathbf{r}\|$ is a Green's function for the Laplace operator.

TABLE A.2 (continued)
Useful Identities and Formulae of Vector/Tensor Calculus

Number	Formula/Identity	Comments
	The following is a list of formulae relating the integral of a scalar or vector function over some finite volume V of three-dimensional space to the integral of the same function over the bounding surface S. Typically, a vector derivative such as the gradient, divergence, or curl of the function within the volume is related to the surface integral of some product of the function with the surface normal unit vector $\mathbf{n}$. These formulae are typically variants of the divergence theorem.	
16	$$\int_V (\nabla w)dV = \int_S (w\,\mathbf{n})dS$$	Gradient theorem. Integral of ∇w over some volume V equals the surface integral of ($w\,\mathbf{n}$) over the bounding surface S.
17	$$\int_V (\nabla \cdot \mathbf{v})dV = \int_S (\mathbf{v}\cdot\mathbf{n})dS$$	Divergence theorem.
18	$$\int_V (\nabla \cdot \underline{\mathbf{A}})dV = \int_S (\underline{\mathbf{A}}\cdot\mathbf{n})dS$$	Divergence theorem also holds for second rank tensors $\underline{\mathbf{A}}$.
19	$$\int_V (\nabla \times \mathbf{v})dV = \int_S (\mathbf{n} \times \mathbf{v})dS$$	Curl theorem: Equivalent of the gradient theorem (16) for the curl operator.
20	$$\int_V (w\nabla^2 g - g\nabla^2 w)dV = \int_S \mathbf{n}\cdot(w\nabla g - g\nabla w)dS$$	Green's theorem. Derived from divergence theorem and widely used in theoretical mechanics.
21	$$\int_V \left[\mathbf{v}\cdot(\nabla\times\nabla\times\mathbf{u}) - \mathbf{u}\cdot(\nabla\times\nabla\times\mathbf{v})\right]dV =$$ $$\int_S \mathbf{n}\cdot(\mathbf{u}\times\nabla\times\mathbf{v} - \mathbf{v}\times\nabla\times\mathbf{u})dS$$	Generalized Green's theorem for the double-curl operator.
	The following formulae relate the integral of either a scalar or a vector function over some surface S embedded in three-dimensional space to the line integral of the same function over the curve C bounding the surface S. In all formulae, $\mathbf{t}$ is a unit vector tangent to the curve C, and $\mathbf{n}$ is a unit vector normal to the surface S. The infinitesimal dl is an infinitely small increment of the curve C, and w is some scalar function in three dimensions. These formulae are typically variants of Stoke's theorem outlined above.	
22	$$\int_S \mathbf{n}\times(\nabla w)dS = \oint_C w\mathbf{t}\,dl$$	Two-dimensional analogue of the gradient theorem in Formula 16.
23	$$\int_S \mathbf{n}\cdot(\nabla\times\mathbf{v})dS = \oint_C \mathbf{v}\cdot\mathbf{t}\,dl$$	Two-dimensional analogue of divergence theorem applied to curl operator.

TABLE A.2 (continued)
Useful Identities and Formulae of Vector/Tensor Calculus

Number	Formula/Identity	Comments
24	$$\int_S (\mathbf{n} \times \nabla) \times \mathbf{v})dS = \oint_C \mathbf{t} \times \mathbf{v}\, dl$$	Vector version of Formula 23.

The following formulae are handy when working in coordinate systems other than the standard rectangular Cartesian (x, y, z) system. Thus, problems exhibiting cylindrical symmetry are far more easily handled in cylindrical coordinates than in Cartesian. Likewise, spherical coordinates are much to be preferred for problems with spherical symmetry. The only fly in the ointment is that the standard vector derivatives of gradient, divergence, curl, and Laplacian have more involved representations in these systems. Standard procedures exist for transforming from Cartesian coordinates to either cylindrical or spherical coordinates, which although fairly straightforward, are nonetheless tedious to derive and subject to a number of slippery errors. It is best to carefully derive these formulae once and tabulate them for future reference as needed.

Vector Operators in Cylindrical Coordinates

Number	Formula/Identity	Comments
25	Definition of cylindrical coordinates in terms of standard cartesian (x, y, z) $$x = \rho\cos\phi$$ $$y = \rho\sin\phi$$ $$z = z$$	ρ = Radial coordinate extending from origin to a point in (x,y) plane. ϕ = Angular sweep of ρ from x axis. z = Identical to standard Cartesian z coordinate.
	Vectors $\mathbf{v}$ and tensors $\underline{\mathbf{A}}$ have the following components in cylindrical coordinates $$\mathbf{v} = \begin{pmatrix} v_\rho \\ v_\phi \\ v_z \end{pmatrix} \quad \underline{\mathbf{A}} = \begin{pmatrix} A_{\rho\rho} & A_{\rho\phi} & A_{\rho z} \\ A_{\phi\rho} & A_{\phi\phi} & A_{\phi z} \\ A_{z\rho} & A_{z\phi} & A_{zz} \end{pmatrix}$$	
26	$(\nabla w)_\rho = \dfrac{\partial w}{\partial \rho}$ Radial Component $(\nabla w)_\phi = \dfrac{1}{\rho}\dfrac{\partial w}{\partial \phi}$ Angular Component $(\nabla w)_z = \dfrac{\partial w}{\partial z}$ z Component	Gradient operator in cylindrical coordinates.

TABLE A.2 (continued)
Useful Identities and Formulae of Vector/Tensor Calculus

Number	Formula/Identity	Comments
27	$$\nabla \cdot \mathbf{v} = \frac{1}{\rho}\frac{\partial}{\partial \rho}(\rho v_\rho) + \frac{1}{\rho}\frac{\partial v_\phi}{\partial \phi} + \frac{\partial v_z}{\partial z}$$	Divergence of vector $\mathbf{v}$ in cylindrical coordinates where v_ρ = Radial component of $\mathbf{v}$. v_ϕ = ϕ component of $\mathbf{v}$. v_z = z component of $\mathbf{v}$.
28	$$(\nabla \times \mathbf{v})_\rho = \frac{1}{\rho}\frac{\partial v_\rho}{\partial \phi} - \frac{\partial v_\phi}{\partial z} \quad \text{Radial Component}$$ $$(\nabla \times \mathbf{v})_\phi = \frac{\partial v_\rho}{\partial z} - \frac{\partial v_z}{\partial r} \quad \text{Angular Component}$$ $$(\nabla \times \mathbf{v})_z = \frac{1}{\rho}\frac{\partial(\rho v_\phi)}{\partial \rho} - \frac{1}{\rho}\frac{\partial v_\rho}{\partial \phi} \quad \text{z Component}$$	Curl of vector $\mathbf{v}$ in cylindrical coordinates where v_ρ = Radial component of $\mathbf{v}$. v_ϕ = ϕ component of $\mathbf{v}$. v_z = z component of $\mathbf{v}$.
29	$$\nabla^2 w = \frac{1}{\rho}\frac{\partial}{\partial \rho}\left(\rho \frac{\partial w}{\partial \rho}\right) + \frac{1}{\rho^2}\frac{\partial^2 w}{\partial \phi^2} + \frac{\partial^2 w}{\partial z^2}$$	Laplacian of scalar function w in cylindrical coordinates.
30	$$(\nabla^2 \mathbf{v})_\rho = \nabla^2 v_\rho - \frac{2}{\rho^2}\frac{\partial v_\phi}{\partial \phi} - \frac{v_\rho}{\rho^2} \quad \text{Radial Component}$$ $$(\nabla^2 \mathbf{v})_\phi = \nabla^2 v_\phi + \frac{2}{\rho^2}\frac{\partial v_\rho}{\partial \phi} - \frac{v_\phi}{\rho^2} \quad \text{Angular Component}$$ $$(\nabla^2 \mathbf{v})_z = \nabla^2 v_z \quad \text{z Component}$$	Laplacian of vector $\mathbf{v}$ in cylindrical coordinates. Note that ∇^2 of the right-hand side is the full Laplace operator as defined in Formula 29.
31	$$(\nabla \cdot \underline{\mathbf{A}})_\rho = \frac{1}{\rho}\frac{\partial(\rho A_{\rho\rho})}{\partial \rho} + \frac{1}{\rho}\frac{\partial A_{\phi\rho}}{\partial \phi} + \frac{\partial A_{z\rho}}{\partial z} - \frac{1}{\rho}A_{\phi\phi}$$ $$(\nabla \cdot \underline{\mathbf{A}})_\phi = \frac{1}{\rho}\frac{\partial(\rho A_{\rho\phi})}{\partial \rho} + \frac{1}{\rho}\frac{\partial A_{\phi\phi}}{\partial \phi} + \frac{\partial A_{z\phi}}{\partial z} + \frac{1}{\rho}A_{\phi\rho}$$ $$(\nabla \cdot \underline{\mathbf{A}})_z = \frac{1}{\rho}\frac{\partial(\rho A_{\rho z})}{\partial \rho} + \frac{1}{\rho}\frac{\partial A_{\phi z}}{\partial \phi} + \frac{\partial A_{zz}}{\partial z}$$	Divergence of a tensor $\underline{\mathbf{A}}$ in cylindrical coordinates. Note that this is a vector quantity.

TABLE A.2 (continued)
Useful Identities and Formulae of Vector/Tensor Calculus

Number	Formula/Identity	Comments
	Vector Operators in Spherical Coordinates	
32	Definition of spherical coordinates in terms of standard Cartesian (x, y, z) $$x = r\cos\theta\cos\phi$$ $$y = r\cos\theta\sin\phi$$ $$z = r\sin\theta$$ Vectors v and tensors $\underline{\mathbf{A}}$ have the following components in spherical coordinates $$v = \begin{pmatrix} v_r \\ v_\theta \\ v_\phi \end{pmatrix} \quad \underline{\mathbf{A}} = \begin{pmatrix} A_{rr} & A_{r\theta} & A_{r\phi} \\ A_{\theta r} & A_{\theta\theta} & A_{\theta\phi} \\ A_{\phi r} & A_{\phi\theta} & A_{\phi\phi} \end{pmatrix}$$	r = Radial coordinate extending from origin to a point in (x, y, z) space. θ = Declination of radius vector from z axis. ϕ = Angular sweep of the projection of r on x,y plane from x axis. Also called azimuthal angle.[10]
33	$(\nabla w)_r = \dfrac{\partial w}{\partial r}$ Radial Component $(\nabla w)_\theta = \dfrac{1}{r}\dfrac{\partial w}{\partial\theta}$ θ Component $(\nabla w)_\phi = \dfrac{1}{r\sin\theta}\dfrac{\partial w}{\partial\phi}$ ϕ Component	Gradient operator in spherical coordinates.
34	$\nabla\cdot\mathbf{v} = \dfrac{1}{r^2}\dfrac{\partial}{\partial r}(r^2 v_r) + \dfrac{1}{r\sin\theta}\dfrac{\partial(v_\theta\sin\theta)}{\partial\theta} + \dfrac{1}{r\sin\theta}\dfrac{\partial v_\phi}{\partial\phi}$	Divergence operator in spherical coordinates where v_r = Radial component of **v**. v_θ = θ component of **v**. v_ϕ = ϕ component of **v**.
35	$(\nabla\times\mathbf{v})_r = \dfrac{1}{r\sin\theta}\dfrac{\partial(v_\phi\sin\theta)}{\partial\theta} - \dfrac{1}{r\sin\theta}\dfrac{\partial v_\theta}{\partial\phi}$ $(\nabla\times\mathbf{v})_\theta = \dfrac{1}{r\sin\theta}\dfrac{\partial v_r}{\partial\phi} - \dfrac{1}{r}\dfrac{\partial(rv_\phi)}{\partial r}$ $(\nabla\times\mathbf{v})_\phi = \dfrac{1}{r}\dfrac{\partial(rv_\theta)}{\partial r} - \dfrac{1}{r}\dfrac{\partial v_r}{\partial\theta}$	Curl operator in spherical coordinates where v_r = Radial component of **v**. v_θ = θ component of **v**. v_ϕ = ϕ component of **v**.

TABLE A.2 (continued)
Useful Identities and Formulae of Vector/Tensor Calculus

Number	Formula/Identity	Comments
36	$$\nabla^2 w = \frac{1}{r^2}\frac{\partial}{\partial r}\left(r^2\frac{\partial w}{\partial r}\right) + \frac{1}{r^2\sin\theta}\frac{\partial}{\partial\theta}\left(\sin\theta\frac{\partial w}{\partial\theta}\right) +$$ $$\frac{1}{r^2\sin^2\theta}\frac{\partial^2 w}{\partial\phi^2}$$	Laplacian of scalar function w in spherical coordinates.
37	$$(\nabla^2\mathbf{v})_r = \nabla^2 v_r - \frac{2}{r^2}\frac{\partial v_\theta}{\partial\theta} - 2\frac{v_r}{r^2} - \frac{2}{r^2\sin\theta}\frac{\partial v_\phi}{\partial\phi}$$ $$(\nabla^2\mathbf{v})_\theta = \nabla^2 v_\theta + \frac{2}{r^2}\frac{\partial v_r}{\partial\theta} - \frac{v_\theta}{r^2\sin^2\theta} - \frac{2\cos\theta}{r^2\sin^2\theta}\frac{\partial v_\phi}{\partial\phi}$$ $$(\nabla^2\mathbf{v})_\phi = \nabla^2 v_\phi - \frac{v_\phi}{r^2\sin^2\theta} + \frac{2}{r^2\sin^2\theta}\frac{\partial v_r}{\partial\phi}$$ $$+ \frac{2\cos\theta}{r^2\sin^2\theta}\frac{\partial v_\theta}{\partial\phi}$$	Laplacian of vector $\mathbf{v}$ in spherical coordinates. Note that ∇^2 of the right-hand side is the full Laplace operator as defined in Formula 36.
38	$$(\nabla\cdot\underline{\mathbf{A}})_r = \frac{1}{r^2}\frac{\partial(r^2 A_{rr})}{\partial r} + \frac{1}{r\sin\theta}\frac{\partial(A_{\theta r}\sin\theta)}{\partial\theta}$$ $$+ \frac{1}{r\sin\theta}\frac{\partial A_{\phi r}}{\partial\phi} - \frac{(A_{\theta\theta} + A_{\phi\phi})}{r}$$ $$(\nabla\cdot\underline{\mathbf{A}})_\theta = \frac{1}{r^2}\frac{\partial(r^2 A_{r\theta})}{\partial r} + \frac{1}{r\sin\theta}\frac{\partial(A_{\theta\theta}\sin\theta)}{\partial\theta}$$ $$+ \frac{1}{r\sin\theta}\frac{\partial A_{\phi\theta}}{\partial\phi} + \frac{1}{r}A_{\phi r} - \frac{\cot\theta}{r}A_{\phi\phi}$$ $$(\Delta\cdot\underline{\mathbf{A}})_\phi = \frac{1}{r^2}\frac{\partial(r^2 A_{r\phi})}{\partial r} + \frac{1}{r\sin\theta}\frac{\partial(A_{\theta\phi}\sin\theta)}{\partial\theta}$$ $$+ \frac{1}{r\sin\theta}\frac{\partial A_{\phi\phi}}{\partial\phi} + \frac{1}{r}A_{\phi r} + \frac{\cot\theta}{r}A_{\phi\theta}$$	Divergence of a tensor in spherical coordinates. Note that this is a vector quantity.

Notes

1. *The Feynman Lectures on Physics*, Vol. 2, R.P. Feynman, R.B. Leighton, and M. Sands (Addison-Wesley, Reading, MA, 1964), Chap. 2 and 3.
2. *Div Grad Curl and All That: An Informal Text on Vector Calculus*, 2nd ed., H. M. Schey (Norton, New York, 1992).
3. *Physics Vade Mecum*, H.L. Anderson, Ed. (American Institute of Physics, New York, 1981).
4. "The Linear Theory of Elasticity," M.E. Gurtin, in *Mechanics of Solids,* Vol. 2, C. Truesdell, Ed. (Springer-Verlag, Berlin, 1984).
5. *Introduction to Linear Elasticity*, P.L. Gould (Springer-Verlag, New York, 1983).

6. *Continuum Mechanics*, Schaum Outline Series, G.E. Mase (McGraw-Hill, New York, 1970).

7. The literature on linear algebra is truly vast. A sampling of vintage texts includes *Introduction to Matrices and Linear Transformations*, D.T. Finkbeiner II (Freeman, San Francisco, 1960); *Elementary Matrix Algebra*, 2nd ed., F.E. Hohn (Macmillan, New York, 1964); *Introduction to Linear Algebra*, M. Marcus and H. Minc (Macmillan, New York, 1965).

8. Those interested in solving systems of linear equations of high order should refer to a dedicated numerical analysis text. One that can be highly recommended is *Computer Methods for Mathematical Computations*, Prentice Hall Series in Automatic Computation, G.E. Forsythe, M.A. Malcom, and C.B. Moler (Prentice-Hall, Englewood Cliffs, NJ, 1977).

9. The reader interested in the numerical analysis aspects of the eigenvalue problem for real symmetric matrices can consult dedicated works such as *The Symmetric Eigenvalue Problem*, Prentice-Hall Series in Computational Mathematics, B.N. Parlett (Prentice-Hall, Englewood Cliffs, NJ, 1980).

10. The reader needs to be aware that, unlike the case of cylindrical coordinates, there is no consistency in the definition of spherical coordinates. Some texts interchange the roles of θ and ϕ, which will cause confusion when trying to compare results. In addition, the declination θ is sometimes replaced by its compliment $(\pi - \theta)$ and relabeled as θ.

Appendix B

Notes on Elementary Strength of Materials (SOM) Theory

From the earliest times when people started to build it was found necessary to have information regarding the strength of structural materials so that rules for determining safe dimensions could be drawn up. No doubt the Egyptians had some empirical rules of this kind, for without them it would have been impossible to erect their great monuments, temples, pyramids, and obelisks, some of which still exist.

Steven P. Timoshenko, *History of Strength of Materials*

As noted in Chapter 3, this volume tries to present the thermal-mechanical aspects of adhesion problems from the point of view of modern continuum theory. The main reason for this is that continuum mechanics covers a much wider range of problems and is essential for the introduction of fracture mechanics concepts that underlie all fully quantitative treatments of adhesion strength. However, the reader should be aware of the concepts and methods of basic strength of materials (SOM) theory for the following reasons:

1. Much of modern continuum theory is based on the earlier concepts derived in SOM theory.
2. Much of the early literature on adhesion measurement is based on SOM theory.
3. A large number of adhesion measurement techniques involve the use of simple beams, and these problems are most conveniently handled using SOM theory.

In what follows a cursory outline of the fundamental concepts and techniques of SOM theory will be given. The reader should note that this treatment is by no means intended to be comprehensive; those interested in a more detailed treatment should consult one of the standard texts on the subject, such as the classic volume by Den Hartog.*

---

* *Strength of Materials*, J.P. Den Hartog (Dover, New York, 1961).

B.1   BASIC CONCEPTS

SOM theory deals with the deformation of solid elastic bodies subjected to forces and torques. The forces and torques are exactly those derived in elementary physics courses and treated in detail in standard texts.* Forces are used for problems that are most conveniently cast in terms of linear Cartesian coordinates. A weight sitting on a table would be a common example. Problems involving the twisting of an object are best treated using angular coordinates; in this case, forces are replaced by torques. Other basic concepts include the notions of deformation, reaction, shear, and moment. Table B.1 gives a summary of the relevant quantities.

B.1.1 NOTION OF STATIC EQUILIBRIUM

The basic program of SOM theory is to determine the deformation of an object subjected to known loading conditions or inversely to determine the forces when the deformations are known. The essential assumption is that a state of equilibrium exists, and nothing is moving.** At equilibrium, the following two conditions must hold:

$$\sum_i F_i = 0 \quad \text{Lateral equilibrium}$$

$$\sum_i M_i = 0 \quad \text{Rotational equilibrium}$$

(B.1)

Equations (B.1) coupled with the definitions in Table B.1 provide the tools required for solving what are called statically determinate problems or, stated more specifically, problems that are completely determined by the conditions of static equilibrium, Equations (B.1). The broader class of SOM problems requires one additional ingredient, which was discussed in some detail in Chapter 3: the constitutive relation for the material body under investigation. In all cases of interest for simple beam-bending problems, the constitutive relation of choice is Hooke's law in its simplest form; that is,

$$\sigma = E\,\varepsilon$$

(B.2)

One should note at this point that, for simple bending problems, the Poisson ratio is assumed to be zero, which is generally consistent with the degree of approximation attainable by the elementary SOM method.

---

* An excellent text to consult for an introduction to elementary point mass mechanics is *Physics*, Parts I and II combined 3rd ed., D. Halliday and R. Resnick (John Wiley and Sons, New York, 1978).
** The case for which movement is occurring but only very slowly compared to any timescale of interest is referred to as a quasi-static equilibrium and is treated in the same manner as truly static equilibrium.

TABLE B.1
Basic Quantities of Strength of Materials Theory

Quantity	Definition	Units	Usual Symbol	Comments
Spatial extent	A coordinate measuring distance from one point to another	Meter (m)	x	This is the usual Cartesian coordinate in one dimension; what one normally measures with a ruler
Deformation	Primitive concept of change in spatial extent and form of a solid body	Meter (m)	u	Everyone is supposed to have an intuitive notion of a primitive concept such as deformation
Force	Primitive concept of the agent that causes deformation	Newton (N)	F	Everyone is supposed to have an intuitive notion of a primitive concept such as force
Reaction	Type of force exerted by a solid body that encounters an imposed external force	Newton (N)	R	A weight on a table exerts a downward force; the table exerts a reaction to counter this force; at equilibrium, nothing moves
Shear	Type of force that gives rise to a shear stress; typically a load applied at an angle to a surface	Newton (N)	S	Shear loads arise from a nonuniform distribution of forces
Torque	Force acting at a right angle to a lever arm	Newton-meter (N-m)	τ	Commonly applied to objects of circular/cylindrical symmetry such as a corkscrew
Moment	A type of torque induced in a body by external forces acting on lever arms	Newton-meter (N-m)	M	Roughly speaking, the moment is to the torque as the reaction is to the force
Stress	Force acting over an element of area	Newton/meter^2 (N/m^2)	σ	Involves both the direction of the force and the surface element and is thus a tensor quantity
Traction	Same as stress	Newton/meter^2 (N/m^2)	T	Tractions are always applied at the surface of a solid and give rise to stresses that exist within the body as a whole

TABLE B.1 (continued)
Basic Quantities of Strength of Materials Theory

Quantity	Definition	Units	Usual Symbol	Comments
Strain	Dimensionless measure of deformation in a solid body; defined as deformed length divided by original length	Dimensionless	ε	Strains exist when neighboring elements of a body are separated by some amount in distinction to a rigid translation or rotation
Modulus	A measure of the tensile stiffness of a solid	Newton/meter^2 (N/m^2); in SI units called a Pascal (Pa)	E	The constant of proportionality in Hooke's famous relation $\sigma = E\varepsilon$; note that 1 Pa is a very small stress, and thus one more often finds it multiplied by 10^6 (MPa) or 10^9 (GPa) to give reasonable numbers
Mass	Primitive property of all solids that couples them to Earth's gravitational field and instills an inherent resistance to any form of acceleration (inertia)	Kilogram (kg)	m	For static problems, the coupling to the Earth's gravity is most important and gives rise to static loads such as a weight sitting on a table
Density	Measure of how a body's mass is distributed over its volume	Kilogram/m^3 (kg/m^3)	ρ	Elementary theory most often assumes a constant uniform density
Moment of inertia	Measure of resistance to rotation or bending	Meter^4 (m^4)	I	Essentially the rotational analog of the mass used mostly for specifying the resistance of a beam to bending

B.1.2 EQUATIONS FOR BENDING OF SIMPLE BEAMS

In what follows the differential equations of equilibrium are derived for the bending deformation of a beam fabricated from a homogeneous isotropic material with an elastic modulus E and a uniform cross section with moment of inertia I. Figure B.1

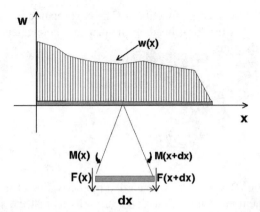

FIGURE B.1 An arbitrary continuous loading function w(x) applied to a beam. Note that w(x) has units force per unit length (N/m). Given this quantity and the beam parameters, one can integrate Equations (B.10) to determine the deflection of the beam and all other ancillary quantities, such as the shear stress and bending moment.

illustrates a uniform beam with some arbitrary loading per unit length $w(x)$.* The bottom of the figure shows a small element of the beam magnified for purposes of illustration. Note that, for the general loading condition portrayed, the loading force at the left end of the element is different from that for the right. This difference gives rise to an incremental shear load dS as follows:

$$dS = F(x+dx) - F(x) \qquad (B.3)$$

The usual procedure now is to use Taylor's theorem to expand $F(x + dx)$ in the incremental quantity dx:

$$dS = F(x) + \frac{dF}{dx}dx - F(x) = \frac{dF}{dx}dx = w(x)dx$$

$$dS = w(x)dx \qquad (B.4)$$

$$\frac{dS}{dx} = w(x)$$

The third relation in Equations (B.4) is the first differential equation for the beam pictured in Figure B.1 and simply states that the first differential of the shear load is given by the load per unit distance function. This can almost be taken as a

---

* The term $w(x)$ is the one-dimensional analog of a surface traction used to specify distributed loads. It has dimension force per unit length or N/m. It is often confused with point loads, which are simple forces with dimension N.

simple definition of the loading function w. The next relation is derived in a similar fashion by noting that the local bending moment on the element dx is simply the shear load times dx; thus,

$$dM = M(x + dx) - M(x) = Sdx$$

$$dM = Sdx \qquad (B.5)$$

$$S = \frac{dM}{dx}$$

From Equation (B.4), we determined that the derivative of the shear gives rise to the loading function w; now, Equation (B.5) tells us that the shear force is just the derivative of the bending moment. The picture is completed by relating the bending moment M to the resulting beam deflection u. This relationship is derived in Chapter 3, Equation (3.59), and is repeated here for convenience:

$$M = \frac{EI}{R} \qquad (B.6)$$

EI is commonly called the bending stiffness of the beam, and R is the local radius of curvature, also derived in Chapter 3, Equation (3.62), as follows:

$$\frac{1}{R} = \frac{d^2u}{dx^2} \qquad (B.7)$$

where u is the beam deflection. With Equations (B.7) and (B.6) in hand, we can now write everything in terms of the beam deflection, starting by substituting Equation (B.7) into Equation (B.6):

$$\frac{d^2u}{dx^2} = \frac{M}{EI} \qquad (B.8)$$

We can get a relationship between the deflection and the shear S by taking the derivative of both sides of Equation (B.8) and using Equation (B.5) to eliminate dM/dx in terms of S; thus,

$$\frac{d^3u}{dx^3} = \frac{S}{EI} \qquad (B.9)$$

Finally, we repeat the above procedure one more time and use Equation (B.4) to eliminate dS/dx.

For future reference, it is useful to collect all the results together:

a) $u = \left(\text{Beam deflection}\right)$

b) $\dfrac{d^2u}{dx^2} = \dfrac{M}{EI} \ \left(\text{Bending moment}\right)$

c) $\dfrac{d^3u}{dx^3} = \dfrac{S}{EI} \ \left(\text{Shear force}\right)$

d) $\dfrac{d^4u}{dx^4} = \dfrac{w}{EI} \ \left(\text{Loading function}\right)$

(B.10)

Equations (B.10) provide a general formulation for determining the deflection of a beam in simple bending in terms of the basic loads and beam parameters.

Appendix C

Material Property Data for Selected Substances

*Some curious and interesting phenomena presented themselves in the experiments —
many of them are anomalous to our preconceived notions of the strength of materials,
and totally different to anything yet exhibited in any previous research. It has invariably
been observed, that in almost every experiment the tubes gave evidence of weakness
in their powers of resistance on the top side, to the forces tending to crush them.*

**Comments of W. Fairbairn in regard to the construction of tubular bridges
in 19th century England (*An Account of the Construction
of the Britannia and Conway Tubular Bridges*)**

This appendix is intended to serve as a handy reference in support of "back-of-the-
envelope" calculations on the stresses and fracture properties of material structures.
It is often useful and timesaving to perform a few rough calculations to estimate the
size of the effect one is trying to understand, such as the stress level in a thin film
or the driving force for fracture behind a given structural defect, before going into
detailed calculations or setting up elaborate experiments. To do this, however, it is
very important to have reasonable estimates of the material properties involved.
When dealing with stress-related problems, the two main quantities of interest are
the Young's modulus and the Poisson ratio. Because thermal stresses are often the
underlying cause of many stress problems, it is also handy to have data on the
thermal expansion coefficient, the density, the specific heat, and the thermal conductivity. Data for this appendix have been gathered from various handbooks, texts,
and journal articles.[1–9] Typical accuracy is not expected to be better than to within
10% of a "true value." Careful work always requires that material properties be
measured directly on the materials of interest and fabricated under conditions as
close as possible to those seen in the actual manufacturing process. As an example
of why caution is always required, we can take the case of electro-deposited copper.
Standard handbook copper has a typical listed Young's modulus of about 110 GPa.
I have come across measured results on electro-deposited copper that gave a value
less than half of the generally accepted value. Close examination later revealed that
the electro-deposited material was highly porous, which explained the anomalously
low modulus data. However, used with appropriate care, the data listed in the

following table can provide some quick and useful insight into stress- and fracture-related problems.

C.1   USEFUL FORMULAE

Table C1 gives the elastic properties of all materials in terms of the common Young's modulus E and Poisson's ratio v. On occasion, it may be more useful to know the values of the following related quantities: shear modulus G, bulk modulus K, and Lame's constant λ. The following formulae can be used to determine these quantities if needed.

$$G = \frac{E}{2(1+v)} \quad \text{Shear modulus}$$

$$K = \frac{E}{3(1-2v)} \quad \text{Bulk modulus}$$

$$\lambda = \frac{vE}{(1+v)(1-2v)} \quad \text{Lame's constant}$$

TABLE C.1
Selected Material Properties of Common Materials at Room Temperature

Material	Young's Modulus (GPa)	Poisson Ratio	Density (kg/m³)	Thermal Expansion Coefficient (ppm/ K)	Specific Heat (J/kg K)	Thermal Conductivity [Watt/(m K)]
			Common Elements			
Aluminum (Al)	70	0.33	2,669	25	896	237
Antimony (Sb)	81.4	0.33	6,620	10.9	204	20.9
Beryllium (Be)	289	0.03	1,830	12.2	1,842	168
Bismuth (Bi)	31.4	0.33	9,780	13.2	125	8.38
Cadmium (Cd)	68.6	0.3	8,650	31	229	92.2
Chrome (Cr)	248	0.31	7,190	6	458	93.9
Cobalt (Co)	207		8,900	13.8	425	67
Copper (Cu)	110	0.36	8,960	16	383	398
Germanium (Ge)	110		5,320	6	304	6.28
Gold (Au)	74	0.42	19,320	14.2	129	315
Hafnium (Hf)	138		13,290	5.9	146	23
Indium (In)	10.8		7,310	33	238	71.1
Iridium (Ir)	518	0.26	22,400	6.7	133	59
Iron (cast) (Fe)	78.5	0.25	7,860	11.6	446	59
Lead (Pb)	15.7	0.43	11,340	29	129	33.5
Magnesium (Mg)	43.1	0.35	1,740	26.4	1,037	151
Molybdenum (Mo)	344	0.33	10,220	5	250	138

TABLE C.1 (continued)
Selected Material Properties of Common Materials

Material	Young's Modulus (GPa)	Poisson Ratio	Density (kg/m³)	Thermal Expansion Coefficient (ppm/ K)	Specific Heat (J/kg K)	Thermal Conductivity [Watt/(m K)]
Nickel (Ni)	206	0.31	890	13	442	90.9
Lead (Pb)	17.9	0.4	11,340	29	129	35.3
Osmium (Os)	552	0.24	22,500	6.2	129	59
Palladium (Pd)	117	0.39	12,020	11.6	242	70.4
Platinum (Pt)	167	0.38	21,400	9.2	133	67
Plutonium (Po)	98.6	0.18	19,800	47.0	142	67.4
Rhodium (Rh)	345	0.26	12,440	8.4	246	88
Ruthenium (Ru)	410	0.3	12,200	6.4	234	117
Silicon (Si)	150	0.06	2,330	3	675	149
Silver (Ag)	72	0.37	10,500	19	237.5	427
Tin (Sn)	41	0.33	7,298	20	225	64.2
Tantalum (Ta)	186	0.35	16,600	6.5	150	57.5
Titanium (Ti)	106	0.33	4,540	8.5	525	21.9
Tungsten (W)	344	0.28	19,300	4.5	133	178
Uranium (U)	165	0.22	18,700	14.9	117	27.6
Vanadium (V)	133	0.26	5,900	7.9	487	30.7
Zinc (Zn)	88.3	0.3	6,920	26.3	387	126
Zirconium (Zr)	100		6,440	5.7	276	22.7
Common Compounds						
Boron nitride (BN) (in plane)	85		2,300	7.51	800–2000	15.1
Boron nitride (BN) (out of plane)	34	0.25	2,300	3.1	800–2000	28.9
Silica (SiO₂)	70	0.14	2,200	0.55	787	1.8
Glass-ceramic	130	0.25	2,560	2.8		5
Granite	30	0.3	2,700	8.4	812	188
Generic glass	60	0.23	2,500	9	750	83
Steel	220	0.28	7,800	11.6	479	46
Polymers						
Natural rubber	10.5×10^{-4}	0.49	910	670 (volume)	1,700	0.153
Polyethylene (low density)	0.2	0.49	920	237 (linear), 710 (volume)	1,550	0.16
Poly(tetrafluoro-ethylene) Teflon	1.13	0.46	2,000	124	1,250	0.4
Nylon	1.9	0.44	1,140	71 (linear)	1,590	0.19
Poly(methyl-methacralate) (PMMA)	3.2	0.4	1,170	580 (volume)	1,370	0.150
Polystyrene	3.4	0.38	1,050	510 (volume)	1,230	0.131

TABLE C.1 (continued)
Selected Material Properties of Common Materials

Material	Young's Modulus (GPa)	Poisson Ratio	Density (kg/m³)	Thermal Expansion Coefficient (ppm/ K)	Specific Heat (J/kg K)	Thermal Conductivity [Watt/(m K)]
Poly(vinyl chloride)	2.96	0.38	1,393	70 (linear)	1352	0.16
Epoxy	2.5	0.4	1,180	7.93	1264	0.682
Polyimide (generic) (pyromellitic-dianhydride-oxydianiline) (PMDA-ODA)	3.0	0.3	1,420	30 (coated thin film)	1090	0.1–0.35
Polyimide (orthotropic low TCE spun-coated film) (BPDA-PDA)	14	0.3 (in plane), 0.08 (out of plane)		6 (in plane), 93 (out of plane)		
Refractory Materials						
Tungsten silicide (WSi$_2$)	296	0.22	9,250	8.41	332	31.1
Alumina (Al$_2$O$_3$)	324	0.24	3,980	6	1,050	25
Sintered alumina (Al$_2$O$_3$)	324	0.24	3,600	6	1,050	25
Diamond	1000	0.07	3,515	1.5	516	700–1600
Titanium diboride (TiB$_2$)	367		4,520	5		24
Zirconium diboride (ZrB$_2$)	400	0.144	6,090	5.5		24
Boron carbide (B$_4$C)	450	0.17	2,520	5	1,000	29–67
Silicon carbide (SiC)	483	0.16	3,200	1.1	708	25.5
Aluminum nitride (AlN)	308	0.21–0.24	3,260	2.7	1,041	180–220
Silicon nitride (Si$_3$N$_4$)	300	0.24	3,190	2.8	708	1.46

Notes

1. *Handbook of Chemistry and Physics*, R.C. Weast, Ed. (CRC Press, Boca Raton, FL). Because this volume originally focused on the properties of organic liquids, it was affectionately known as the "Rubber Bible" among nonchemists. The recently published 86th edition, edited by David R. Lide; is a virtual treasure trove of physical property data on practically all materials.
2. *The Nature and Properties of Engineering Materials*, Z.D. Jastrzebski (John Wiley & Sons, New York, 1976).
3. *Engineering Materials Handbook*, C. Mantel (McGraw-Hill, New York, 1958).
4. *Physical Properties of Crystals*, J.F. Nye (Oxford Press, London, 1967).
5. *Handbook of Material Science*, D.F. Miner and J.B. Seastone (John Wiley & Sons, New York, 1955).
6. "Calculated Elastic Constants for Stress Problems," W.A. Brantley, *Journal Applied Physics*, 44, 1 (1973).
7. *Properties of Polymers: Their Estimation and Correlation with Chemical Structure*, D.W. van Krevelen and P.J. Hoftyzer (Elsevier, Amsterdam, 1976).
8. *Handbook of Physical Calculations*, J.J. Tuma (McGraw-Hill, New York, 1983).
9. *Physics Vade Mecum*, H.L. Anderson, Ed. (American Institute of Physics, New York, 1981).
10. *Polymer Handbook*, J. Brandrup and E.H. Immergut, Eds. (John Wiley and Sons, New York, 1975).

Appendix D
Driving Force Formulae for a Variety of Laminate Structures

He had a singular distaste for mathematical studies, and never made himself acquainted with the elements of geometry; so remarkable indeed was this peculiarity, that when we had occasion to recommend to him a young friend as a neophyte in his office, and founded our recommendation on his having distinguished himself in mathematics, he did not hesitate to say that he considered such acquirements as rather disqualifying than fitting him for the situation.

Comments of D. Brewster on Thomas Telford, 18th–19th century English bridge builder, many of whose structures stand to this day

Brewster's comments on the attitude of pioneer bridge builder Thomas Telford, although archaic in terms of modern engineering practice, still have relevance to those would try to anticipate delamination and fracture phenomena solely on the basis of a mathematical analysis. Professor J. E. Gordon goes on to explain that

We must be clear that what Telford and his colleagues were objecting to was not a numerate approach as such—they were at least as anxious as anybody else to know what forces were acting on their materials—but rather the means of arriving at these figures. They felt that theoreticians were too often blinded by the elegance of their methods to the neglect of their assumptions, so that they produced the right answer to the wrong sum.[1]

With this caveat in mind, we give a table of formulae that can be useful in estimating the driving force for various failure modes that are common in laminate structures. For coatings in general, there can be a wide range of failure phenomena, including surface cracks, channeling, blistering, spalling, edge delamination, and so on, all caused by an excessive level of residual stress in the coating. In an extensive review article, Hutchinson and Suo[2] pointed out that, for many of the failure modes, the strain energy release rate G (also called the driving force) has the following simplified form in terms of the stress level, the coating thickness h, and the coating elastic properties (E and v):

$$G = Z \frac{(1-v)\sigma^2 h}{E} \qquad (D.1)$$

The right-hand side of Equation (D.1) is proportional to the elastic strain energy per unit area stored in the coating. Different failure modes are distinguished by the numerical factor Z, which can be easily tabulated. If we know the critical value of the driving force G_c* at which the failure mode in question starts to propagate, then Equation (D.1) implies that, for a given value of the film stress and coating elastic properties, the maximum stable coating thickness will be given by

$$h_c = \frac{G_c E}{Z(1-v)\sigma^2} \qquad (D.2)$$

If one has data on G_c gathered by any of the quantitative methods outlined in Chapter 1, then Equation (D.2) can be used to obtain a quick estimate of the maximum stable film thickness that can be attained for a given level of film stress. Table D.1 gives a list of Z factors for a variety of common coating failure processes.

---

* In the literature, when referring to delamination processes G_c is commonly called the *surface fracture energy* and given the symbol γ.

TABLE D.1
Numerical Factors Z for Estimating the Driving Force G for a Variety of Coating/Substrate Failure Modes Where G Is Given by the Formula

$$G = Z \frac{(1-\nu)\sigma^2 h}{E}$$

	Z Factor	Graphic Configuration	Comments
1	0.5		Edge delamination; substrate assumed perfectly rigid and crack length at steady state limit
2	$(1 + \nu)/2$		Remaining undelaminated ligament in superlayer test[3]; require b/h >> 1
3	$1.2\pi a/h$		Through fracture; coating and substrate have close elastic properties; steady-state conditions assumed[4]
4	3.951		Surface cracking; substrate assumed perfectly rigid[2]

TABLE D.1 (continued)
Numerical Factors Z for Estimating the Driving Force G for a Variety of Coating/Substrate Failure Modes Where G Is Given by the Formula

$$G = Z\frac{(1-v)\sigma^2 h}{E}$$

	Z Factor	Graphic Configuration	Comments
5	1.976		Channeling behavior[2]
6	3.951		Substrate penetration[2]
7	0.343		Delamination followed by subsequent substrate penetration and lifting; spalling[2]
8	1.028 (initiation) 0.5 (steady state)		Edge delamination from a long cut; no substrate penetration[2]
9	0.83		Delamination kinking into substrate; elastic properties of coating and substrate assumed close and long delamination length

TABLE D.1 (continued)
Numerical Factors Z for Estimating the Driving Force G for a Variety of Coating/Substrate Failure Modes Where G Is Given by the Formula

$$G = Z \frac{(1-v)\sigma^2 h}{E}$$

	Z Factor	Graphic Configuration	Comments
10	0.5 (initiation)		Fracture in substrate below interface; very short crack initiation assumed
11	$\dfrac{2(1+v)}{\left[1+v+\left(\dfrac{a}{R}\right)^2 (1-v)\right]^2}$		Axisymmetric edge delamination[5]
12	$\dfrac{2(1+v)}{\left[1+v+\left(\dfrac{a}{R}\right)^2 (1-v)\right]^2}$		Axisymmetric delamination from circular hole[4]
13	$\dfrac{\pi}{4(1-v)}$; $\quad \dfrac{a}{h} \to \infty$ $\dfrac{2}{\pi(1-v)}$; $\quad \dfrac{a}{h} = 1$		Flaw of length a in adhesive layer of thickness h between two rigid substrates under conditions of biaxial plane stress σ^2

TABLE D.1 (continued)
Numerical Factors Z for Estimating the Driving Force G for a Variety
of Coating/Substrate Failure Modes Where G Is Given by the Formula

$$G = Z \frac{(1-\nu)\sigma^2 h}{E}$$

	Z Factor	Graphic Configuration	Comments
14	$\left[1 - \left(\dfrac{\sigma_G}{\sigma}\right)\right]\left[1 + 3\left(\dfrac{\sigma_c}{\sigma}\right)\right]\dfrac{(1+\nu)}{2}$ $\sigma_c = \dfrac{\pi^2 E}{12(1-\nu^2)}\left(\dfrac{h}{b}\right)^2$		Long strip buckle delamination caused by compressive stress σ; buckled region extends into page for long distance compared to width b; σ_c is the Euler buckling stress for coating of thickness h^2
15	$\dfrac{1 - (\sigma_c/\sigma)^2}{1 + 0.9021(1-\nu)}$ $\sigma_c = \dfrac{1.2235E}{(1-\nu^2)}\left(\dfrac{h}{R}\right)^2$		Circular buckle in biaxially compressed film; formula is asymptotic for the case $(\sigma/\sigma_c) \gg 1$; σ_c is the Euler buckling stress for coating of thickness h^2

Notes

1. *Structures or Why Things Don't Fall Down*, J.E. Gordon (DaCapo Press, New York, 1978).
2. "Mixed Mode Cracking in Layered Materials," J.W. Hutchinson and Z. Suo, *Advances in Applied Mechanics*, 29, 63 (1991).
3. Details of this test are discussed in a review by A.G. Evans, "Interface Adhesion: Measurement and Analysis," in *Adhesion Measurement of Films and Coatings*, Vol. 2, K.L. Mittal, Ed. (VSP, Utrecht, The Netherlands, 2001), p. 9.
4. "The Decohesion of Thin Brittle Films From Brittle Substrates," M.S. Hu, M.D. Thouless, and A.G. Evans, *Acta Metallica*, 36, 1301 (1988).
5. "Decohesion of Films With Axisymmetric Geometries," M.D. Thouless, *Acta Metallica*, 36, 3131 (1988).

Appendix E

Selected References and Commentary on Adhesion Measurement and Film Stress Literature

E.1   GENERAL REFERENCES

1. *Adhesion Aspects of Polymeric Coatings*, K.L. Mittal, Ed. (Plenum Press, New York, 1983). Good general overview of adhesion problems as they arise in the general area of coating technology, including paints and photoresists. Broad discussion given of issues concerning bond durability, coupling agents, and adhesion measurement methods.

2. "Selected Bibliography on Adhesion Measurement of Films and Coatings," K.L. Mittal, *Journal of Adhesion Science and Technology*, 1, 247 (1987). Comprehensive list of publications on adhesion measurement dating from 1934 to 1987. Divided into books, review articles, and original research articles, it covers nearly all of the first five decades of work in this field. It is fairly safe to say that if it is not in this bibliography, it probably was not published in the open literature.

3. *Adhesion Measurement of Thin Films, Thick Films and Bulk Coatings*, ASTM 640, K.L. Mittal, Ed. (American Society for Testing and Materials, Philadelphia, PA, 1978). Records the proceedings of what was perhaps the first international symposium on adhesion measurement testing. Most of the articles deal with problems of the then rapidly growing microelectronics industry; however, the techniques discussed have wide applicability to the general field of thin films and coatings. This work is as relevant and useful today as it was over 20 years ago.

4. *Adhesion Measurement of Films and Coatings*, K.L. Mittal, Ed. (VSP, Utrecht, The Netherlands, 1995). This volume is the logical follow-up to the 1978 ASTM meeting and could in a sense be called the "Second International Meeting" on adhesion measurement methods. As usual, many problems of direct interest to the microelectronics industry are covered. By this time, microchips had shrunk to the size at which one could say that at least half of the structure consisted of surfaces and interfaces. However, many new topics emerged that did not exist at the time of the first meeting, including treatment of diamond and diamondlike films, thermodynamics of peel testing, finite element stress analysis applied to adhesion problems, laser spallation, acoustic wave sensors, and internal friction measurements for nondestructive adhesion testing.

5. *Adhesion Measurement of Films and Coatings*, Vol. 2, K.L. Mittal, Ed. (VSP, Utrecht, The Netherlands, 2001). This is the third volume in the series documenting the international symposia on adhesion measurement, which began with the 1978 ASTM publication. In this volume, the ever-popular scratch test receives extensive coverage. Applications involving microelectronic structures also continue to dominate the proceedings.

6. *Handbook of Adhesion*, 2nd ed., D.E. Packham, Ed. (John Wiley & Sons, Chichester, U.K., 2005). This is a handy encyclopedia covering nearly all aspects of adhesion measurement plus several articles on adhesives and the phenomenon of adhesion in general. Depth of coverage is limited as would be expected given the broad score of the volume, but it is a good place to start if looking for a given topic on adhesion.

7. *Molecular Adhesion and Its Applications: The Stickey Universe*, K. Kendall (Kluwer Academic/Plenum Publishers, New York, 2001). This is a truly eclectic work covering nearly the entire range of adhesion phenomena, including particle and colloid adhesion, bioadhesion, thin film adhesion, laminates, adhesive joints, and composites. The treatment is more on the intuitive and phenomenological level with a minimum of mathematics. Although this work does not treat the problem of adhesion measurement in much detail, it is nonetheless an excellent overview of the field of adhesion in general and would be well recommended to the beginning student. The author has done significant original work in the field of particle adhesion and represents the K in the JKR theory of particle adhesion.

8. *Adhesion and Adhesives Technology*, A.V. Pocius (Carl Hanser Verlag, Munich, 2002). This volume is similar to the work of Kendall but more heavily slanted toward industrial applications. Again, the mathematical treatment is quite limited and at an elementary level. The coverage is fairly broad, with chapters covering material properties, adhesion measurement, surface science, and surface modification methods. A fairly in-depth treatment of adhesive technology is covered. A good volume for the novice to start, especially if the interest is more toward practical problems of industry.

9. *The Mechanics of Adhesion*, D.A. Dillard and A.V. Pocius, Eds. (Elsevier, Amsterdam, 2002). This volume is a general collection of articles dealing with a number of the more important problems related to the mechanics of adhesion and adhesion measurement. Topics covered include fracture mechanics, adhesive joints, mechanics of coatings, rheological issues, composites, and bond inspection methods. The authors are all distinguished investigators in the field of adhesion and mechanics. The level of presentation is definitely advanced and would be recommended for the graduate student contemplating original thesis work.

E.2 SELECTED REFERENCES ON ADHESION MEASUREMENT METHODS

E.2.1 BLISTER TEST

10. "Adhesion (Fracture Energy) of Electropolymerized Poly(n-octyl maleimide-co-styrene) on Copper Substrates Using a Constrained Blister Test," J.-L. Liang, J.P. Bell, and A. Mehta, in *Adhesion of Films and Coatings*, K.L. Mittal, Ed. (VSP, Utrecht, The Netherlands, 1995), p. 249. Illustrates the use of the constrained blister test on polymer adhesion to various copper substrates.

11. "A Study of the Fracture Efficiency Parameter of Blister Tests for Films and Coatings," Y.-H. Lai and D. Dillard, in *Adhesion Measurement of Films and Coatings*, K.L.

Mittal, Ed. (VSP, Utrecht, The Netherlands, 1995). Gives an extensive study of the mechanics of the blister test, covering the standard blister and the Island blister tests.

E.2.2 SCRATCH TEST

The literature on the scratch test starts officially with the work of Heavens:

12. "Some Factors Influencing the Adhesion of Films Produced by Vacuum Evaporation," O.S. Heavens, *Journal of Physics Radium*, 11, 355 (1950).

and of Heavens and Collins:

13. "L'Epitaxie dans les Lames Polycrystallines," O.S. Heavens and L.E. Collins, *Journal of Physics Radium*, 13, 658 (1952).

These authors performed some of the first studies on the adhesion of evaporated metals to glass and the effect of chromium interlayers. The first critical analysis of the scratch test is universally attributed to Benjamin and Weaver:

14. "Measurement of Adhesion of Thin Film," P. Benjamin and C. Weaver, *Proceedings of the Royal Society (London)*, 254A, 163 (1960).

These authors concluded that the critical load required to cleanly scratch the coating off the substrate or to create a so-called clear channel was an accurate indicator of the strength of the adhesion of the coating to the substrate.

Subsequent workers found that this was true only under rare circumstances. In particular:

15. "Hardness and Adhesion Filmed Structures as Determined by the Scratch Technique," J. Ahn, K.L. Mittal, and R.H. MacQueen, *Adhesion Measurement of Thin Films, Thick Coatings and Bulk Coatings*, ASTM Special Technical Publication 640, K.L. Mittal, Ed. (American Society for Testing and Materials, Philadelphia, PA, 1978), p. 134.

Ahn et al. concluded that the complexities of the scratch test preclude any simple correlation between the critical load and the coating adhesion. They did show, however, how the scratch test can be used for a useful qualitative evaluation of coating adhesion. In addition, Oroshnik and Croll

16. "Threshold Adhesion Failure: An Approach to Aluminum Thin-Film Adhesion Measurement Using the Stylus Method," J. Oroshnik and W.K. Croll, *Adhesion Measurement of Thin Films, Thick Coatings and Bulk Coatings*, ASTM Special Technical Publication 640, K.L. Mittal, Ed. (American Society for Testing and Materials, Philadelphia, PA, 1978), p. 158

modified the Benjamin and Weaver approach by looking at what they called the threshold adhesion failure (TAF) concept. These authors noted that before one could clearly observe complete removal of the coating, local spots of apparent clean delamination appeared. The appearance of these isolated spots of clean coating removal was taken as an indication of the threshold of adhesion failure. They further observed that the load at which this phenomenon occurred could be used as a consistent semiquantitative measure of adhesion failure.

More recently, Ollivier and Matthews

17. "Adhesion of Diamond-Like Carbon Films on Polymers: An Assessment of the Validity of the Scratch Test Technique Applied to Flexible Substrates," B. Ollivier and A. Matthews, in *Adhesion Measurement of Films and Coatings*, K.L. Mittal, Ed. (VSP, Utrecht, The Netherlands, 1995), p. 103.

applied the scratch test to the study of the adhesion of diamondlike carbon films to flexible polymer substrates. The last sentence of their article seems to sum up the situation: "The scratch test appears to be a good tool for semiquantitative comparisons if both the critical load and the surface 'hardness' of the sample are taken into account for the calculation of interfacial shear strength values."

More advanced applications of the scratch test combine a standard apparatus with acoustic emission equipment as in the following:

18. "Scratch Indentation, a Simple Adhesion Test Method for Thin Films on Polymeric Supports," G.D. Vaughn, B.G. Frushour, and W.C. Dale, in *Adhesion Measurement of Films and Coatings*, K.L. Mittal, Ed. (VSP, Utrecht, The Netherlands, 1995), p. 127

E.2.3   INDENTATION DEBONDING TEST

A close relative of the scratch test is the indentation experiment. Much the same type of experimental setup is used as for the scratch test except that only a simple indentation is made, and there is no need for the stylus to be moved over the sample surface. An estimate of the coating adhesion can be made by the elementary analysis given in

19. "Indentation-Debonding Test for Adhered Thin Polymer Layers," P.A. Engle and G.C. Pedroza, in *Adhesion Aspects of Polymeric Coatings*, K.L. Mittal, Ed. (Plenum Press, New York, 1983), p. 583.

The following authors used the indentation test to study the adhesion of epoxy to copper in laminated circuit boards. The fact that the indentation area can be made quite small makes this test amenable to real-time inspection during the manufacturing process.

20. "Use of Microindentation Technique for Determining Interfacial Fracture Energy," L.G. Rosenfeld, J.E. Ritter, T.J. Lardner, and M.R. Lin, *Journal of Applied Physics*, 67, 3291 (1990). This article is essentially an application of the indentation technique to epoxy on glass. The use of a microindenter is a technical innovation allowing one to deal with very small indentation spots. A nice feature of this article is that it gives a comparison of results among the microindentation, double cantilevered beam, and four-point flexure tests.

21. "Mechanics of the Indentation Test and Its Use to Assess the Adhesion of Polymeric Coatings," R. Jaychandran, M.C. Boyce, and A.S. Argon, in *Adhesion Measurement of Films and Coatings*, K.L. Mittal, Ed. (VSP, Utrecht, The Netherlands, 1995), p. 189. These authors gave a fairly heavy-duty analysis of the indentation test using the high-power tools of continuum theory and finite element analysis. An excellent reference for anyone who really wants to delve into the mechanics of the indentation process. This is not recommended for beginners.

E.2.4 Scotch Tape Test

The Scotch tape test has been useful in assessing the adhesion of inks to paper. A nice treatment is given by

22. "Quantifying the Tape Adhesion Test," G.V. Calder, F.C. Hansen, and A. Parra, in *Adhesion Aspects of Polymeric Coatings*, K.L. Mittal, Ed. (Plenum Press, New York, 1983), p. 569.

These authors introduced the use of the spectrophotometer and advanced statistical methods to eliminate operator bias in evaluating the results of the test. The method proved effective in quantitatively comparing the adhesion of different ink formulations.

E.2.5 Laser Spallation

23. "Recent Developments in the Laser Spallation Technique to Measure the Interface Strength and Its Relationship to Interface Toughness With Applications to Metal/Ceramic, Ceramic/Ceramic and Ceramic/Polymer Interfaces," V. Gupta, J. Yuan, and A. Pronin, in *Adhesion Measurement of Films and Coatings*, K.L. Mittal, Ed. (VSP, Utrecht, The Netherlands, 1995). This chapter gives a detailed study of the laser spallation technique, including the innovative use of laser interferometry to monitor in real time the deformation of the sample surface caused by the laser pulse. Extensive results are presented on systems of practical interest.

E.2.6 Selected References on Mechanics of Peel Test

24. "Theory of Peeling Through a Hookean Solid," J.J. Bikerman, *Journal of Applied Physics*, 28, 1484 (1957). One of the classic articles in peel test mechanics using simple linear elasticity theory. Unfortunately. few if any real systems are perfectly elastic; however, this short and precise article provides a good starting point for later work.
25. "Theory and Analysis of Peel Adhesion: Mechanisms and Mechanics," D.H. Kaelble, *Transactions of the Society of Rheology*, 3, 161 (1959). Another of the classic works on the mechanics of peel testing, as with the work of Bikerman, the analysis is essentially elastic, avoiding the difficult complications of viscous and plastic flow.
26. "Film Structure and Adhesion," T.R. Bullett, *Journal of Adhesion*, 4, 73 (1972). This work studied the effect of residual stress buildup in paint films because of solvent loss. It points out the effect of film formation and the development of mechanical properties on stress buildup and adhesion strength.
27. "Shrinkage and Peel Strength of Adhesive Joints," K. Kendall, *Journal of Physics, D: Applied Physics*, 6, 1782 (1973). This is one of the earliest articles to discuss the effect of residual shrinkage stress on peel strength for the rubber-on-glass system.

28. "Peel Mechanics for an Elastic-Plastic Adherend," A.N. Gent and G.R. Hamed, *Journal of Applied Polymer Science*, 21, 2817 (1977). One of the earliest studies of peel mechanics to address the problem of elastic-plastic behavior of the peel strip, it studied the effect of peel angle and film thickness on peel strength for Mylar films. It gives elementary but detailed analysis of the stress distribution and yielding behavior in the peel bend region.

29. "An Experimental Partitioning of the Mechanical Energy Expended During Peel Testing," R.J. Farris and J.L. Goldfarb, *Journal of Adhesion Science and Technology*, 7, 853 (1993). This work was based on a unique deformation calorimeter that can follow heat flow while simultaneously deforming the sample. Applied to the peel test, it gives the only data of which I am aware that shows how the work expended by the tensile test machine is divided among irreversible elastoplastic/viscoelastic deformation, internal strain energy of the peel strip, and the work of adhesion required to remove the coating from the substrate.

E.2.7 Nondestructive Methods

30. "Nondestructive Dynamic Evaluation of Thin NiTi Film Adhesion," Q. Su, S.Z. Hua, and M. Wuttig, in *Adhesion Measurement of Films and Coatings*, K.L. Mittal, Ed. (VSP, Utrecht, The Netherlands, 1995). This chapter gives an account of a unique approach to coating adhesion that involves the shift in dynamic modulus of a coated reed sample caused by weakness in the adhesion of the coating to the underlying reed.

31. "Adhesive Joint Characterization by Ultrasonic Surface and Interface Waves," S.I Rokhlin, in *Adhesive Joints: Formation, Characterization, and Testing*, K.L. Mittal, Ed. (Plenum Press, New York, 1984). This is one of the earliest works to apply the full range of surface acoustic wave (SAW) phenomena to the problem of evaluating adhesion behavior. The chapter explores the use of surface acoustic waves, leaky waves, and Raleigh waves to evaluate the adhesion of coatings and the viscoelastic properties of adhesives used in adhesive joints. This is an important and interesting work because viscoelastic effects dominate the measured strength of most adhesive joints.

32. "Ultrasonic Methods," R.B. Thompson, W.-Y. Lu, and A.V. Clarke, Jr., in *Handbook of Measurement of Residual Streses*, J. Lu, Ed. (Society for Experimental Mechanics, published by Fairmont Press, Lilburn, GA, 1996).

33. "Adhesion Studies of Polyimide Films Using a Surface Acoustic Wave Sensor," D.W. Galipeau, J.F. Vetelino, and C. Feger, in *Adhesion Measurement of Films and Coatings*, K.L. Mittal, Ed. (VSP, Utrecht, The Netherlands, 1995). This chapter applied the surface acoustic wave method to polyamide films on silica glass and emphasized the effect of relative humidity and adhesion promoters.

E.3 SELECTED REFERENCES ON STRESSES IN LAMINATE STRUCTURES AND COATINGS

34. "Film Formation and Mudcracking in Latex Coatings," G.P. Bierwagen, *Journal of Coatings Technology*, 51, 117 (1979). This article took a detailed look at stress buildup in latex films caused by the solvent drying process and was one of the first articles to look at the time dependence of stress buildup during the drying process.

35. "Theory of Unsymmetric Laminated Plates," T.S. Chow, *Journal of Applied Physics*, 46, 219 (1975). Gives an analytical treatment of the problem of multilevel laminates. On reading this article, you will appreciate why this type of calculation is now done numerically on computers.

36. "Strains Induced Upon Drying Thin Films Solvent-Cast on Flexible Substrates," T.S. Chow, C.A. Liu, and R.C. Penwell, *Journal of Polymer Science: Polymer Physics Edition*, 14, 1311 (1976). This treatment of solvent drying stresses for a coating on a flexible substrate is of significant interest in xerography technology.

E.4 SELECTED REFERENCES ON FRACTURE MECHANICS AS RELATED TO PROBLEMS OF ADHESION OF FILMS AND COATINGS

37. "A Path Independent Integral and the Approximate Analysis of Strain Concentration by Notches and Cracks," J.R. Rice, *Journal of Applied Mechanics*, 35, 379 (1968). This is one of the seminal articles in the modern theory of fracture mechanics and introduced for the first time the mysterious J integral. It is not recommended for beginners or those whose interests are purely practical. However, it is nice to know where these things come from and who to blame for the subsequent obscurity of the fracture mechanics literature.

38. *Elementary Engineering Fracture Mechanics*, 3rd rev. ed., D. Broek (Martinus Nijhoff, Dordrecht, The Netherlands, 1982). As the title implies, this is a text for the practicing engineer. It is a good place to start if you are looking for answers to practical questions; it devotes an entire chapter to the phenomenology of material fracture processes in addition to the usual introductory theory.

39. *Advanced Fracture Mechanics*, M.F. Kannien and C.H. Popelar (Oxford University Press, Oxford, UK, 1985). This is the next step up from Broek's text for those who wish to push further and deeper into the arcana of fracture mechanics. Nonlinear phenomena such as viscoplasticity receive extensive treatment. Also, the very difficult field of dynamic fracture, the time evolution of crack propagation, is treated in some detail. Fortunately, for most practical purposes the realm of dynamic fracture is not of that much interest because the main goal is to prevent a crack from starting in the first place rather than worry about its time evolution.

40. "Mixed Mode Cracking in Layered Materials," J.W. Hutchinson, *Advances in Applied Mechanics*, 29, 63 (1991). A monumental and useful reference in the literature on stress-related failure of coatings either by delamination or cracking. Hutchinson unified the entire field of stress-related coating failure within the formalism of fracture mechanics. Although this is definitely not a work for beginners, the practical engineer with some background in fracture mechanics will find a wealth of useful formulae for estimating the driving force for cracking or delamination for the most common coating failure modes.

E.5 SELECTED REFERENCES ON STRESSES IN SOLIDS

41. "*Structures or Why Things Don't Fall Down*," J.E. Gordon (Da Capo Press, New York, 1978). An excellent reference for the beginner who would like to gain some understanding of stresses in solids and their consequences for structures made out of those solids, it is well written, understandable, and entertaining to boot. The discussion alone of the problem of shear stress in wings that led to severe stability problems in early aircraft is worth the "price of admission."

42. *The New Science of Strong Materials or Why You Don't Fall Through the Floor*, J.E. Gordon (Princeton University Press, Princeton, NJ, 1984). Another fine introductory book by Prof. Gordon, the same comments apply here as above. In this book, he

gives a fascinating account of the British *Comet* aircraft disaster. The British *Comet* was the first commercial jet aircraft. However, because of a failure to fully comprehend the nature of the material stresses in the structure, the wings fell off during a maiden flight, leading to the untimely demise of what were perhaps the first jet-setters. This incident set the British aircraft industry back by several years, thus allowing American companies to take the lead. If ever there were a practical consequence of not properly understanding the stresses in a structure, this has to be it.

43. "Beams and Plates on Elastic Foundations And Related Problems," M. Hetényi, *Applied Mechanics Reviews*, 19, 95 (1966). A classic review article on the problem of elastic foundations, the ubiquitous "Winkler foundation" is covered, which is important for subsequent analysis of the peel test. It is recommended only for those who want to probe deeper into the mechanics of peel testing.

44. *Mechanics of Solids: Volume I: The Experimental Foundations of Solid Mechanics; Volume II: Linear Theories of Elasticity and Thermoelasticity — Linear and Nonlinear Theories of Rods, Plates and Shells; Volume III: Theory of Viscoelasticity, Plasticity, Elastic Waves and Elastic Stability; Volume IV: Waves in Elastic and Viscoelastic Solids (Theory and Experiment)*, C. Truesdell, Ed. (Springer-Verlag Berlin, 1984). This four-volume work is clearly intended only for those who are ready to do some very heavy lifting in the field of theoretical and applied mechanics. We list it here only for the truly dedicated who want to go deeper into the foundations of continuum theory.

Appendix F
General Adhesion Measurement References

F.1   REVIEW ARTICLES

1. "Selected Bibliography on Adhesion Measurement of Films and Coatings," K.L. Mittal, *JAST*, 1, 247 (1987).
2. "A Review of Adhesion Test Methods for Thin Hard Coatings," J. Valli, *J. Vac. Sci. Technol.*, A4, 3007 (November/December 1986).
3. "Assessment of Coating Adhesion," J. Valli, *Surface Eng.*, 2(1), 49 (1986).
4. "Adhesion: A Review of Various Test Methods," John E. Fitzwater, Jr., *J Waterborne Coatings*, 8, 2 (1985).
5. "Practical Adhesion Testing," J.B. Mohler, *Metal Finishing*, 71 (September 1983).
6. "Adhesion Testing," G.A. DiBari, *Metal Finishing Guidebook*, 370 (1983).
7. "Simple Adhesion Tests," B. Hantschke, *Mappe,* 103(12), 31 (1983).
8. "Survey of Methods for Measuring Adhesion," M.N. Sathyanarayana, *J. Colour Soc.*, 21, 11 (1982).
9. "Adhesion," H.K. Pilker, A.J. Perry, and R. Berger, *Surface Technol.*, 14, 25 (1981).
10. "Measurement of Adhesion of Organic Coatings," M.N. Sathyanarayana, P.S. Sampathkumaran, and M.A. Sivasamban, *Paintindia, Annu.*, 96 (1981).
11. "Techniques for Quantitatively Measuring Adhesion of Coatings," J.W. Dini and H.R. Johnson, *Metal Finishing*, 42 (March 1977) and 48 (April 1977).
12. "Adhesion Measurement of Thin Films," K.L. Mittal, *Electrocomponent Sci. Technol.*, 3, 21 (1976).
13. J. Savage, in *Handbook of Thick Film Technology*, P.J. Holmes and R.G. Loasby, Eds. (Electrochemical Publications, Ayr, Scotland, 1976), p. 108.
14. "A Critical Appraisal of the Methods for Measuring Adhesion of Electrodeposited Coatings," K.L. Mittal, in *Properties of Electrodeposits: Their Measurement and Significance*, R. Sard, H. Leidheiser, Jr., and F. Ogburn, Eds. (Electrochemical Society, Princeton, NJ, 1975), p. 273.
15. "Adhesion of Thin Films," C. Weaver, *J. Vac. Sci. Technol.*, 12, 18 (1975).
16. "Thin Film Adhesion," B.N. Chapman, *J. Vac. Sci Technol.*, 11, 106 (1974).
17. "Measurement of Adhesion," F.H. Reid, in *Gold Plated Technology*, F.H. Reid and W.W. Goldie, Eds. (Electrochemical Publications, Ayr, Scotland, 1974), p. 422.
18. "Existing State of Testing Methods for Adhesion of Organic Coatings," S. Minami, *Shikizai Kyokaishi*, 45, 370 (1972).
19. "Adhesion of Metals to Polymers," C. Weaver, *Faraday Special Discussion*, 2, 18 (1972).

20. "The Measurement of Adhesion," T.R. Bullet and J.L. Prosser, *Prog. Org. Coatings*, 1, 45 (1972).

21. "Adhesion," E.M. Corcorn, in *Paint Testing Manual*, 13th ed., G.G. Sward, Ed. (American Society for Testing and Materials, Philadelphia, PA, 1972), Chap. 5.3, p. 314.

22. "Thick Film Conductor Adhesion Testing," R.P. Anjard, *Microelectronics and Reliability*, 10(4), 269 (1971).

23. "Mechanical Properties of Thin Films," D.S. Campbell, in *Handbook of Thin Film Technology*, L.I. Maissel and R. Glang, Eds., (McGraw Hill, New York, 1970), Chap. 12.

24. "Adhesion of Coatings," A.F. Lewis and L.J. Forrestal, in *Treatise on Coatings*, Vol. 2, Part 1, R.R. Myers and J.S. Long, Eds., (Marcel Dekker, New York, 1969), p. 57.

25. "The Adhesion of Metal Films to Surfaces," C. Weaver, *Chem. Ind.*, 370 (February 27, 1965).

26. "Adhesion of Electrodeposits," A.T. Vagramyan and Z.A. Soloveva, *Electroplat. Metal Finish.*, 15, 84 (1962).

27. "Adhesion of Thin Films," C. Weaver, in *Proc. of the First International Conference on Vacuum Techniques,* Namur, Belgium, 1958, Vol. 2, (Pergamon Press, London, 1960), p. 734.

F.2   GENERAL ADHESION BOOKS/PAPERS

28. *Adhesion and Adhesives Technology*, A.V. Pocius (Carl Hanser Verlag, Munich, 2002), p. 186.

29. "Adhesion Strength of Indium to Glass and a Method of Measurement of It," A.M. Magomedov, *Strength Materials* [English translation], 18(3), 369 (1986).

30. "Practical Measurements of Adhesion and Strain for Improved Optical Coatings," S.D. Jacobs, J.E. Hayden, and A.L. Hrycin, *Proc. SPIE*, 678, 66 (1986).

31. "Testing Adhesion of Coating Materials According to DIN55991, Part 1," F. Rissel and D.J.H. Scheff, *Farbe Lack*, 92, 582 (1986).

32. "Assessment of the Adhesion of Plasma Sprayed Coatings," C.C. Berndt and R. McPherson, *Surfacing J. Int.*, 1, 49 (1986).

33. "About Testing Coating Adhesion," L. Durney, *Prod. Finish.*, 50(12), 74 (1986).

34. "Semiquantitative Method for Thin Film Adhesion Measurement," B.-Y. Ting, W.O. Winer, and S. Ramalingham, *J. Tribology Trans. AIME*, 107(4), 472 (1985).

35. "Experimental Determination of Adhesion: Factors Influencing Measurement," A.A. Roche, *Couches Minces*, 40, 511 (1985).

36. "Method for Determining the Adhesive Strength of Coatings Produced by Electrodeposition," V.M. Antonov, V.K. Gorshkov, and A.L. Sibirev, *Lakorus. Mater. Ikh Primen*, p. 44 (1985).

37. "A Device for Determining the Adhesion Strength of Films on Elastic Substrates," V.S. Shkoylar, A.I. Razinkov, and A.V. Korneev, *Ind. Lab.*, 51(9), 863 (1985).

38. "Assessing Measurement Standards for Coating Adhesion to Plastics," K.N. Gray, S.E. Buckley, and G.L. Nelson, *Modern Paint Coatings*, 75, 160 (1985).

39. "An Acceleration Method for Measuring the Adherence of Plasma-Sprayed Coatings at Elevated Temperatures," S.D. Brown and K.M. Ferber, *J. Vac. Sci. Technol.*, A3(6), 2506 (1985).

40. "Rigs for Testing the Adhesion Strength Between Coatings and Metal," V.E. Ved, Y.A. Gusev, A.I. Skripka, and F.Y. Kharitonov, *Industrial Lab.*, 51(2), 188 (1985).

41. "Apparatus for Testing Antireflection Coating Hardness and Contact Metallization Adhesion," R.K. Dhingra, H.L. Chhatwal, and T.R.S. Reddy, *Research and Industry*, 30(4), 366 (1985).

42. "Evaluating the Adhesion of Glow-discharge Plasma Polymerized Films by a Novel Voltage Cyclic Technique," M.F. Nichols, A.W. Hahn, W.J. James, A.K. Sharma, and H. Yasuda, *J. Appl. Polymer Sci. Appl. Polymer Symp.*, 38, 21 (1984).

43. "Effect of Soldering Techniques and Pad Geometry," P.J. Whalen and J.B. Blum, *Intl. J. Hybrid Microelectronics*, 7(3), 72 (1984).

44. "Adhesion: Theories and Experimental Measurements," A.A. Roche and M.J. Romand, *Double Liason-Chim. Peint.*, 31, 503 (1984).

45. "Universal Test for Adhesion of Thin Films," Y.N. Borisenko, V.T. Gritsyna, N.A. Kasatkina, and N.I. Polyakov, *Industrial Lab.*, 49(12), 1274 (1983).

46. "Adhesion Testing of Paints," E.E. Michaelis, *Paint and Resin*, 53(3), 30 (1983).

47. "New Insights Into the Process of Adhesion Measurement and the Interactions at Polymer/Substrate Interfaces," U. Zoril, *J. Oil Colour Chem. Assoc.*, 66(7), 193 (1983).

48. "Measurement of the Adhesion of Evaporated Metal Film," A. Kikuchi, S. Baba, and A. Kinbara, *J. Vac. Sci. Technol.*, 25(4), 258 (1982).

49. "New Knowledge on Film Formation and Adhesion of Paints," W. Brushwell, *Farbe+Lack (Stuttgart)*, 88(5), 375 (1982).

50. "Rational Testing of Coatings," H. Schmidt, *Metalloberflaeche (Munchen)*, 36(6), 275 (1982).

51. "Coating Update VI. Review of Literature on Adhesion," W. Brushwell, *Mod. Paints Coatings*, 66(4), 61 (1982).

52. "Ways of Checking the Adherence of Electrodeposits," S. Sekowski, *Mechanik*, 55, 295 (1982).

53. "Adhesion Testing," J.B. Mohler, *Metal Finishing*, 51 (November 1981).

54. "A Plating Adhesion Testing Procedure," S. Murata, *Industrial Finishing*, 28 (January 1980).

55. "Some Aspects of Measurement of Adhesion," A.N. Gent, *Rubber World*, 180(3), 33 (1979).

56. "Testing the Adhesion of Thin Films Using Real-Time Holographic Interferometry: A Simple Method," B.S. Ramprasad and T.S. Radha, *Applied Optics*, 17, 2670 (1978).

57. "Experimental Aspects of Adhesion Testing. An IUPAC Study," J. Sickfeld, *J. Oil Colour Chem. Assoc.*, 61, 292 (1978).

58. "Method for Measuring Chalk Adhesion of Latex Paints," O.E. Brown and K.L. Hoy, *J. Coatings Technol.*, 37 (1977).

59. "A Fracture Approach to Thick Film Adhesion Measurement," W.D. Bascom and J.L. Bitner, *J. Mat. Sci.*, 12, 1401 (1977).

60. "The Adhesion of Thin Carbon Films to Metallic Substrates," H.S. Shim, N.K. Agarwal, and A.D. Haubold, *J. Bioengineering*, 1, 45 (1976).

61. "Methods and Apparatus for the Study of Adhesion of Polymers," I. Novak and A. Romanov, *Chemicke Listy*, 70, 250 (1976).

62. "The Adhesion of Metal Films in Glass and Magnesium Oxide in Tangential Shear," D.S. Lin, *J. Phys. (D): Appl. Phys.*, 4, 1977 (1971).

63. "The Continuum Interpretation for Fracture and Adhesion," M.L. Williams, *J. Appl. Polymer Sci.*, 13, 29 (1969).

F.3 ACOUSTIC EMISSION/ULTRASONIC METHODS

64. Ultrasonic Characterization of the Interface Between a Die Attach Adhesive and a Copper Leadframe in IC Packaging," N. Guo, J. Abdul, H. Du, and B.S. Wong, *J. Adhesion Sci. Technol.*, 16(9), 1261 (2002).

65. "Evaluation of the Polymer-Nonpolymer Adhesion in Particle Filled Polymers With the Acoustic Emission Method," S. Minko, A. Karl, A. Voronov, V. Senkovskii, T. Pomper, W. Wilke, H. Malz, and J. Pionteck, *J. Adhesion Sci. Technol.*, 14(8), 999 (2000).

66. "The Influence of the Bevel Angle on the Micro-mechanical Behavior of Bonded Scarf Joints," A. Objois, B. Fargette, and Y. Gilbert, *J. Adhesion Sci. Technol.*, 14(8), 1057 (2000) (acoustic emission, strain gauge measurement).

67. "Ultrasonic Evaluation of Adhesive Bond Degradation by Detection of the Onset of Nonlinear Behavior," Z. Tang, A. Cheng, and J.D. Achenbach, *J. Adhesion Sci. Technol.*, 13(7), 837 (1999).

68. "Use of Acoustic Emission Analysis in Studying the Adhesion of Variously Pigmented Laquers," G.M. Leibinger and H.G. Mosle, *Metalloberflaeche*, 39, 257 (1985).

69. "Adhesion Measurement of Thin Metal Films on Plastics," R.E. Van de Leest, *Thin Solid Films*, 124, 335 (1985).

70. "Monitoring of Adhesion Strength Between Materials by Photo-acoustic Pulse Generation at or Near the Interface," H. Coufal, S. Guerriero, S. Lam, and A. Tam, *IBM Technical Disclosure Bull.*, 28(2), 487 (1985).

71. "Testing of the Adhesive Strength of Metal Coatings by Using an Ultrasonic Coupling Oscillator," K. Stallman, F.W. Firth, and H. Speckhardt, *Z. Werkstofftech*, 15(7), 250 (1984).

72. "Ultrasonic Inspection Potential for Polymeric Coatings," M.S. Good, J.B. Nestleroth, and J.L. Rose, in *Adhesion Aspects of Polymeric Coatings*, K.L. Mittal, Ed. (Plenum Press, New York, 1983), p. 623.

73. "Application of Acoustic Emission Analysis to Adhesion Tests of Paints on Steel. II. Polyester and Acrylate Coatings," H. Hansmann and H.G. Mosle, *Adhäsion*, 26(8-9), 18 (1982).

74. "Film Adhesion Studies With the Acoustic Microscope," R.C. Bray, C.F. Quate, J. Calhoun, and R. Koch, *Thin Solid Films*, 74, 295 (1980).

75. "Determination of Adhesion of Metallic Films on Elastic Polymer Base," V. Kadek, T. Rudik, Z. Rubess, A. Lusis, J. Sokolov, and L. Lepin, *Latv. PSR Zinat. Akad.*, 5, 569 (1976).

76. "Ultrasonic Determination of the Adherence in Porcelain Enamel," D.E. Scherperell, N.F. Fiore, and R.A. Kettler, *Mat. Evaluation*, 32(11), 235 (1974).

77. "The Ultrasonic Pulse-Echo Technique as Applied to Adhesion Testing," H.G. Tattersall, *J. Phys. (D): Appl. Phys.*, 6, 819 (1973).

F.4   BEND TEST

78. "Three Dimensional Finite Element Analysis of Single Lap Adhesive Joints Subjected to Impact Bending Moments," I. Higuchi, T. Sawa, and H. Suga, *J. Adhesion Sci. Technol.*, 16(10), 1327 (2002) (three-point bend on single lap shear sample).

79. "The Effect of Moisture on the Failure Locus and Fracture Energy of an Epoxy-Steel Interface," W.K. Loh, A.D. Crocombe, M.M. Abdel Wahab, J.F. Watts, and I.A. Ashcroft, *J. Adhesion Sci. Technol.*, 16(11), 1407 (2002) (three-point bend).

80. "The Role of a Metallic Interlayer on the Mechanical Response of SiC Coatings on Stainless Steel Substrates," M. Andrieux, M. Ignat, M. Ducarroir, B. Feltz, and G. Farges, *J. Adhesion Sci. Technol.*, 16(1), 81 (2002) (three-point bend).

81. "The Role of the Polymer/Metal Interphase and Its Residual Stresses in the Critical Strain Energy Release Rate (G_c) Determined Using a Three Point Flexure Test," J. Bouchet, A.A. Roche, and E. Jacquelin, *J. Adhesion Sci. Technol.*, 15(3), 345 (2001) (three-point bend, double cantilevered beam).

82. "Interface Adhesion: Measurement and Analysis," A.G. Evans, in *Adhesion Measurement of Films and Coatings,* Vol. 2, K.L. Mittal, Ed. (VSP, Utrecht, The Netherlands, 2001), p. 1.

83. "Influence of Glass Mechanical Strengthening on the Adhesion Properties of Poly(vinyl butyral) Films to a Float Glass Surface," A.V. Gorokhovsky and K.N. Matazov, *J. Adhesion Sci. Technol.*, 14(13), 1657 (2000) (three-point bend).

84. "Influence of Silane Coupling Agents on the Surface Energetics of Glass Fibers and Mechanical Interfacial Properties of Glass Fiber Reinforced Composites," S.-J. Park, J.-S. Jin, and J.-R. Lee, *J. Adhesion Sci. Technol.*, 14(13), 1677 (2000) (interfacial shear stress, three-point bend).

85. "Stress Analysis and Strength Evaluation of Single Lap Band Adhesive Joints of Dissimilar Adherends Subjected to External Bending Moments," J. Liu and T. Sawa, *J. Adhesion Sci. Technol.*, 14(1), 67 (2000) (four-point bend test on single lap shear joints).

86. "Analysis of Adhesion and Interface Debonding in Laminated Safety Glass," Y. Sha, C.Y. Hui, E.J. Kramer, P.D. Garret, and J.W. Knapczyk, *J. Adhesion Sci. Technol.*, 11(1), 49 (1997) (three-point bend, flexure adhesion test, FEM analysis).

87. "Practical Adhesion Measurement in Adhering Systems: A Phase Boundary Sensitive Test," A.A. Roche, M.J. Romand, and F. Sidoroff, in *Adhesive Joints: Formation, Characteristics and Testing*, K.L. Mittal, Ed., (Plenum Press, New York, 1984), p. 19.

88. "The Three Point Bend Test for Adhesive Joints," N.T. McDevitt and W.L. Baun, in *Adhesive Joints: Formation, Characteristics and Testing*, K.L. Mittal, Ed. (Plenum Press, New York, 1984), p. 381.

89. "Use of Fracture Mechanics Concepts in Testing of Film Adhesion," W.D. Bascom, P.F. Becher, J.L. Bitner, and J.S. Murday, in *Adhesion Measurement of Thin Films, Thick Films and Bulk Coatings*, ASTM STP 640, K. . Mittal, Ed. (American Society for Testing and Materials, Philadelphia, 1978), p. 63.

90. "Fracture Design Criteria for Structural Adhesive Bonding — Promise and Problems," W.D. Bascom, R.L. Cottington, and C.O. Timmons, *Naval Engineers Journal*, 73 (August 1976).

F.5 BLISTER TEST

91. "Effects of the Lamination Temperature on the Properties of Poly(ethylene terephthalate-*co*-isophthalate) in Polyester Laminated Tin Free Steel Can. II. Adhesion Mechanism of Poly Poly(ethylene terephthalate-*co*-isophthalate) to TFS," C.F. Cho, J.D. Kim, K. Cho, C.E. Park, S.W. Lee, and M. Ree, *J. Adhesion Sci. Technol.*, 14(9), 1145 (2000).

92. "Effect of the Environment on the Measured Values of the Adhesion of Polyimide to Silicon," S.-H. Paek, K.-W. Lee, and C.J. Durning, *J. Adhesion Sci. Technol.*, 13(4), 423 (1999) (blister test using variety of liquids, correlate with 90° peel test).

93. "Study of Adherence Loss for Paint Films Coated Onto Polymeric Substrates by Induced Blistering," F. Touyeras, B. George, R. Cabala, A. Chambaudet, and J. Vebrel, *J. Adhesion Sci. Technol.*, 11(2), 263 (1997).

94. "The Use of Stimulated Boundary Gas Release for Determination of the Adhesion of Thin Films by the Blister Test," Y. N. Borisenko, V.T. Gritsyna, and T.V. Ivko, *J. Adhesion Sci. Technol.*, 9(11), 1413 (1995) (novel use of radiation to create gas pressure under coating).

95. "A Study of the Fracture Efficiency Parameter of Blister Tests for Films and Coatings," Y.-H. Lai and D.A. Dillard, in *Adhesion Measurement of Films and Coatings*, K.L. Mittal, Ed. (VSP, Utrecht, The Netherlands, 1995), p. 231.

96. "Adhesion (Fracture Energy) of Electropolymerized Poly(*n*-octyl maleimide-*co*-styrene) on Copper Substrates Using a Constrained Blister Test," J.-L. Liang, J.P. Bell, and A. Mehta, in *Adhesion of Films and Coatings*, K.L. Mittal, Ed. (VSP, Utrecht, The Netherlands, 1995), p. 249.

97. "The Adhesion of Poly(urethane) to Rough Counterfaces: The Influence of Weak Boundary Layers," B.J. Briscoe and S.S. Panesar, *J. Adhesion Sci. Technol.*, 8(12), 1485 (1994).

98. "Effects of Residual Stresses in the Blister Test," H.M. Jensen and M.D. Thouless, *Intl. J. Solids Struct.*, 30, 779 (1993).

99. "Growth and Configurational Stability of Circular Buckling-Driven Film Delaminations," J.W. Hutchinson, M.D. Thouless, and E.G. Liniger, *Acta Metall. Mater.*, 40, 295 (1992).

100. "Plane-Strain, Buckling-Driven Delamination of Thin Films," M.D. Thouless, J.W. Hutchinson, and E.G. Liniger, *Acta Metall. Mater.*, 40, 2639 (1992).

101. "Thin Film Cracking and the Roles of Substrate and Interface," T. Ye and A.G. Evans, *Intl. J. Solids Struct.*, 29, 2639 (1992).

102. "Cracking and Delamination of Coatings," M.D. Thouless, *J. Vac. Sci. Technol.*, A9, 2510 (1991).

103. "Numerical Analysis of the Constrained BlisterTest," Y.H. Lai and D.A. Dillard, *J. Adhesion*, 33, 63 (1990).

104. "An Elementary Plate Theory Prediction for Strain Energy Release Rate of the Constrained Blister Test," Y.H. Lai and D.A. Dillard, *J. Adhesion*, 31, 177 (1990).

105. "Some Mechanics for the Adhesion of Thin Films," M.D. Thouless, *Thin Solid Films*, 181, 397 (1989).

106. M.G. Allen and S.D. Senturia, *J. Adhesion*, 29, 219 (1989).

107. "The Constrained Blister — A Nearly Constant Strain Energy Release Rate Test for Adhesives," Y.S. Chang, Y.H. Lai, and D.A. Dillard, *J. Adhesion*, 27, 197 (1989).

108. "The Constrained Blister Test for the Energy of Interfacial Adhesion," M.J. Napolitano, A. Chudnovsky, and A. Moet, *J. Adhesion Sci. Technol.*, 2(4), 311 (1988).

109. "Analysis of Critical Debonding Pressures of Stressed Thin Films in the Blister Test," M.G. Allen and S.D. Senturia, *J. Adhesion*, 25, 303 (1988).

110. M.J. Napolitano, A. Chudnovsky, and A. Moet, *Proc. ACS, Div. Polym. Mater. Sci. Eng.*, 57, 755 (1987).

111. "Microfabricated Structures for the Measurement of Adhesion and Mechanical Properties of Polymer of Polymer Films," M.G. Allen and S. Senturia, *ACS Polymeric Materials Science and Engineering*, 56, 735 (April 1987).

112. "Hydrodynamic Method of Determination of Strength of the Adhesional Bond and Cohesional Strength of Polymeric Coatings," A.S. Sabev, K.K. Moroz, and T.A. Sabaeva, *Vysokomol. Soedin*, 27(11), 827 (1985).

113. "The Adhesion of Chlorinated Rubber to Mild Steel Using a Blister Technique," A. Elbasir, J.D. Scantlebury, and L.M. Callow, *J. Oil. Colour Chem. Assoc.*, 67(11), 282 (1985).

114. "Hydrodynamic Method of Determining the Adhesion Strength of Polymer Coatings," A.S. Sabev and K.K. Moroz, *R. Zh. Mash.*, 3B636 (1983) [Russian].

115. "A. Blister Test for Adhesion of Polymer Films to SiO$_2$," J.A. Hinkley, *J. Adhesion*, 16, 115 (1983).

116. "Hydraulic Adhesiometer," L.I. Pyatykin and N.V. Chinilina, *Zavod Lab.*, 48(6), 82 (1981 or 1982).

117. "The Influence of Pressure on Blister Growth," L.A. Van der Meer-Lerk and P.M. Heertjes, *J. Oil Colour Chem. Assoc.*, 64, 30 (1981).

118. "Measurement of Adhesion by a Blister Method," H. Dannenberg, *J. Appl. Polymer Sci.*, 5, 125 (1961).
119. "Measurement of Adhesion of Paint Films," E. Hoffman and O. Geogoussis, *J. Oil Colour Chem. Assoc.*, 42, 267 (1959).

F.6   DOUBLE CANTILEVERED BEAM TEST

120. "Tapered Double Cantilever Beam Test Used as a Practical Adhesion Test for Metal/Adhesive/Metal Systems," M. Meiller, A.A. Roche, and H. Sautereau, *J. Adhesion Sci. Technol.*, 13(7), 773 (1999).
121. "Effect of Modified Silicone Elastomer Particles on the Toughening of Epoxy Resin," T. Okamatsu, M. Kitajima, H. Hanazawa, and M. Ochi, *J. Adhesion Sci. Technol.*, 12(8), 813 (1998).
122. "Environmental Aging of the Ti-6Al-4V/FM-5 Polyimide Adhesive Bonded System: Implications of Physical and Chemical Aging on Durability," H. Parvatareddy, J.G. Dillard, J.E. McGrath, and D.A. Dillard, *J. Adhesion Sci. Technol.*, 12(6), 615 (1998).
123. "Modelling Crack Propagation in Structural Adhesives Under External Creep Loading," A.D. Crocombe and G. Wang, *J. Adhesion Sci. Technol.*, 12(6), 655 (1998) (double cantilevered beam compact tension specimen, FEM analysis).
124. "An Atomic Force Microscopy Study of the Effect of Surface Roughness on the Fracture Energy of Adhesively Bonded Aluminum," Y.L. Zhang and G.M. Spinks, *J. Adhesion Sci. Technol.*, 11(2), 207 (1997) (double cantilevered beam and single lap shear tests).
125. "Effect of the Addition of Aramid-CTBN Block Copolymer on Adhesive Properties of Cured Epoxy Resins Modified With CTBN," M. Ochi, M. Yamauchi, O. Kiyohara, and T. Tagami, *J. Adhesion Sci. Technol.*, 9(12), 1559 (1995) (double cantilevered beam and dynamic mechanical tests).

F.7   CENTRIFUGAL LOADING TEST

126. "Centrifugal Determination of Adhesion of Protective Coatings to a Steel Substrate," V.M. Ponizovskii, G.P. Spelkov, and N.V. Voronova, *Protection of Metals*, 19(5), 686 (1983).
127. "Determination of the Adhesion Strength of Electrolytic Zinc Coatings to Steel by a Centrifugal Method," V.S. Kolevatova, V.M. Ponizovskii, and G.P. Spelkov, *Ind. Lab. (U.S.S.R.)*, 42(5), 811 (1976).
128. "Determination of the Strength of the Adherence of Chromium Coatings With a Steel Base by a High Centrifugal Field Method," V.M. Ponizovskii and G.P. Spelkov, *Zavod Lab.*, 40(1), 107 (1974).
129. "Determination of the Adhesion of Coatings to Steel Based on the Application of Large Centrifugal Force," V.M. Ponizovskii, G.P. Spelkov, and N.F. Shishkina, *Lakokras. Mater. Ikh Primen.*, No. 5, 52 (1972).
130. "Determination of Adhesion of Copper and Nickel Coatings on a Steel Base by Means of Large Centrifugal Fields," V.M. Ponizovskii, Y.G. Svetlov, and G.V. Chirkov, *Zasch Metal*, 3(4), 515 (1967).
131. "Adhesion of Chromium and Other Deposits to Steel," W.H. Dancy, Jr., and A. Zavarella, *Plating*, 52, 1009 (1965).
132. "Ultracentrifugal Measurement of Adhesion: Some Limitations," J.R. Huntsberger, *Official Digest*, 33, 635 (1961).

133. "Influence of Selective Adsorption of Adhesion," J.R. Huntsberger, *J. Polymer Sci.*, 43, 581 (1960).
134. "Ultracentrifugal Measurement of the Adhesion of Epoxy Polymer," H. Alter and W. Soller, *Ind. Eng. Chem.*, 50, 922 (1958).
135. "The Ultracentrifuge as an Instrument of Testing Adhesion," R.L. Patrick, W.A. Vaughan, and C.M. Doede, *J. Polymer Sci.*, 28, 11 (1958).
136. "Study of Adhesion by the High Speed Rotor Technique," J.W. Beams, *Tech. Proc. Amer. Electroplaters Soc.*, 43, 211 (1956).
137. "Mechanical Strength of Thin Films of Metal," J.W. Beams, J. Breazeale, and W.L. Bart, *Phys. Rev.*, 100, 1657 (1955).
138. "Production and Use of High Centrifugal Fields," J.W. Beams, *Science*, 120, 619, (1954).
139. "Evaluation of Adhesion of Organic Coatings by Ultracentrifugal and Other Methods — Part 1," A.M. Malloy, W. Soller, and A.G. Roberts, *Paint Oil Chem. Rev.*, 14 (August 27, 1953).
140. "Bond Testing of Silver Plated Engine Bearings," F.D. Hallworth, *Automotive and Aviation Ind.*, 95(2), 30 (1946).

F.8   ELECTROMAGNETIC TEST

141. "Electromagnetic Tensile Adhesion Test Method," S. Krongelb, in *Adhesion Measurement of Thin Films, Thick Films and Bulk Coatings*, ASTM STP 640, K.L. Mittal, Ed. (American Society for Testing and Materials, Philadelphia, 1978), p. 107.
142. "Adhesion of Aluminum Films on Polyimide by the Electromagnetic Tensile Test," B.P. Baranski and J.H. Nevin, *J. Electronic. Mat.*, 16(1), 39 (1987).
143. "Dielectrometric Method for Evaluating the Adhesion Properties of Coatings and Adhesives to Metal," F.M. Smekhov, E.A. Andryushchenko, and I.S. Shvartsman, *Lakokrus. Mater. Ikh. Primen.*, 4, 45 (1986).

F.9   FRACTURE MECHANICS STUDIES

144. "Fracture Mechanics Characterization of Mixed-Mode Toughness of Thermoplastic/ Glass Interfaces (Brittle/Ductile Interfacial Mixed-Mode Fracture)," J. Dollhofer, W. Beckert, B. Lauke, and K. Schneider, *J. Adhesion Sci. Technol.*, 15(13), 1559 (2001).
145. "Fracture Mechanics Testing of the Bond Between Composite Overlays and a Concrete Substrate," V. Giurgiutiu, J. Lyons, M. Petrou, D. Laub, and S. Whitley, *J. Adhesion Sci. Technol.*, 15(11), 1351 (2001).
146. "A Stress Singularity Apprach to Failure Initiation in a Bonded Joint With Varying Bondline Thickness," D.M. Gleich, M.J.L. van Tooren, and A. Beukers, *J. Adhesion Sci. Technol.*, 15(10), 1247 (2001) (analysis of single lap joint).
147. "Analysis and Evaluation of Bondline Thickness Effects on Failure Load in Adhesively Bonded Structures," D.M. Gleich, M.J.L. van Tooren and A. Beukers, *J. Adhesion Sci. Technol.*, 15(9), 1091 (2001) (analysis of single lap joint).
148. "Prediction of Fatigue Thresholds in Adhesively Bonded Joints Using Damage Mechanics and Fracture Mechanics," M.M. Abdel Wahab, I.A. Ashcroft, A.D. Crocombe, and S.J. Shaw, *J. Adhesion Sci. Technol.*, 15(7), 763 (2001) (double lap joint, lap strap joint).
149. "Deformation and Fracture of Adhesive Layers Constrained by Plastically Deforming Adherends," M.S. Kafkalidis, M.D. Thouless, Q.D. Yang, and S.M. Ward, *J. Adhesion Sci. Technol.*, 14(13), 1593 (2000).

150. "On the Use of Fracture Mechanics in Designing a Single Lap Adhesive Joint," M.M. Abdel Wahab, *J. Adhesion Sci. Technol.*, 14(6), 851 (2000).

151. "Numerical and Experimental Analyses of a Fracture Mechanics Test for Adhesively Bonded Joints," *J. Adhesion Sci. Technol.*, 13(8), 931 (1999) (double cantilevered beam, FEM analysis).

152. "Relevance of Perturbation Methods in the Mechanics of Adhesion," J.F. Ganghoffer and J. Schultz, *J. Adhesion Sci. Technol.*, 10(9), 775 (1996) (theoretical analysis).

153. "A Fracture Mechanics Study of Natural Rubber to Metal Bond Failure," A.H. Muhr, A.G. Thomas, and J.K. Varkey, *J. Adhesion Sci. Technol.*, 10(7), 593 (1996) (peel test, quadruple lap shear, fiber pullout).

154. "Fracture Mechanics for Thin Film Adhesion," M.D. Thouless, *IBM J. Res. Dev.*, 38, 376 (1994).

155. "Mixed-Mode Fracture of a Lubricated Interface," M.D. Thouless, *Acta Metall. Mater.*, 40, 1281 (1992).

156. "Axisymmetric Shielding in Interfacial Fracture Under In-Plane Shear," K.M. Liechti and Y.-S. Chai, *J. Appl. Mech.*, 59, 295 (1992).

157. "Cracking of Brittle Films on an Elastic Substrate," M.D. Thouless, E. Olsson, and A. Gupta, *Acta Metall. Mater.*, 40, 1287 (1992).

158. "Cracking of Thin Bonded Films in Residual Tension," J.L. Beuth, Jr., *Intl J. Solids Struct.*, 29, 1657 (1992).

159. "Mixed Mode Cracking in Layered Materials," J.W. Hutchinson and Z. Suo, in *Advances in Applied Mechanics*, Vol. 29, J.W. Hutchinson and T.Y. Wu, Eds. (Academic Press, New York, 1991), p. 63.

160. "Decohesion of a Cut Prestressed Film on a Substrate," H.M. Jensen, J.W. Hutchinson, and K.S. Kim, *Intl. J. Solids Struct.*, 26, 1099 (1990).

161. "Interface Crack Between Two Elastic Layers," Z. Suo and J.W. Hutchinson, *Intl. J. Fract.*, 43, 1 (1990).

162. "Fracture of a Model Interface Under Mixed-Mode Loading," M.D. Thouless, *Acta Metall. Mater.*, 38, 1135 (1990).

163. "Crack Spacing in Brittle Films on Elastic Substrates," M.D. Thouless, *J. Amer. Ceram. Soc.*, 73, 2144 (1990).

164. "Steady State Cracking in Brittle Substrates Beneath Adherent Films," Z. Suo and J.W. Hutchinson, *Intl. J. Solids Struct.*, 25, 1337 (1989).

165. "Kinking of a Crack Out of an Interface," M.Y. He and J.W. Hutchinson, *J. Appl. Mechs.*, 56, 270 (1989).

166. "Delamination from Surface Cracks in Composite Materials," M.D. Thouless, H.C. Cao, and P.A. Mataga, *J. Mater. Sci.*, 24, 1406 (1989).

167. "Experimental Study of the Fracture Resistance of Bimaterial Interfaces," H.C. Cao and A.G. Evans, *Mech. Mater.*, 7, 295 (1989).

168. "Effects of Non-Planarity on the Mixed-Mode Fracture Resistance of Bimaterial Interfaces," A.G. Evans and J.W. Hutchinson, *Acta Metall.*, 37, 909 (1989).

169. "Decohesion of Films With Axisymmetric Geometries," M.D. Thouless, *Acta Metall.*, 36, 3131 (1988).

170. "The Decohesion of Thin Films from Brittle Substrates," M.S. Hu, M.D. Thouless, and A.G. Evans, *Acta Metall.*, 36, 1301 (1988).

171. "On the Decohesion of Residually Stressed Thin Films," M.D. Drory, M.D. Thouless, and A.G. Evans, *Acta Metall.*, 36, 2019 (1988).

172. "The Edge Cracking and Spalling of Brittle Plates," M.D. Thouless, A.G. Evans, M.F. Ashby, and J.W. Hutchinson, *Acta Metall.*, 35, 1333 (1987).

173. "Subcritical Crack Growth Along Ceramic-Metal Interfaces," T.S. Oh, R.M. Cannon, and R.O. Ritchie, *J. Amer. Ceram. Soc.*, 70, C352 (1987).
174. "Decohesion of Thin Films From Ceramic Substrates," R.M. Cannon, R.M. Fisher, and A.G. Evans, *Mater. Res. Soc. Proc.*, 54, 799 (1986).
175. "Strength of Thin Films and Coatings," G. Gille, in *Current Topics in Materials Science*, Vol. 12, E. Kaldis, Ed. (North-Holland, Amsterdam, 1985), Chap. 7, p. 420.
176. "Closure and Repropagation of Healed Cracks in Silicate Glass," T.A. Michalske and E.R. Fuller, Jr., *J. Am. Ceram. Soc.*, 68(11), 586 (1985).
177. "Stress Relief Forms of Diamond-Like Carbon Thin Films under Internal Compressive Stress," D. Nir, *Thin Solid Films*, 112, 41 (1984).
178. "Mathematical Analysis of the Mechanics of Fracture," J.R. Rice, in *Fracture*, Vol. 2, H. Liebowitz, Ed. (Academic Press, New York, 1968), Chap. 3, p. 191.
179. "A Simulated Crack Experiment Illustrating the Energy Balance Criterion," S.J. Burns and B.R. Lawn, *Intl. J. Fract. Mechs.*, 4, 339 (1968).
180. "Influence of Water Vapor on Crack Propagation in Soda-Lime Glass," S.M. Weiderhorn, *J. Amer. Ceram. Soc.*, 50, 407 (1967).

F.10   INDENTATION TEST

181. "Effect of Surface Oxygen Content and Roughness on Interfacial Adhesion in Carbon Fiber-Polycarbonate Composites," V.K. Raghavendran, L.T. Drzal, and P. Askeland, *J. Adhesion Sci. Technol.*, 16(10), 1283 (2002) (microindentation).
182. "The Effect of Polyol OH Number on the Bond Strength of Rigid Polyurethane on an Aluminum Substrate," J. Kim and E. Ryba, *J. Adhesion Sci. Technol.*, 15(14), 1747 (2002).
183. "Prospect of Nanoscale Interphase Evaluation to Predict Composite Properties," E. Mäder and S. Gao, *J. Adhesion Sci. Technol.*, 15(9), 1015 (2001) (atomic force microscopy nanoindentation).
184. "Mechanics of the Indentation Test and Its Use to Assess the Adhesion of Polymeric Coatings," R. Jayachandran, M.C. Boyce, and A.S. Argon, in *Adhesion Measurement of Films and Coatings*, K.L. Mittal, Ed. (VSP, Utrecht, The Netherlands, 1995), p. 189.
185. "Observations and Simple Fracture Mechanics Analysis of Indentation Fracture Delamination of TiN Films on Silicon," E.R. Weppelmann, X.-Z. Hu, and M.V. Swain, in *Adhesion Measurement of Films and Coatings*, K.L. Mittal, Ed. (VSP, Utrecht, The Netherlands, 1995), p. 217.
186. "Adhesion Measurement of Thin Films by Indentation," M.J. Matthewson, *Appl. Phys. Lett.*, 49(21), 1426 (1986).
187. "Measurement of Adherence of Residually Stressed Thin Films by Indentation, I. Mechanics of Interface Delamination," D.B. Marshall and A.G. Evans, *J. Appl. Phys.*, 56(10), 2632 (1984).
188. "Measurements of Adherence of Residually Stressed Thin Films by Indentation. II. Experiments With ZnO/Si," C. Rossington, A.G. Evans, D.B. Marshall, and B.T. Khuri-Yakub, *J. Appl. Phys.*, 56(10), 2639 (1984).
189. "Indentation-Debonding Test for Adhered Thin Polymer Layers," P.A. Engel and G.C. Pedroza, in *Adhesion Aspects of Polymeric Coatings*, K.L. Mittal, Ed. (Plenum Press, New York, 1983), p. 583.
190. "Measurement of Adhesion Energy of Thin Films," E.A. Nesmelov, A.S. Nikitin, A.G. Gusev, and O.N. Ivanov, *Sov. J. Optical Technol.*, 49(10), 640 (1982).

191. "A Simple Method for Adhesion Measurements," S.S. Chiang, D.B. Marshall, and A.G. Evans, in *Surfaces and Interfaces in Ceramic-Metal Systems*, J. Pask and A. Evans, Eds. (Plenum Press, New York, 1981), p. 603.
192. "Indentation Tests for Plastic to Copper Bond Strength," P.A. Engel, D.D. Roshon, and D.A. Thorne, *Insulation/Circuits*, 35 (November 1978).

F.11 INTERNAL FRICTION

193. "The Design and Application of Decrement 'Q' Meter as a Nondestructive Adhesion Test for Coatings," J.A. Muaddi, M. Izzard, J.K. Dennis, and J.F.W. Bell, *Trans. Inst. Metal Finishing*, 62(4), 134 (1985).
194. "Mechanical Properties of Thin Polyimide Films," R.H. Lacombe and J. Greenblatt, in *Polyimides,* Vol. 2, K.L. Mittal, Ed. (Plenum, New York, 1984), p. 647.
195. "Nondestructive Dynamic Evaluation of Thin NiTi Film Adhesion," Q. Su, S.Z. Hua, and M. Wuttig, in *Adhesion Measurement of Films and Coatings*, K.L. Mittal, Ed. (VSP, The Netherlands, 1995), p. 357.

F.12 IMPACT METHODS

196. "Measurement of Thick Film Adhesion by an Impact Separation Technique," W.E. Snowden and I.A. Aksay, in *Surfaces and Interfaces in Ceramic-Metal Systems*, J. Pask and A. Evans, Eds. (Plenum Press, New York, 1981), p. 651.

F.13 LAP SHEAR TEST

197. "Moisture Durability of Four Moisture Cure Urethane Adhesives," J. Verhoff, K. Ramani, N. Blank, and S. Rosenberg, *J. Adhesion Sci. Technol.*, 16(4), 373 (2002).
198. "Adhesion of Underfill and Components in Flip Chip Encapsulation," L. Fan, K.-S. Moon, and C.P. Wong, *J. Adhesion Sci. Technol.*, 16(2), 213 (2002).
199. "Liquid Crystalline Vinyl Ester Resins for Structural Adhesives," V. Ambrogi, C. Carfagna, M. Giamberini, E. Amendola, and E.P. Douglas, *J. Adhesion Sci. Technol.*, 16(1), 15 (2002).
200. "Thermal Characteristics of Tubular Single Lap Adhesive Joints Under Axial Loads," J.K. Kim and D.G. Lee, *J. Adhesion Sci. Technol.*, 15(12), 1511 (2002) (lap shear on tubular specimens).
201. "Stress Analysis and Strength Evaluation of Single-Lap Adhesive Joints Combined With Rivets Under External Bending Moments," J. Liu and T. Sawa, *J. Adhesion Sci. Technol.*, 15(1), 43 (2001).
202. "Prediction of the Tensile Load Bearing Capacity of a Co-cured Single Lap Joint Considering Residual Thermal Stresses," K.C. Shin and J.J. Lee, *J. Adhesion Sci. Technol.*, 14(13), 1691 (2000).
203. "Tensile Load Bearing Capacity of Co-cured Double Lap Joints," K.C. Shin and J.J. Lee, *J. Adhesion Sci. Technol.*, 14(12), 1539 (2000).
204. "Optimization of Adhesively Bonded Single Lap Joints by Adherend Notching," E. Sancaktar and S.R. Simmons, *J. Adhesion Sci. Technol.*, 14(11), 1363 (2000).
205. "Selective Use of Rubber Toughening to Optimize Lap-Joint Strength," E. Sancaktar and S. Kumar, *J. Adhesion Sci. Technol.*, 14(10), 1265 (2000) (single lap shear, stress analysis, FEA).

206. "Optimum Design of Co-cured Steel Composite Tubular Single Lap Joints Under Axial Load," D.H. Cho and D.G. Lee, *J. Adhesion Sci. Technol.*, 14(7), 939 (2000) (tubular lap shear, FEM stress analysis).

207. "Shear and Peel Stress Analysis of an Adhesively Bonded Scarf Joint," D.M. Gleich, M.J.L. van Tooren, and P.A.J. de Haan, *J. Adhesion Sci. Technol.*, 14(6), 879 (2000).

208. "A Study on the Lap Shear Strength of a Co-cured Single Lap Joint," K.C. Shin, J.J. Lee, and D.G. Lee, *J. Adhesion Sci. Technol.*, 14(1), 123 (2000).

209. "Effects of Adhesive Fillers on the Strength of Tubular Single Lap Adhesive Joints," *J. Adhesion Sci. Technol.*, 13(11), 1343 (1999) (tubular lap shear).

210. "Stress Analysis and Strength Evaluation of Single Lap Band Adhesive Joints Subjected to External Bending Moments," J. Liu and T. Sawa, *J. Adhesion Sci. Technol.*, 13(6), 729 (1999).

211. "Quinone-Amine Polyurethanes: Novel Corrosion Inhibiting Coupling Agents for Bonding Epoxy to Steel," K. Vaideeswaran, J.P. Bell, and D.E. Nikles, *J. Adhesion Sci. Technol.*, 13(4), 477 (1999) (tubular lap shear, XPS analysis).

212. "Adhesion of ABS Resin to Metals Treated With Triazine Trithiol Monosodium Aqueous Solution," H. Sasaki, I. Kobayashi, S. Sai, H. Hirahara, Y. Oishi, and K. Mori, *J. Adhesion Sci. Technol.*, 13(4), 523 (1999) (lap shear).

213. "Effect of Phase Structure on the Adhesion Properties of the Dimethoxysilyl-Terminated Polypropylene Oxide/Epoxy Resin System," Y. Okamatsu, M. Kitajima, H. Hanazawa, and M. Ochi, *J. Adhesion Sci. Technol.*, 13(1), 109 (1999) (lap shear).

214. "Surface Modification of High Density Polyethylene and Polypropylene by DC Glow Discharge and Adhesive Bonding to Steel," S. Bhowmik, P.K. Ghosh, S. Ray, and S.K. Barthwal, *J. Adhesion Sci. Technol.*, 12(11), 1181 (1998) (lap shear).

215. "Hygrothermal Effects on the Strength of Adhesively Bonded Joints," D.G. Lee, J.W. Kwon, and D.H. Cho, *J. Adhesion Sci. Technol.*, 12(11), 1253 (1998) (tubular lap shear).

216. "A Two-Dimensional Stress Analysis of Single Lap Adhesive Joints Subjected to External Bending Moments," J. Liu, T. Sawa, and H. Toratani, *J. Adhesion Sci. Technol.*, 12(8), 795 (1998) (single lap shear, theoretical analysis, FEM analysis).

217. "The Effect of Surface Roughness on the Adhesion Properties of Ceramic/Hybrid FRP Adhesively Bonded Systems," R. Park and J. Jang, *J. Adhesion Sci. Technol.*, 12(7), 713 (1998) (double lap shear).

218. "Structure and Adhesive Properties of Epoxy Resins Modified With Core/Shell Acrylic Particles," T. Ashida, M. Ochi, and K. Handa, *J. Adhesion Sci. Technol.*, 12(7), 749 (1998) (lap shear and T peel tests).

219. "Stress Analysis of Adhesive Lap Joints of Dissimilar Hollow Shafts Subjected to an Axial Load," Y. Nakano, M. Kawawaki, and T. Sawa, *J. Adhesion Sci. Technol.*, 12(1), 1 (1998) (tubular lap shear).

220. "Surface Pretreatment of Glass Fiber Reinforced Liquid Crystalline Polymer Composites by Eximer Laser for Adhesive Bonding," H.C. Man, M. Li, and T.M. Yue, *J. Adhesion Sci. Technol.*, 11(2), 183 (1997) (single lap shear and butt joint).

221. "Studies on Wood to Wood Bonding Adhesives Based on Natural Rubber Latex," N. John and R. Joseph, *J. Adhesion Sci. Technol.*, 11(2), 225 (1997).

222. "Study of a Single Lap Compression Shear Test for Brittle Substrates Bonded With a Structural Adhesive," D. Amara, F. Levallois, Y. Baziard, and J.A. Petit, *J. Adhesion Sci. Technol.*, 10(11), 1153 (1996) (novel apparatus for single lap shear under compression).

223. "The Role of an Anaerobic Accelerator in Dental Adhesives," I. Eppelbaum, H. Dodiuk, S. Kenig, B. Zalsman, A. Valdman, and R. Pilo, *J. Adhesion Sci. Technol.*, 10(10), 1075 (1996) (single lap shear).

224. "Strength Prediction of Adhesively Bonded Carbon/Epoxy Joints," K. Haruna, H. Hamada, and Z.-I. Maekawa, *J. Adhesion Sci. Technol.*, 10(10), 1089 (1996) (single lap shear, butt joint, torsion, three-point bend, FEM analysis).

225. "The Effect of Surface Characteristics of Polymeric Materials an the Strength of Bonded Joints," P.I.F. Niem, T.L. Lau, and K.M. Kwan, *J. Adhesion Sci. Technol.*, 10(4), 361 (1996) (single lap shear).

226. "An Innovative Approach to Fatigue Disbond Propagation in Adhesive Joints," H. Aglan and Z. Abdo, *J. Adhesion Sci. Technol.*, 10(3), 183 (1996) (Single lap shear).

227. "Adhesion Between Ni/Fe Lead Frame and Epoxy Molding Compounds in IC Packages," S. Asai, T. Ando, and M. Tobita, *J. Adhesion Sci. Technol.*, 10(1), 1 (1996) (single lap shear, contact angle, dynamic mechanical).

228. "Ammonia Plasma Treatment of Polyolefins for Adhesive Bonding With a Cyanoacrylate Adhesive," D.Y. Wu, W. (V.) S. Gutowski, S. Li, and H. Griesser, *J. Adhesion Sci. Technol.*, 9(4), 501 (1995) (single lap shear).

229. "Interfacial Chemistry of Spontaneous Disbonding in Stress Durability Testing of Adhesively Bonded Galvanized Steel," R.A. Dickie, M.A. Debolt, L.P. Haak, and J.E. Devries, *J. Adhesion Sci. Technol.*, 8(12), 1413 (1994) (single lap shear, XPS analysis).

230. "Surface Modification of Copper for Adhesion Promotion of Polybenzimidazole," G. Xue, J. Dong, X. Gu, Y. Qian, W. Sheng, and G.H. Wang, *J. Adhesion Sci. Technol.*, 8(9), 971 (1994) (single lap shear).

F.14 LASER/ELECTRON SPALLATION

231. "Effect of Primer Curing Conditions on Basecoat-Primer Adhesion — A LIDS Study," J.S. Meth, in *Adhesion Measurement of Films and Coatings*, Vol. 2, K.L. Mittal, Ed. (VSP, Utrecht, The Netherlands, 2001), p. 255.

232. "Measurement of Interfacial Bond Strength by Laser Spallation," A.W. Stephens and J.L. Vossen, *J. Vac. Sci. Technol.*, 13, 38 (1976); *J. Adhesion Sci. Technol.*, 11(10), 1249 (1998) (single lap shear, XPS analysis).

233. "Recent Developments in the Laser Spallation Technique to Measure the Interface Strength and Its Relationship to Interface Toughness With Applications to Metal/Ceramic, Ceramic/Ceramic and Ceramic/Polymer Interfaces," V. Gupta, J. Yuan, and A. Pronin, in *Adhesion Measurement of Films and Coatings*, K.L. Mittal, Ed. (VSP, Utrecht, The Netherlands, 1995), p. 367.

234. "Measurement of Interface Strength by a Laser Spallation Technique," V. Gupta, A.S. Argon, D.M. Parks, and J.A. Cornie, *J. Mech. Phys. Solids*, 40, 141 (1992).

235. "Electron Beam Defined Delamination and Ablation of Carbon Diamond Thin Films on Silicon," G.A.J. Amaratunga and M.E. Welland, *J. Appl. Phys.*, 68, 5140 (1990).

236. "Measurement of Film-Substrate Bond Strength by Laser Spallation," J.L. Vossen, in *Adhesion Measurement of Thin Films, Thick Films and Bulk Coatings*, ASTM STP 640, K.L. Mittal, Ed. (American Society for Testing and Materials, Philadelphia, 1978), p. 122.

F.15 MISCELLANEOUS METHODS

237. "The Prediction of Adhesion Between Polymer Matrices and Silane Treated Glass Surfaces in Filled Composites," A.C. Miller and J.C. Berg, *J. Adhesion Sci. Technol.*, 16(5), 495 (2002) (tensile test with single imbedded glass particle).

238. "Fracture Toughness Curves for Epoxy Molding Compound/Leadframe Interfaces," H.-Y. Lee, *J. Adhesion Sci. Technol.*, 16(5), 565 (2002) (Brazil nut, mixed-mode fracture study).

239. "Testing the Rolling Tack of Pressure Sensitive Adhesive Materials, Part II: Effect of Adherend Surface Roughness," O. Ben-Zion and A. Nussinovitch, *J. Adhesion Sci. Technol.*, 16(5), 597 (2002) (rolling drum method).

240. "Testing the Rolling Tack of Pressure-Sensitive Adhesive Materials. Part I. Novel Method and Apparatus," O. Ben-Zion and A. Nussinovitch, *J. Adhesion Sci. Technol.*, 16(3), 227 (2002) (rolling drum method).

241. "Tensile Stress-Strain Behavior of a Structural bonding Tape," F. Kadioglu and R.D. Adams, *J. Adhesion Sci. Technol.*, 16(2), 179 (2002).

242. "Stress Analysis and Strength Evaluation of Bonded Shrink Fitted Joints Subjected to Torsional Loads," T. Sawa, M. Yoneno, and Y. Motegi, *J. Adhesion Sci. Technol.*, 15(1), 23 (2001).

243. "The use of UNIFAC for the Estimation of Adhesion Enhancement Between Polymers and Mineral Surfaces Treated With Silane Coupling Agents," A.C. Miller, M.T. Knowlton, and J.C. Berg, *J. Adhesion Sci. Technol.*, 14(12), 1471 (2000) (embedded bead tensile test).

244. "Direct Adhesion Measurements of Pharmaceutical Particles to Gelatin Capsule Surfaces," T.H. Ibrahim, T.R. Burk, F.M. Etzler, and R.D. Neuman, *J. Adhesion Sci. Technol.*, 14(10), 1225 (2000) (atomic force microscopy, scanning probe microscopy).

245. "Fiber Stretching Test: A New Technique for Characterizing the Fiber-Matrix Interface Using Direct Observation of Crack Initiation and Propagation," S. Zhandarov, E. Pisanova, and K. Schneider, *J. Adhesion Sci. Technol.*, 14(3), 381 (2000) (interfacial shear strength).

246. "Adhesion Characterization of Protective Organic Coatings by Electrochemical Impedance Spectroscopy," F. Deflorian and L. Fedrizzi, *J. Adhesion Sci. Technol.*, 13(5), 629 (1999) (correlated with pull off test).

247. "Influence of Creep Phenomena and Surface Roughness on the Adhesion of *Xenopus laevis* Eggs to Different Substrates," N. Kampf and A. Nussinovitch, *J. Adhesion Sci. Technol.*, 13(4), 453 (1999) (custom peel and tensile testing).

248. "Improved Latex Film — Glass Adhesion Under Wet Environments by Using an Aluminum Polyphosphate Filler," E.F. Souza, M. Do Carmo, V.M. Da Silva, and F. Galembeck, *J. Adhesion Sci. Technol.*, 13(3), 357 (1999) (immersion float off test).

249. "Determination of the Adhesion Properties of an Alkyd Pigmented Coating by Electrochemical Impedence Spectroscopy," J.E.G. Gonzalez and J.C. Mirza Rosca, *J. Adhesion Sci. Technol.*, 13(3), 379 (1999) (electrochemical impedence spectroscopy [EIS]).

250. "Evaluation of the Bond Strength of Hard Coatings by the Contact Fatigue Test," K. Xu, N. Hu, and J. He, *J. Adhesion Sci. Technol.*, 12(10), 1055 (1998) (rolling sphere/cylinder test).

251. "Shear Bond Strength of Repair Resin Using an Intraoral Tribochemical Coating on Ceramic, Metal, and Resin Surfaces," P. Proaño, P. Pfeiffer, I. Nergiz, and W. Niedermeier, *J. Adhesion Sci. Technol.*, 12(10), 1121 (1998) (simple shear test, denture veneers).

252. "Evaluation of the Adhesion Strength in DLC Film Coated Systems Using the Film Cracking Technique," J.H. Jeong and D. Kwon, *J. Adhesion Sci. Technol.*, 12(1), 29 (1998) (sample coating supported on underlying tensile test carrier).

253. "The Characterization of Interfacial Strength Using Single Particle Composites," P.H. Harding and J.C. Berg, *J. Adhesion Sci. Technol.*, 11(8), 1063 (1997) (single glass sphere embedded in tensile specimen).

254. "The Nail Solution: Adhesion at Interfaces," R.P. Wool, D.M. Bailey, and A.D. Friend, *J. Adhesion Sci. Technol.*, 10(4), 305 (1996) (analysis of literature data).

255. "Butt Joint Strength: Effect of Residual Stress and Stress Relaxation," E.D. Reedy, Jr., and T.R. Guess, *J. Adhesion Sci. Technol.*, 10(1), 33 (1996) (double cantilevered beam apparatus for nail pullout test).

256. "Substrate Treatments/Modifications to Improve the Adhesion of Diamond Coated Cutting Inserts," M. Murakawa and S. Takeuchi, *J. Adhesion Sci. Technol.*, 9(6), 695 (1995) (scraping blade test).

257. "Relationship Between Interlayer Hardness and Adhesion and Pin-on-Disc Behavior for Fast Atom Beam Source Diamond Like Carbon Films," B. Olliver and A. Matthews, *J. Adhesion Sci. Technol.*, 9(6), 725 (1995) (pin-on-disk test).

258. "Salt Bath Test for Assessing the Adhesion of Silver to Poly(ethylene terephthalate) Web," J.M. Grace, V. Botticelle, D.R. Freeman, W. Kosel, and R.G. Spahn, in *Adhesion Measurement of Films and Coatings*, K.L. Mittal, Ed. (VSP, Utrecht, The Netherlands, 1995), p. 423.

259. "A Compression Test for Adhesion Evaluation of Diamond Coatings," R. Rozbicki, V.L. Rabinovich, and V.K. Sarin, *J. Adhesion Sci. Technol.*, 9(6), 737 (1995) (three-point bend, four-point bend, compression test).

260. "Adhesion Assessment of DLC Films on PET Using a Simple Tensile Tester: Comparison of Different Theories," *J. Adhesion Sci. Technol.*, 9(6), 769 (1995) (tensile test on precut coating).

261. "The Effect of Substrate Roughness Characteristics on Wettability and on the Mechanical Properties of Adhesive Joints Loaded at High Strain Rates," C. Keisler and J.L. Lataillade, *J. Adhesion Sci. Technol.*, 9(4), 395 (1995) (dynamic shear wave lap shear test).

262. "Fracture and Fatigue Behavior of Serim Cloth Adhesively Bonded Joints With and Without Rivet Holes," H. Aglan, Z. Abdo, and S. Shroff, *J. Adhesion Sci. Technol.*, 9(2), 177 (1995) (single lap shear test).

263. "Direct Measurement of Molecular Level Adhesion Between Poly(ethylene terephthalate) and Polyethylene Films: Determination of Surface and Interfacial Energies," V. Mangipudi, M. Tirrell, and A.V. Pocius, *J. Adhesion Sci. Technol.*, 8(11), 1251 (1994) (surface force analysis).

264. "Buckles in Adhering Elastic Films and a Test Method for Adhesion Based on the Elastica," A.N. Gent, *J. Adhesion Sci. Technol.*, 8(7), 807 (1994) (theory and data for highly novel adhesion measurement method).

265. "Experimental Determination of Interfacial toughness Curves Using Brazil-Nut Sandwitches," J.S. Wang and Z. Suo, *Acta Metall. Mater.*, 38, 1279 (1990).

266. *Adhesion and Adhesives*, A.J. Kinloch (Chapman and Hall, London, 1987).

267. "Adhesion of Dielectric and Encapsulant Layers Examined by Means of the Silver Migration Effect," B. Szczezytko, S. Achmatowicz, and D. Szymanski, *Electronika (Poland)*, 25, 20 (1984).

268. "Adhesiometer," O.A. Marveev, L.I. Pyatykhin, S.N. Pokalitsyn, and N.V. Chinilina, *Industrial Laboratory*, 50(10), 989 (1984).

269. "Measurement of Adhesion of Protective Paint Coatings," W.D. Kaiser, *Korrosion (Dresden)*, 15(2), 71 (1984).

270. "Method of Testing the Adhesion of Lithographic Materials," A.S. Bergendahl, S.E. Greco, A.F. Scaduto, and M.A. Zaitz, *IBM Technical Disclosure Bull.*, 27(5), 2973 (1984).

271. "Evaluation of the Adhesion of Polymer Films by Means of Their Surface Electric Charge," N.N. Morgunov, E.A. Kokhanova, and A.N. Kozhevnikov, *Plast. Massy (Russian)*, 3, 56 (1984).

272. "Combined Mode Fracture via the Cracked Brazilian Disk Test," C. Atkinson, R.E. Smelser, and J. Sanchez, *Int. J. Fracture*, 18, 279 (1982).

273. "Testing Adhesion of Coatings Using Limited Dome Height Method," A. Mertens et al., *Mem. Sci. Rev. Metall.*, 77(4), 579 (1980).

274. "Adhesion Measurement of Polymeric Films and Coatings With Special Reference to Photoresist Materials," K.L. Mittal and R.O. Lussow, in *Adhesion and Adsorption of Polymers*, Part B, L.H. Lee, Ed. (Plenum Press, New York, 1980), p. 503.

275. "Novel Spin Test for Adherence, Wear Resistance and Thermal Shock of Thin Films," R.R. Riegert, *J. Vac. Sci. Technol.*, 15(2), 789 (1978).

276. "'DID' Tester for Measuring the Presence of Ultra Adherence in Thin Films," R.R. Riegert, *J. Vac. Sci Technol.*, 15(2), 790 (1978).

277. "Problems in Adhesion Testing by the Tear Method," W. Dietrich, *Farbe Lack*, 82(7), 589 (1976).

278. "Adhesive Strength of Thin Organic Polymer Layers," M. Starke, K. Vohland, and M. Wroblowski, *Plastic Kautsch*, 22(10), 799 (1975) (piston bonding).

279. "Flyer Plate Techniques for Quantitatively Measuring the Adhesion of Plated Coatings Under Dynamic Conditions," J.W. Dini, H.R. Johnson, and R.S. Jacobson, in *Properties of Electrodeposites: Their Measurement and Significance*, R. Sard, H. Leidheiser, Jr., and F. Ogburn, Eds. (Electrochemical Society, Princeton, NJ, 1975), p. 307.

280. "Flyer Plate Adhesion Test for Copper and Nickel Plated A286 Stainless Steel," J.W. Dini and H.R. Johnson, *Rev. Sci. Instrum.*, 46, 1706 (1975).

281. "The Brittle-to-Ductile Transition in Pre-Cleaved Silicon Single Crystals," C. St. John, *Philos. Mag.*, 32, 1193 (1975).

282. "Ring Shear Adhesion Test - Various Deposit/Substrate Combinations," J.W. Dini and H.R. Johnson, *Metal Finishing*, 44 (August 1974).

283. "Nickel Plated Uranium: Bond Strength," J.W. Dini, H.R. Johnson, and J.R. Helms, *Plating*, 53 (January 1974) (ring shear).

284. "Device for Determining the Adhesion of Polymers by Tangential Shear," Y.I. Krasnov et al., *Zavod Lab.*, 38(4), 506 (1972).

285. "Ring Shear Test for Quantitatively Measuring Adhesion of Metal Deposites," J.W. Dini, J.R. Helms, and H.R. Johnson, *Electroplating Metal Finish.*, 25, 5 (1972).

286. "Adhesion of Evaporated Metal Films," F. Schossberger and K.D. Franson, *Vacuum*, 9, 28 (1959) (abrasion).

F.16 PEEL TEST

287. "Effect of Plasma Treatment of Aluminum on the Bonding Characteristics of Aluminum — CFRP Composite Joints," K.Y. Rhee, N.-S. Choi, and S.-J. Park, *J. Adhesion Sci. Technol.*, 16(11), 1487 (2002) (T peel and lap joint).

288. "A Kinetic Approach to Study the Hydrolytic Stability of Polymer-Metal Adhesion," A. Namkanisorn, A. Ghatak, M.K. Chaudhury, and D.H. Berry, *J. Adhesion Sci. Technol.*, 15(14) 1725 (2001) (hydrothermal peel test).

289. "Adhesion Property of Novel Polyimides Containing Fluorine and Phosphine Oxide Moieties," K.U. Jeong, Y.J. Jo, and T.H. Yoon, *J. Adhesion Sci. Technol.*, 15(14), 1787 (2001) (T peel test).

290. "Tack and Green Strength of Brominated Isobutylene-*co-p*-methylstyrene and Its Blends," B. Kumar, P.P. De, D.G. Peiffer, and A.K. Bhowmick, *J. Adhesion Sci. Technol.*, 15(10), 1145 (2001) (T peel test).

291. "Peel Testing of Adhesively Bonded Thin Metallic Plates With Control of Substrate Plastic Straining and of Debonding Mode Mixity," J.-Y. Sener, F. van Dooren, and F. Delannay, *J. Adhesion Sci. Technol.*, 15(10),1165 (2001) (T peel test with rollers).

292. "Adhesion Strength and Mechanism of Poly(imide-siloxane) to Alloy 42 Leadframe," J.H. Kang, K. Cho, and C.E. Park, *J. Adhesion Sci. Technol.*, 15(8), 913 (2001) (90° peel test).

293. "Adhesion Strength and Fracture Behavior Between Carbon Black Filled Rubber Sheets," C. Nah, J.M. Rhee, J.-H. Lee, and S. Kaang, *J. Adhesion Sci. Technol.*, 15(5), 583 (2001) (T peel test, trouser tear).

294. "Effect of the Physical and Mechanical Properties of Epoxy Resins on the Adhesion Behavior of Epoxy/Copper Leadframe Joints," K. Cho, E.C. Cho, and C.E. Park, *J. Adhesion Sci. Technol.*, 15(4), 439 (2001) (90° peel test).

295. "Peel Strength and Peel Angle Measured by the T-Peel Test on Cr/BPDA-PDA Interfaces," J.W. Choi and T.S. Oh, *J. Adhesion Sci. Technol.*, 15(2), 139 (2001).

296. "Properties of Polyurethane Adhesives Containing Natural Calcium Carbonate + Fumed Silica Mixtures," J. Sepulcre-Guilabert, T.P. Ferrándiz-Gómez, and J. Miguel Martín-Martínez, *J. Adhesion Sci. Technol.*, 15(2), 187 (2001) (T peel test).

297. "Synthesis and Characterization of New Thermoplastic Polyurethane Adhesives Containing Rosin Resin as an Internal Tackifier," F. Arán-Aís, A.M. Torró-Palau, A.C. Orgilés-Barceló, and J.M. Martín-Martínez, *J. Adhesion Sci. Technol.*, 14(12), 1557 (2000) (T peel test).

298. "Effect of the Microstructure of Copper Oxide on the Adhesion Behavior of Epoxy/Copper Leadframe Joints," K. Cho and E.C. Cho, *J. Adhesion Sci. Technol.*, 14(11), 1333 (2000).

299. "Surface Modification of Poly(tetrafluoroethylene) Films by Graft Copolymerization for Adhesion Enhancement With Electrolessly Deposited Copper," S. Wu, E.T. Kang, and K.G. Neoh, *J. Adhesion Sci. Technol.*, 14(11), 1451 (2000).

300. "The Aging of Repeated Oxygen Plasma Treatment on the Surface Rearrangement and Adhesion of LDPE to Aluminum," C.K. Cho, B.K. Kim, K. Cho, and C.E. Park, *J. Adhesion Sci. Technol.*, 14(8), 1071 (2000) (90° peel test).

301. "Theoretical and Experimental Studies for the Failure Criterion of Adhesively Bonded Joints," K.Y. Lee and B.S. Kong, *J. Adhesion Sci. Technol.*, 14(6), 817 (2000) (T peel test, single lap joint, FEM analysis).

302. "Influence of Block Molecular Weight on the Adhesion Properties of Segmented Polyamides," S. Ghosh, D. Khastgir, and A.K. Bhowmick, *J. Adhesion Sci. Technol.*, 14(46), 529 (2000) (180° peel test).

303. "Chlorination of Vulcanized Styrene-Butadiene Rubber Using Solutions of Trichloroisocyanuric Acid in Different Solvents," M.M. Pastor-Blas, T.P. Ferrándiz-Gómez, and J.M. Martín-Martínez, *J. Adhesion Sci. Technol.*, 14(4), 561 (2000) (T peel test).

304. "Effect of the Rheological Properties of Industrial Hot-Melt and Pressure Sensitive Adhesives on the Peel Behavior," F.X. Gibert, A. Allal, G. Marin, and C. Derail, *J. Adhesion Sci. Technol.*, 13(9), 1029 (1999) (floating roller peel test).

305. "Effects of Coupling Agent Thickness on Residual Stress in Polyimide/-APS/Silicon Wafer Joints," D.I. Kong, C.E. Park, S.T. Hong, H.C. Yang, and K.T. Kim, *J. Adhesion Sci. Technol.*, 13(7), 805 (1999) (90° peel test, correlate residual stress, FEM analysis).

306. "Interfacial Interaction and Peel Adhesion Between Polyamide and Acrylate Rubber in Thermoplastic Elastomeric Blends," A. Jha, A.K. Bhowmick, R. Fujitsuka, and T. Inoe, *J. Adhesion Sci. Technol.*, 13(6), 649 (1999) (T peel test).

307. "Adhesion of Ethylene-Acrylic Acid Copolymers and Their Blends With Paraffin to Aluminum," X. Lion and G. Geuskens, *J. Adhesion Sci. Technol.*, 13(6), 669 (1999) (T peel test).

308. "Studies of Ionomeric Polyolefins as Adhesives in Metal-Metal Bonding," P. Antony, S.K. De, and A.K. Bhowmick, *J. Adhesion Sci. Technol.*, 13(5), 561 (1999) (180° peel test).

309. "Adhesion of Latex Films. Part II. Loci of Failure," J.Y. Charmeau, A. Sartre, L. Vovelle, and Y. Holl, *J. Adhesion Sci. Technol.*, 13(5), 593 (1999) (180° peel test, XPS analysis, contact angle).

310. "Surface Graft Copolymerization of Poly(tetrafluoroethylene) Film With Simultaneous Lamination to Copper Foil," E.T. Kang, Y.X. Liu, K.G. Neoh, K.L. Tan, C.Q. Cui, and T.B. Lim, *J. Adhesion Sci. Technol.*, 13(3), 293 (1999) (T peel test, lap shear test).

311. "Adhesion of Latex Films. Part III. Surfactant Effects at Various Peel Rates," J.Y. Charmeau, P.A. Gerin, L. Vovelle, R. Schirrer, and Y. Holl, *J. Adhesion Sci. Technol.*, 13(2), 203 (1999) (180° peel test).

312. "Adhesion of Latex Films. Part IV. Dominating Interfacial Effect of the Surfactant," P.A. Gerin, Y. Grohens, R. Schirrer, and Y. Holl, *J. Adhesion Sci. Technol.*, 13(2), 217 (1999) (180 degree peel test).

313. "XPS/AFM Study of the PET Surface Modified by Oxygen and Carbon Dioxide Plasmas: Al/PET Adhesion," Q.T. Le, J.J. Pireaux, R. Caudana, P. Leclere, and R. Lazzaroni, *J. Adhesion Sci. Technol.*, 13(7), 773 (1999).

314. "Adhesion Improvement of a Poly(tetrafluoroethylene)-Copper Laminate by Thermal Graft Copolymerization," J. Zhang, C.Q. Cui, T.B. Lim, E.-T. Kang, and K.G. Neoh, *J. Adhesion Sci. Technol.*, 12(11), 1091 (1998) (T peel test).

315. "Effects of the Surface Modification by Remote Hydrogen Plasma on Adhesion in the Electroless Copper/Tetrafluoroethylene — Hexafluoropropylene Copolymer (FEP) System," M. Inagaki, S. Tasaka, and Y.W. Park, *J. Adhesion Sci. Technol.*, 12(11), 1105 (1998) (T peel test).

316. "Influence of a Monofunctional Acid-Base Environment on Peel Mechanics," R.A. Sinicki and J.C. Berg, *J. Adhesion Sci. Technol.*, 12(10), 1091 (1998) (180° peel test).

317. "Low Temperature Thermal Graft Copolymerization of 1-Vinyl Imidazole on Poly-imide Films With Simultaneous Lamination to Copper Foils," A.K.S. Ang, E.T. Kang, K.G. Neoh, K.L. Tan, C.Q. Cui, and T.B. Lim, *J. Adhesion Sci. Technol.*, 12(8), 889 (1998) (T peel test with single lap shear test).

318. "Ta and TaN Adhesion to High-Temperature Fluorinated Polyimides. Surface and Interface Chemistry," K.-W. Lee, E. Simonyi, and C. Jahnes, *J. Adhesion Sci. Technol.*, 12(7), 773 (1998) (90° peel test with XPS analysis).

319. "Effect of Curing Temperature on the Adhesion Strength of Polyamideimide/Copper Joints," J.H. Cho, D.I. Kong, C.E. Park, and M.Y. Jin, *J. Adhesion Sci. Technol.*, 12(5), 507 (1998) (90° peel test, effect of residual stress).

320. "Adhesion Improvement of Epoxy Resin/Copper Lead Frame Joints by Azole Com-pounds," S.M. Song, C.E. Park, H.K. Yun, C.S. Hwang, S.Y. Oh, and J.M. Park, *J. Adhesion Sci. Technol.*, 12(5), 541 (1998) (90° peel test).

321. "Experimental Determination of the Peel Adhesion Strength for Metallic Foils," E. Breslauer and T. Troczynski, *J. Adhesion Sci. Technol.*, 12(4), 367 (1998) (90° peel test, with and without roller).

322. "Effects of the Alkaline Permanganate Etching of Epoxy on the Peel Adhesion of Electrolessly Plated Copper on a Fibre Reinforced Epoxy Composite," J. Kirmann, X. Roizard, J. Pagetti, and J. Halut, *J. Adhesion Sci. Technol.*, 12(4), 383 (1998) (90° peel test).

323. "Adhesion Behavior of PDMS Containing Polyimide to Glass," K. Cho, D. Lee, T.O. Ahn, K.H. Seo, and H.M. Jeong, *J. Adhesion Sci. Technol.*, 12(3), 253 (1998) (90° peel test, locus of failure).

324. "Relationship Between Interfacial Phenomena and Viscoelastic Behavior of Laminated Materials," P. Cuillery, J. Tatibouët, and M. Mantel, *J. Adhesion Sci. Technol.*, 12(3), 271 (1998) (T peel test).

325. "Rubber to Steel Bonding Studies Using a Rubber Compound Strip Adhesive System," N.J. and R. Joseph, *J. Adhesion Sci. Technol.*, 12(1), 59 (1998) (180° peel test).

326. "Rate Dependence of the Peel force in Peel Apart Imaging Films," L. Kogan, C.Y. Hui, E.J. Kramer, and E. Wallace, Jr., *J. Adhesion Sci. Technol.*, 12(1), 71 (1998) (180° peel test, micromechanical model, comparison with experiment).

327. "Development of a Surface Treatment Method for AlN Substrates to Improve Adhesion With Thick Film Conductors," Y. Kuromitsu, T. Nagase, H. Yoshida, and K. Morinaga, *J. Adhesion Sci. Technol.*, 12(1), 105 (1998) (90° wirebond peel test).

328. "Direct Adhesion of Fluorinated Rubbers to Nickel Plated Steel and Nitrile Rubber During Curing Using the Tetrabutyl Ammonium Salt of 1,3,5-Triazine-2,4,6-trithiol," *J. Adhesion Sci. Technol.*, H. Hirahara, K. Mori, and Y. Oishi, *J. Adhesion Sci. Technol.*, 11(11), 1459 (1997) (T peel test).

329. "Fracture Mechanism and Oil Absorption of an Oil Accommodating Adhesive," T. Ogawa and M. Hongo, *J. Adhesion Sci. Technol.*, 11(9), 1197 (1997) (T peel test).

330. "Adhesion at Copper-Polyamide 11-Copper and Aluminum-Polyamide 11-Aluminum Laminate Interfaces: Influence of Metal Surface Oxidation," M. Gundjian and B. Fisa, *J. Adhesion Sci. Technol.*, 11(6), 759 (1997) (T peel test, XPS analysis).

331. "Adhesion Improvement of Epoxy Resin to Alloy 42 Lead Frame by Silane Coupling Agents," S.M. Song, C.E. Park, H.K. Yun, S.Y. Oh and J.M. Park, *J. Adhesion Sci. Technol.*, 11(6), 797 (1997) (90° peel test, XPS analysis).

332. "Surface Properties of Chemically Modified Polyimide Films," I. Ghosh, J. Konar, and A. Bhowmick, *J. Adhesion Sci. Technol.*, 11(6), 877 (1997) (180° peel test, XPS analysis).

333. "Dynamic Measurement of Humidity Attack on Polymer/Glass Interfaces Under Stress," H. Wu, J.T. Dickinson, and S.C. Langford, *J. Adhesion Sci. Technol.*, 11(5), 695 (1997) (180° peel with custom laser monitored cantilever strain gague).

334. "Relationship Between Surface Rearrangement and Adhesion Strength of Poly(ethylene-*co*-acrylic acid)/LDPE Blends to Aluminum," C.K. Cho, K. Cho, and C.E. Park, *J. Adhesion Sci. Technol.*, 11(4), 433 (1997) (90° peel with contact angle measurements).

335. "Influence of the Nature and Formulation of Styrene-Butadiene Rubber on the Effects Produced by Surface Treatment With Trichloroisocyanuric Acid," M.M. Pastor-Blas, J.M. Martín-Martínez, and J.G. Dillard, *J. Adhesion Sci. Technol.*, 11(4), 447 (1997) (T peel with contact angle measurements, XPS analysis).

336. "Adhesion Properties of Butyl Rubber Coated Polyester Frabric," S.N. Lawandy, A.F. Younan, N.A. Daewish, F. Yousef, and A. Mounir, *J. Adhesion Sci. Technol.*, 11(3), 317 (1997) (T peel correlated with dielectric measurements).

337. "Effect of Carbon Black Type on the Adhesion Between Butyl Rubber and Polyester Fabric," S.N. Lawandy, A.F. Younan, N.A. Daewish, F. Yousef, and A. Mounir, *J. Adhesion Sci. Technol.*, 11(2), 141 (1997) (90° peel test, roller assisted).

338. "Plastic Work in the Peeling of Work Hardening Foils," M. Sexsmith, T. Troczynski, and E. Breslauer, *J. Adhesion Sci. Technol.*, 11(3), 327 (1997) (T peel correlated with dielectric measurements).

339. "Effects of Plasma Treatment of Polyimide on the Adhesion Strength of Epoxy Resin/Polyimide Joints," H.K. Yun, K. Cho, J.K. Kim, C.E. Park, S.M. Sim, S.Y. Oh, and J.M. Park, *J. Adhesion Sci. Technol.*, 11(1), 95 (1997) (90° peel, contact angle, XPS analysis).

340. "Adhesion Behavior of Cu-Cr Alloy/Polyimide Films Under 350°C and 85°C/85% RH Environments," E.C. Ahn, J. Yu, I.S. Park, and W.J. Lee, *J. Adhesion Sci. Technol.*, 10(12), 1343 (1996) (T peel, contact angle, auger analysis).

341. "Adhesion of Poly(arylene ether benzimidazole) to Copper and Polyimides," K.-W. Lee, A. Viehbeck, G.F. Walker, S. Cohen, P. Zucco, R. Chen, and M. Ree, *J. Adhesion Sci. Technol.*, 10(9), 807 (1996) (90° peel, residual stress on thermal cycling).

342. "Adhesion of Latex Films. Part I. Poly(2-ethyl-hexyl methacrylate) on Glass)," E. Kientz, J.Y. Carmeau, Y. Holl, and G. Nanse, *J. Adhesion Sci. Technol.*, 10(8), 745 (1996) (180° peel, XPS analysis, locus of failure).

343. "Plasma Treatment and Adhesion Properties of a Rubber-Modified Polypropylene," A. Nihlstrand, T. Hjertberg, H.P. Schreiber, and J.E. Klemberg-Sapieha, *J. Adhesion Sci. Technol.*, 10(7), 651 (1996) (180° peel, XPS analysis, contact angle).

344. "Plasma Surface Modification of Polyimide for Improving Adhesion to Electroless Copper Coatings," G. Rozovskis, J. Vinkevicius and J. Jaciauskiene, *J. Adhesion Sci. Technol.*, 10(5), 399 (1996) (90° peel, XPS analysis).

345. "Surface Modification of Silicone Rubber by Gas Plasma Treatment," J.Y. Lai, Y.Y. Lin, Y.L. Denq, S.S. Shyu, and J.K. Chen, *J. Adhesion Sci. Technol.*, 10(3), 231 (1996) (180° peel, XPS analysis, contact angle).

346. "Improvement in the Adhesion Between Copper Metal and Polyimide Substrate by Plasma Polymer Deposition of Cyano Compounds Onto Polyimide," N. Inagaki, S. Tasaka, H. Ohmori, and S. Mibu, *J. Adhesion Sci. Technol.*, 10(3), 243 (1996) (T peel, XPS analysis).

347. "An Experimental Partitioning of the Mechanical Energy Expended During Peel Testing," R.J. Farris and J.L. Goldfarb, in *Adhesion Measurement of Films and Coatings*, K.L. Mittal, Ed. (VSP, Utrecht, The Netherlands, 1995), p. 265 (deformation calorimetry, peel test, thermodynamic analysis).

348. "Plasma Polymerization of Allyl Alcohol on Polyolefin Based Materials: Characterization and Adhesion Properties," A. Nihlstrand, T. Hjertberg, and K. Johansson, *J. Adhesion Sci. Technol.*, 10(2), 123 (1996) (90° peel, XPS analysis, contact angle, Fourier transform infrared [FTIR]).

349. "Effect of Ion Irradiation and Heat Treatment on Adhesion in the Cu/Teflon System," L. Wang, N. Angert, C. Trautmann, and J. Vetter, *J. Adhesion Sci. Technol.*, 9(12), 1523 (1995) (90° peel, XPS analysis).

350. "Peel Strength Improvement of Silicone Rubber Film by Plasma Pretreatment Followed by Graft Copolymerization," J.Y. Lai, Y.L. Denq, J.K. Chen, L.Y. Yan, Y.Y. Lin, and S.S. Shyu, *J. Adhesion Sci. Technol.*, 9(7), 813 (1995) (180° peel).

351. "The Adhesive Properties of Chlorinated Ultra-high Molecular Weight Polyethylene," H.N.A.M. Steenvakkers-Menting, P.E.L. Voets, and P.J. Lemstra, *J. Adhesion Sci. Technol.*, 9(7), 889 (1995) (asymmetric T peel).

352. "Role of Helium Plasma Pretreatment in the Stability of the Wettability, Adhesion and Mechanical Properties of Ammonia Plasma Treated Polymers. Application to the Al-Polypropylene System," M. Tatoulian, F. Arefi-Khonsari, I. Mabille-Rouger, J. Amouroux, M. Gheorgiu, and D. Bouchier, *J. Adhesion Sci. Technol.*, 9(7), 923 (1995) (180° tape peel).

353. "Electrical Transients Generated by the Peel of a Pressure Sensitive Adhesive Tape From a Copper Substrate. Part I: Initial Observations," S. Lee, L.C. Jensen, S.C. Langford, and J.T. Dickinson, *J. Adhesion Sci. Technol.*, 9(1), 1 (1995) (90° tape peel, triboluminescent study).

354. "Electrical Transients Generated by the Peel of a Pressure Sensitive Adhesive Tape From a Copper Substrate. Part II: Analysis of Fluctuations — Evidence for Chaos," L. Scudiero, J.T. Dickinson, L.C. Jensen, and S.C. Langford, *J. Adhesion Sci. Technol.*, 9(1), 27 (1995) (180° tape peel, triboluminescent study, fractal analysis).

355. "'Surface Photografting' Onto Polymers — A New Method for Adhesion Control," B. Rånby, *J. Adhesion Sci. Technol.*, 9(5), 599 (1995) (90° peel, lap shear, single-fiber pullout).

356. "Chromium and Tantalum Adhesion to Plasma Deposited Silicon Dioxide and Silicon Nitride," L.P. Buchwalter, *J. Adhesion Sci. Technol.*, 9(1), 97 (1995) (90° peel, XPS analysis).

357. "Effects of the Imidization Temperature of Poly(imide-siloxane) on the Adhesion Strength of Epoxy Resin/Poly(imide-siloxane) Joints," H.K. Yun, K. Cho, J.H. An, C.E. Park, S.M. Sim, and J.M. Park, *J. Adhesion Sci. Technol.*, 8(12), 1395 (1994) (90° peel, contact angle).

358. "Fracto-Emission and Electrical Transients Due to Interfacial Failure," J.T. Dickinson, L.C. Jensen, S. Lee, L. Scudiero, and S.C. Langford, *J. Adhesion Sci. Technol.*, 8(11), 1285 (1994) (90° peel, triboluminescent study).

359. "Modification of Polyimide Surface Morphology: Relationship Between Modification Depth and Adhesion Strength," K.-W. Lee, *J. Adhesion Sci. Technol.*, 8(10), 1077 (1994) (90° peel, XPS analysis).

360. "Surface Modification of Synthetic Vulcanized Rubber," M.M. Pastor-Blas, M.S. Sánchez-Adsuar, and J.M. Martín-Martínez," *J. Adhesion Sci. Technol.*, 8(10), 1039 (1994) (T degree peel, XPS analysis, infrared [IR] analysis).

361. "Tantalum and Chromium Adhesion to Polyimide, Part 2: Peel Test and Locus of Failure Analyses," L.P. Buchwalter, *J. Adhesion Sci. Technol.*, 7, 948 (1993).

362. "Elastic Fracture Mechanics of the Peel-Test Geometry," M.D. Thouless and H.M. Jensen, *J. Adhesion*, 38, 185 (1992).

363. "Elasto-Plastic Analysis of the Peel Test for Thin Film Adhesion," K.-S. Kim and J. Kim, *J. Eng. Mater. Technol.*, 110, 266 (1988).

364. "Mechanics of the Peel Test for Thin Film Adhesion," K.S. Kim, *Mat. Res. Soc. Symp. Proc.*, 119, 31 (1988).

365. "Elastoplastic Analysis of the Peel Test," K.S. Kim and N. Aravas, *Intl. J. Solids Struct.*, 24, 417 (1988).

366. "Elasto-Plastic Analysis of the Peel Test for Thin Film Adhesion," K.S. Kim and J. Kim, *J. Eng. Mater. Technol.*, 110, 266 (1988).

367. "On the Mechanics of Adhesion Testing of Flexible Films," N. Aravas, K.S. Kim, and M.J. Loukis, *Materials Science and Engineering*, A107, 159 (1989).

368. "Mechanical Effects in Peel Adhesion Test," J. Kim, K.S. Kim, and Y.H. Kim, *J. Adhesion Sci. Technol.*, 3, 175 (1989).

369. "A Bond Strength Test for Electroplated Metals," W.M. Robertson and F. Mansfield, *Plating Surface Finishing*, 59 (June 1980).

370. "The Adherence of Chromium and Titanium Films Deposited on Alumina, Quartz and Glass — Testing and Improvement of Electron-Beam Deposition Techniques," M. Caulton, W.L. Sked, and F.S. Wozniak, *RCA Rev.*, 40(2), 115 (June 1979).

371. "Adhesion of Chromium films to Soda Lime Glass," N.M. Poley and H.L. Whitaker, *J. Vac. Sci. Technol.*, 11, 114 (1974).

372. "Mechanics of Film Adhesion: Elastic and Elastic-Plastic Behavior," W.T. Chen and T.F. Flavin, *IBM J.Res. Dev.*, 16(3), 203 (1972).

373. "The Effect of Surface Roughness and Substrate Toughness on the Peel Value Obtained With Electroplated Plastics," A. Rantell, *Trans. Inst. Metal Finish.*, 50(4), 146 (1972).

374. "Adhesion of Viscoelastic Materials to Rigid Substrates," A.N. Gent and R.P. Petrich, *Proc. Roy. Soc.*, 310A, 433 (1969).

375. "Failure Mechanisms in Peeling of Pressure-Sensitive Adhesive Tape," D.W. Aubrey, G.N. Welding, and T. Wong, *Journal of Applied Polymer Science*, 13, 2193 (1969).

376. "Peel Adhesion and Pressure Sensitive Tapes," D. Satas and F. Egan, *Adhesives Age*, 22 (August 1966).

F.17 PULL TEST

377. "Adhesion of Electrolessly Deposited Copper to Photosensitive Epoxy," J. Ge, R. Tuominen, and J.K. Kivilahti, *J. Adhesion Sci. Technol.*, 15(10), 1133 (2001).

378. "The Investigation of Sized Cellulose Surfaces With Scanning Probe Microscopy Techniques," L.E. Dickson and J.C. Berg, *J. Adhesion Sci. Technol.*, 15(2), 171 (2001) (pull off with atomic force microscopy).

379. "On the Theory of Pull-off of a Viscoelastic Sphere from a Flat Surface," V.M. Muller, *J. Adhesion Sci. Technol.*, 13(9), 999 (1999).

380. "A Study of the Adhesion Between Hydraulic Mortars and Concrete," J. Barroso de Aguiar and M.D. Cruz, *J. Adhesion Sci. Technol.*, 12(11), 1243 (1998).

381. "Adhesion of an Acrylic Paint Coating to a Concrete Substrate — Some Observations," V. Saraswathy and N.S. Rengaswamy, *J. Adhesion Sci. Technol.*, 12(7), 681 (1998).

382. "Atomic Force Microscopy Investigation of the Adhesion Between a Single Polymer Sphere and a Flat Surface," S. Biggs and G. Spinks, *J. Adhesion Sci. Technol.*, 12(5), 461 (1998) (pull off test with AFM probe).

383. "Effect of Bonded Metal Substrate Area and Its Thickness on the Strength and Durability of Adhesively Bonded Joints," P. Pfeiffer and M. Abd el Salam Shakal, *J. Adhesion Sci. Technol.*, 12(3), 339 (1998) (symmetric pull off test).

384. "Effect of Different Surface Preparations on the Tensile Strength of Adhesively Bonded Metal Joints," M. Abd el Salam Shakal and P. Pfeiffer, *J. Adhesion Sci. Technol.*, 12(3), 349 (1998) (symmetric pull off test).

385. "Effect of Wax on the Adhesion Strength Between Acrylic Urethane Film and Chromated Steel Sheet and the Friction Coefficient of the Film," C.K. Cho, C.E. Park, S.K. Noh, S.H. Cho, C.S. Park, and Y.S. Jin, *J. Adhesion Sci. Technol.*, 11(11), 1365 (1997) (asymmetric pull off test).

386. "Measurement of the Adhesion Strength of Metal Oxide and Metal Nitride Thin Films Sputtered Onto Glass by the Direct Pull-off Method," S. Suzuki, Y. Hayashi, K. Suzuki, and E. Ando, *J. Adhesion Sci. Technol.*, 11(8), 1137 (1997).

387. "A Two Dimensional Stress Analysis and Strength Evaluation of Band Adhesive Butt Joints Subjected to Tensile Loads," T. Sawa and H. Uchida, *J. Adhesion Sci. Technol.*, 11(6), 811 (1997) (symmetric pull test, theoretical stress analysis, comparison with experiment).

388. "Study of the Interfacial Chemistry of Poly(vinyl Chloride) Paint on Steel Exposed to the Ultraviolet-Water Condensation Test," S. Feliu, Jr., J.L.G. Fierro, C. Maffiotte, B. Chico, and M. Morcillo, *J. Adhesion Sci. Technol.*, 11(4), 591 (1997) (pull test, XPS analysis).

389. "Surface Force Induced Bonding at Metal Multiasperity Contacts," J. Maniks and A. Simanovskis, *J. Adhesion Sci. Technol.*, 10(6), 541 (1996) (symmetric pull test, creation of atomically clean surfaces).

390. "An Advanced Multipurpose Dental Adhesive System," I. Eppelbaum, H. Dodiuk, S. Kenig, B. Zalsman, and A. Valdman, *J. Adhesion Sci. Technol.*, 9(10), 1357 (1995) (asymmetric pull test).

391. "Adhesion of Diamond Coatings on Tungsten Substrates," M. Alam, Feng He, D.E. Peebles, J.A. Ohlhausen, and D.R. Tallant, *J. Adhesion Sci. Technol.*, 9(6), 653 (1995) (asymmetric pull test).

392. "Analysis of Pull Tests for Determining the Effects of Ion Implantation on the Adhesion of Iron Films to Sapphire Substrates," J.E. Pawel and C.J. McHargue, in *Adhesion Measurement of Films and Coatings*, K.L. Mittal, Ed. (VSP, Utrecht, The Netherlands, 1995), p. 323.

393. "Measurement of the Adhesion of Diamond Films on Tungsten and Correlations With Processing Parameters," M. Alam, D.E. Peebles, and J.A. Ohlhausen, in *Adhesion Measurement of Films and Coatings*, K.L. Mittal, Ed. (VSP, Utrecht, The Netherlands, 1995), p. 331.

394. "Adhesion of Silver Films to Ion Bombarded Silicon Carbide: Ion Dose Dependence and Weibull Statistical Analysis of Pull Test Results," R.A. Erck, B.A. Lauer, and F.A. Nichols, *J. Adhesion Sci. Technol.*, 8(8), 885 (1994).

395. "Practical Adhesion of an Ag-Pd Thick film Conductor: An Acoustic Emission Study of Pull Tests — Part I Monitoring System and Initial Adhesion," P.J. Whalen and J.B. Blum, *IEEE Trans. CHMT*, CHMT-9, 161 (1986).

396. "Improved Adhesion Tester," F.I. Azimov, Y.S. Orevkov, A.E. Antipov, and E.K. Zagidullin, *Lakokras. Mater. Ikh Primen*, 5, 43 (1984).

397. "Practical Adhesion of an Ag-Pd Thick film Conductor: An Acoustic Emission Study of Pull Tests — Part II Aged Adhesion," P.J. Whalen and J.B. Blum, *IEEE Trans. CHMT*, CHMT-9, 168 (1986).

398. "The Direct Pull-off Test," P.C. Hopman, *J. Oil Colour Chem. Assoc.*, 67(7), 179 (1984).

399. "Pull-off Test, an Internationally Standardized Method for Adhesion Testing — Assessment of the Relevance of Test Results," J. Sickfeld, in *Adhesion Aspects of Polymeric Coatings*, K.L. Mittal, Ed. (Plenum Press, New York, 1983), p. 543.

400. "Adhesion Failure: Environmental Effect on a Two-Pack Polyurethane Coating," E.K. Rajan, B.K.G. Murthy, and M.A. Sivasamban, *Paintindia*, 29(10), 5 (1979).

401. "Sandwitch Pull-off Method for Evaluation of Adhesion of Coatings," M.N. Sathyanarayan, P.S. Sampathkumaran, and M.A. Sivasamban, *J. Colour Soc.*, 19, 98 (1980).

402. "Measurement of the Adhesion of Thin Films," R. Jacobson, *Thin Solid Films*, 34, 191 (1976).

403. "Adhesion of Evaporated Metal Film to fused Silica," L.D. Yurk, *J. Vac. Sci. Technol.*, 13, 102 (1976).

404. "Measurement of Adhesion of Thin Evaporated Films on Glass Substrates by Means of Direct Pull Off Method," R. Jacobson and B. Kruse, *Thin Solid Films*, 15, 71 (1973).

405. "Measurement of Bond Strength of Flame Sprayed Deposits," J.A. Catherall and J.E. Kortegas, *Japanese Inst. Metals Trans.*, 12, 218 (1972).

406. "Effect of Ion Pump Evacuation on the Adhesion of Evaporated Thin Films," K. Kuwahara, T. Nakagawa, and K. Kuramasu, *Metal. Const. British Weld. J.*, 4(1), 11 (1971).

407. "Measuring Electrodeposit Adhesion," S.D. Cramer, V.A. Cammarota, Jr., and D. Schlain, *Metal Finishing*, 45 (March 1970).
408. "High Vacuum Adhesion Test Apparatus," M.J. Hardon and M.A. Wright, *Rev. Sci. Instrum.*, 40, 1017 (1969).
409. "Effect of Ion Bombardment on the Adhesion of Aluminum Films on Glass," L.E. Collins, J.G. Perkins, and P.T. Stroud, *Thin Solid Films*, 4, 41 (1969).
410. "Interface and Adhesion Studies on Evaporated Selenium on Oxide Surfaces," Y. Chiang and S.W. Ing, *J. Vac. Sci. Technol.*, 6, 809 (1969).
411. "The Adhesion of Electrodeposits to Plastics," E.B. Saubestre, L.J. Durney, J. Hajdu, and E. Bastenbeck, *Plating*, 52, 982 (1965).
412. "Adhesion of Vapor Deposited Aluminum to Lime-Soda Glass," J.R. Frederick and K.C. Ludema, *J. Appl. Phys.*, 35, 256 (1964).
413. "Simple Rapid Sputtering Apparatus," R.B. Belser and W.H. Hicklin, *Rev. Sci. Instrum.*, 27, 293 (1956).

F.18   PULLOUT TEST

414. "Pull-Out Behavior of Oxidized Copper Leadframes From Epoxy Molding Compounds," H.-Y. Lee and S.-R. Kim, *J. Adhesion Sci. Technol.*, 16(6), 621 (2002).
415. "Plasma Treatment of Sisal Fibres and Its Effects on Tensile Strength and Interfacial Bonding," X. Yuan, K. Jayaraman, and D. Bhattacharyya, *J. Adhesion Sci. Technol.*, 16(6), 703 (2002).
416. "Enhanced Adhesion of Steel Filaments to Rubber via Plasma Etching and Plasma-Polymerized Coatings," H.M. Kang, K.H. Chung, S. Kaang, and T.H. Yoon, *J. Adhesion Sci. Technol.*, 15(4), 467 (2001).
417. "Investigation of Load Transfer Between the Fiber and the Matrix in Pull-out Tests With Fibers Having Different Diameters," S. Zhandarov, E. Pisanova, E. Mäder, and J.A. Nairn, *J. Adhesion Sci. Technol.*, 15(2), 205 (2001).
418. "Enhanced Interfacial Adhesion of Carbon Fibers to Vinyl Ester Resin Using Poly(arylene ether phosphine oxide) Coatings as Adhesion Promoters," I.-C. Kim and Y.-H. Yoon, *J. Adhesion Sci. Technol.*, 14(4), 545 (2000) (interfacial shear strength, microbond pullout test).
419. "Acid-Base Interactions and Covalent Bonding at a Fiber-Matrix Interface: Contribution to the Work of Adhesion and Measured Adhesion Strength," E. Pisanova and E. Mäder, *J. Adhesion Sci. Technol.*, 14(3), 415 (2000) (fiber pullout test, interfacial shear strength).
420. "Comparison of the Stress Transfer in Single and Multifiber Pullout Tests," S.-Y. Fu and B. Lauke, *J. Adhesion Sci. Technol.*, 14(3), 437 (2000) (fiber pullout test, interfacial shear strength).
421. "A Failure Criterion for Debonding Between Encapsulants and Leadframes in Plastic IC Packages," S. Yi, L. Shen, J.K. Kim, and C.Y. Yue, *J. Adhesion Sci. Technol.*, 14(1), 93 (2000) (lead frame pullout test, FEM analysis).
422. "Improvement of the Adhesion of Kevlar Fiber to Bismaleimide Resin by Surface Chemical Modification," T.K. Lin, B.H. Kuo, S.S. Shyu, and S.H. Hsiao, *J. Adhesion Sci. Technol.*, 13(5), 545 (1999) (microbond pullout, interfacial shear strength).
423. "Effects of a Pulsed XeCl Eximer Laser on Ultra High Strength Polyethylene Fiber and Its Interface With Epoxy Resin," Q. Song and A.N. Netravali, *J. Adhesion Sci. Technol.*, 13(4), 501 (1999) (single fiber pullout, interfacial shear strength).

424. "The Single Fiber Pull-out Test Analysis: Influence of a Compliant Coating on the Stresses at Bonded Interfaces," M.Y. Quek, *J. Adhesion Sci. Technol.*, 12(12), 1391 (1998) (single fiber pullout, interfacial shear strength).

425. "Eximer Laser Surface Modification of Ultra High Strength Polyethylene Fibers for Enhanced Adhesion With Epoxy Resins. Part I. Effect of Laser Operating Parameters," Q. Song and A.N. Netravli, *J. Adhesion Sci. Technol.*, 12(9), 957 (1998) (single-fiber pullout, interfacial shear strength).

426. "Eximer Laser Surface Modification of Ultra High Strength Polyethylene Fibers for Enhanced Adhesion With Epoxy Resins. Part 2. Effect of Treatment Environment," Q. Song and A.N. Netravli, *J. Adhesion Sci. Technol.*, 12(9), 983 (1998) (single-fiber pullout, interfacial shear strength).

427. "Modification of the Carbon Fiber/Matrix Interface Using Gas Plasma Treatment With Acetylene and Oxygen," S. Feih and P. Schwartz, *J. Adhesion Sci. Technol.*, 12(5), 523 (1998) (single-fiber pullout, microbond test, interfacial shear strength).

428. "Work of Adhesion and Local Bond Strength in Glass Fibre-Thermoplastic Polymer Systems," E. Pisanova, V. Dutschk, and B. Lauke, *J. Adhesion Sci. Technol.*, 12(3), 305 (1998) (single-fiber pullout, microbond test, interfacial shear strength).

429. "Thermostable Adhesives Based on Monomers With CN-multiple Bonds," Y.A. Gorbatkina, V.A. Pankratov, T.S.M. Frenkel, A.G. Chernyshova, and A.E. Shvorak, *J. Adhesion Sci. Technol.*, 11(1), 113 (1997) (single-fiber pullout).

430. "Adhesion Between High Strength and High Modulus Polyethylene Fibers by Use of Polyethylene Gel as an Adhesive," K. Takagi, H. Fujimatsu, H. Usami, and S. Ogasawara, *J. Adhesion Sci. Technol.*, 10(9), 869 (1996) (single-fiber pullout).

431. "Surface Modification of Kevlar 149 Fibers by Gas Plasma Treatment. Part II. Improved Interfacial Adhesion to Epoxy Resin," G.S. Sheu and S.S. Shyu, *J. Adhesion Sci. Technol.*, 10(9), 869 (1996) (microbond pullout, contact angle).

F.19 PUSH OUT TEST

432. "Adhesion and Reactivity in the Copper-Alumina System: Influence of Oxygen and Silver," A. Kara-Slimane, B. Mbongo, and D. Tréheux, *J. Adhesion Sci. Technol.*, 13(1), 35 (1999) (push out test).

433. "A New Tensile Rupture Test for the Mechanical Characterization of an Adhesively Bonded Structural Ceramic Assembly," D. Amara, B. Hassoune, R. el Abdi, Y. Baziard, and J.A. Petit, *J. Adhesion Sci. Technol.*, 12(10), 1029 (1998) (push out test, ceramic to ceramic, FEM analysis).

F.20 RESIDUAL STRESS SELF-LOADING TEST

434. "Cracking and Loss of Adhesion of Si_3N_4 and SiO_2:P Films Deposited on Al Substrates," P. Scafidi and M. Ignat, *J. Adhesion Sci. Technol.*, 12(11), 1219 (1998).

435. "A Self-Delamination Method of Measuring the Surface Energy of Adhesion of Coatings," R.J. Farris and C.L. Bauer, *J. Adhesion*, 26, 293 (1988).

436. "Mechanical Properties of Thin and Thick Coatings Applied to Various Substrates. Part I. An Elastic Analysis of Residual Stresses Within Coating Materials," M. Benabdi and A.A. Roche, *J. Adhesion Sci. Technol.*, 11(2), 281 (1997).

437. "Measurement of Interfacial Fracture Energy in Microelectronic Multifilm Applications," J.C. Hay, E. Lininger, and X.H. Liu, in *Adhesion Measurement of films and Coatings*, Vol. 2, K.L. Mittal, Ed. (VSP, Utrecht, The Netherlands, 2001), p. 205.

F.21 SCRATCH TEST

438. "Can the Scratch Adhesion Test Ever Be Quantitative," S.J. Bull, in *Adhesion Measurement of Films and Coatings*, Vol. 2, K.L. Mittal, Ed. (VSP, Utrecht, The Netherlands, 2001), p. 107.

439. "Effect of Annealing on the Cohesion, Adhesion and Tribological Behavior of Amorphous Silicon Containing Diamond-like Carbon (Si-DLC) Coatings on Steel," T. Kattamis, S. Skolianos, and C.G. Fountzoulas, *J. Adhesion Sci. Technol.*, 14(6), 805 (2000) (scratch test).

440. "Adhesion of Gold and Copper Thin Films Deposited on Alumina and Magnesium Oxide," A. Assaban and M. Gillet, *J. Adhesion Sci. Technol.*, 13(8), 871 (1999) (scratch test with pull test).

441. "Adhesion Strength of Plasma Assisted CVD B(C,N) Film to Silicon Substrate," S. Nakamura, N. Saito, S. Yoshioka, I. Nakaaki, and Y. Suzaki, *J. Adhesion Sci. Technol.*, 13(5), 615 (1999) (scratch test with microindentaion test).

442. "The Effect of Thermal Annealing on the Adherence of Al_2O_3 Films Deposited by LP-MOCVD on Several High Alloy Steels," V.A.C. Haanappel, A.H.J. van den Berg, H.D. van Corbach, T. Fransen, and P.J. Gellings, *J. Adhesion Sci. Technol.*, 11(7), 905 (1997) (scratch test with auger electron spectroscopy).

443. "Principle and Application of the Constant Strain Rate Tensile Test With In Situ Scanning Electron Mocroscopy for Evaluation of the Adhesion of Films," L. Yu. J.W. Liu, S.X. Dai, and Z.J. Jin, *J. Adhesion Sci. Technol.*, 11(1), 127 (1997) (tensile test on prescratched coatings).

444. "Residual Stress and Adhesion in Ion-Assisted Hafnia Coatings on Glass," L. Caneve, S. Scaglione, D. Flori, M.C. Cesile, and S. Martelli, *J. Adhesion Sci. Technol.*, 10(12), 1333 (1996).

445. "Adhesion Study of SiO_x/PET Films: Comparison Between Scratch Test and Fragmentation Test," Y. Pitton, S.D. Hamm, F.-R. Lang, H.J. Mathieu, Y. Leterrier, and J.-A.E. Månson, *J. Adhesion Sci. Technol.*, 10(10), 1047 (1996).

446. "Microstructure, Adhesion and Tribological Properties of Conventional Plasma Sprayed Coatings on Steel Substrate," T.Z. Kattamis, M. Chen, R. Huie, J. Kelly, C. Fountzoulas, and M. Levy, *J. Adhesion Sci. Technol.*, 9(7), 907 (1995) (scratch test with acoustic emission).

447. "Scratch Indentation, a Simple Adhesion Test Method for Thin Films on Polymeric Supports," G.D. Vaughn, B.G. Frushour, and W.C. Dale, in *Adhesion Measurement of Films and Coatings*, K.L. Mittal, Ed. (VSP, Utrecht, The Netherlands, 1995), p. 127.

448. "Scratch Test Failure Modes and Performance of Organic Coatings for Marine Applications," S.J. Bull, K. Horvathova, L.P. Gilbert, D. Mitchell, R.I. Davidson, and J.R. White, in *Adhesion Measurement of Films and Coatings*, K.L. Mittal, Ed. (VSP, Utrecht, The Netherlands, 1995), p. 175.

449. "Micro-scratch Test for Adhesion Evaluation of Thin Films," V.K. Sarin, in *Adhesion Measurement of Films and Coatings*, K.L. Mittal, Ed. (VSP, Utrecht, The Netherlands, 1995), p. 175.

450. "Adhesion Studies of Radio-Frequency Sputtered AlN Films on XC 70 Steel Substrates. Effect of Ageing Treatment After Deposition," A. Lahmar and M. Cailler, *J. Adhesion Sci. Technol.*, 8(9), 981 (1994) (scratch test with auger analysis).

451. "Improved Adhesion of Aluminum Layers Deposited on Plasma and Thermally Treated Poly(paraphenylene-vinylene) Films Substrates," T.P. Nguyen, K. Amgaad, M. Cailler, and V.H. Tran, *J. Adhesion Sci. Technol.*, 8(7), 821 (1994) (scratch test with XPS analysis).

452. "Applications of the Scratch Test Method for Coating Adhesion Assessment," J. Valli and U. Makela, *Wear*, 115(1-2), 215 (March 1987).

453. "Scratch Adhesion Testing: A Critique," A.J. Perry, *Surface Engng.*, 2(3), 183 (1986).

454. "Groove Adhesion Test for Electrodeposited Chromium," S.K. Pan and E.S. Chen, *J. Vac. Sci. Technol.*, A4(6), 3019 (1986).

455. "A Microtribometer for Measurement of Friction and Adhesion of Coatings," S. Baba, A. Kikuchi, and A. Kinbara, *J. Vac. Sci. Technol.*, A4(6), 3015 (1986).

456. "Scratch Test of Reactively Sputtered TiN Coatings on a Soft Substrate," J.H. Je, E. Gyarmati, and N. Naoumidis, *Thin Solid Films*, 136, 57 (1986).

457. "TiN Coating Adhesion Studies Using the Scratch Test Method," J. Valli, U. Makela, A. Matthews, and V. Murawa, *J. Vac. Sci. Technol.*, A3(6), 2411 (1985).

458. "An Energy Approach to the Adhesion of Coatings Using the Scratch Test," M.T. Laugier, *Thin Solid Films*, 117, 243 (1984).

459. "Measurements of the Adhesion of Evaporated Metal Films," A. Kikuchi, S. Baba, and A. Kinbara, *Shinku*, 27(6), 448 (1984).

460. "The Scratch Test as a Measure of Adhesion," M. Cailler, O.S. Ghee, and D. Poptin, *Vide, Couches Minces*, 39(220), 55 (1984).

461. "Scratch Adhesion Test Applied to Thin and Hard Metallurgical Coatings," P.A. Steinmann, P. Laeng, and H. Hintermann, *Vide, Couches Minces*, 39(220), 87 (1984).

462. "Scratch-Profiles Study in Thin Films Using SEM and EDS," M. El-Shabasy, L. Pogany, G. Konczos, E. Hajto, and B. Szikora, *Electrocomponent Sci. Technol.*, 11, 237 (1984).

463. "Scratch Adhesion Testing of Hard Coatings," A.J. Perry, *Thin Solid Films*, 107, 167 (1983).

464. "Investigation of Multilayer Systems by the Scratch Method," M. El-Shabasy, B. Szikora, G. Peto, I. Szabo, and K.L. Mittal, *Thin Solid Films*, 109, 127 (1983).

465. "Scratch Test of TiC Deposited Industrially by Chemical Vapor Deposition on Steel," B. Hammer, A.J. Perry, P. Laeng, and P.A. Steinmann, *Thin Solid Films*, 96, 45 (1982).

466. "The Adhesion of Chemically Vapor-Deposited Hard Coatings to Steel — The Scratch Test," A.J. Perry, *Thin Solid Films*, 78, 77 (1981).

467. "The Development of the Scratch Test Technique for the Determination of the Adhesion of Coatings," M. Laugier, *Thin Solid Films*, 76, 289 (1981).

468. "Adhesion and Internal Stress in Thin films of Aluminum," M. Laugier, *Thin Solid Films*, 79, 15 (1981).

469. "Adhesion of Chromium Films to Various Glasses Used in Photomasks," Y. Nakajima et al., *SPIE Developments in Semiconductor Microlithography III*, 135, 164 (1978).

470. "Threshold Adhesion Failure: an Approach to Aluminum Thin-film Adhesion Measurement Using the Stylus Method," J. Oroshnik and W.K. Croll, in *Adhesion Measurement of Thin Films, Thick Films and Bulk Coatings*, ASTM STP 640, K.L. Mittal, Ed. (American Society for Testing and Materials, Philadelphia, 1978), p. 158.

471. "Hardness and Adhesion of Filmed Structures as Determined by the Scratch Technique," J. Ahn, K.L. Mittal, and R.H. MacQueen, in *Adhesion Measurement of Thin Films, Thick Films and Bulk Coatings*, ASTM STP 640, K.L. Mittal, Ed. (American Society for Testing and Materials, Philadelphia, 1978), p. 134.

472. "Adhesion of Sputter Deposited Carbide Films to Steel Substrates," J.E. Green and M. Pestes, *Thin Solid Films*, 37, 373 (1976).

473. "Evaluation of Metal Film Adhesion to Flexible Substrates," L.F. Goldstein and T.J. Bertone, *J. Vac. Sci. Technol.*, 12, 1423 (1975).

474. "Quantitative Adhesion Tests of Vacuum Deposited Thin Films," O. Berendsohn, *J. Testing Evaluation*, 1, 139 (1973).

475. "Stability of Conductor Metallization in Corrosive Environments," A.T. English and P.A. Turner, *J. Electronic Mat.*, 1, 1 (1972).
476. "The Stylus of Scratch Method for Thin Film Adhesion Measurement: Some Observations and Comments," D.W. Butler, C.T.H. Stoddart, and P.R. Stuart, *J. Phys. (D): Appl. Phys.*, 3, 877 (1970).
477. "Diffusion in Gold-Aluminum," C. Weaver and D.J. Parkinson, *Phil. Mag.*, 22, 377 (1970).
478. "Adhesion of High Energy Surfaces," C. Weaver, in *Adhesion Fundamentals and Practice* (Maclaren and Sons, London, 1969), p. 46.
479. *Thin Film Phenomena*, K.L. Chopra (McGraw-Hill, New York, 1969), p. 313.
480. "Scratch Test for Measuring Adherence of Thin Films to Oxide Substrates," M.M. Karnowsky and W.B. Estill, *Rev. Sci. Instrum.*, 35, 1324 (1964).
481. "The Adhesion of Metals to Crystal Faces," P. Benjamin and C. Weaver, *Proc. Roy. Soc.*, 274A, 267 (1963).
482. "The Adhesion of Evaporated Metal Film on Glass," P. Benjamin and C. Weaver, *Proc. Roy, Soc.*, 261A, 516 (1961).
483. "Adhesion of Metal Films to Glass," P. Benjamin and C. Weaver, *Proc. Roy, Soc.*, 254A, 177 (1960).
484. "Measurement of Adhesion of Thin Film," P. Benjamin and C. Weaver, *Proc. Roy, Soc.*, 254A, 163 (1960).
485. "Adhesion of Evaporated Aluminum Films," C. Weaver and R.M. Hill, *Phil. Mag.*, 3, 1402 (1958).
486. "L'Epitaxie dans les Lames Polycrystallines," O.S. Heavens and L.E. Collins, *J. Phys. Radium*, 13, 658 (1952).
487. "Some Factors Influencing the Adhesion of Films Produced by Vacuum Evaporation," O.S. Heavens, *J. Phys. Radium*, 11, 355 (1950).

F.22   TAPE TEST

488. "Adhesion Measurement of Thin Films to a Porous Low Dielectric Constant Film Using a Modified Tape Test," L.L. N. Goh, S.L. Toh, S.Y.M. Chooi, and T.E. Tay, *J. Adhesion Sci. Technol.*, 16(6), 729 (2002).
489. "Surface Analysis and Printability Studies on Electron Beam Irradiated Thermoplastic Elastomeric Films From LDPE and EVA Blends," S. Chattopadhyay, R.N. Ghosh, T.K. Chaki, and A.K. Bhowmick, *J. Adhesion Sci. Technol.*, 15(3), 303 (2001).
490. "Determination of the Acid-Base Properties of Metal Oxide Films and of Polymers by Contact Angle Measurements," E. McCafferty and J.P. Wightman, *J. Adhesion Sci. Technol.*, 13(12), 1415 (1999) (tape test correlated with acid-base and contact angle).
491. "Effects of Nitrogen Plasma Treatment of Pressure Sensitive Adhesive Layer Surfaces on Their Peel Adhesion Behavior," M. Kawabe, S. Tasaka, and N. Inagaki, *J. Adhesion Sci. Technol.*, 13(5), 573 (1999) (180° tape test).
492. "Effect of the Functional Groups of Polystyrene on Its Adhesion Improvement and Corrosion Resistance," R. Kurbanova, A. Okudan, R. Mirzaolu, S. Kurbanov, I. Karata, M. Ersöz, E. Özcan, G. Ahmedova, and V. Pamuk, *J. Adhesion Sci. Technol.*, 12(9), 947 (1998) (tape pull on lattice of squares).
493. "Structure Effect on the Peel Strength of Polyurethane," S.L. Huang, S.J. Yu, and J.Y. Lai, *J. Adhesion Sci. Technol.*, 12(4), 445 (1998) (180° tape peel).

494. "Complex Dynamic Behavior in Adhesive Tape Peeling," M.C. Gandur, M.U. Kleinke, and F. Galembeck, *J. Adhesion Sci. Technol.*, 11(1), 11 (1997) (novel peeling dynamics study).

495. "Silicone Pressure Sensitive Adhesives With Selective Adhesion Characteristics," S.B. Lin, *J. Adhesion Sci. Technol.*, 10(6), 559 (1996) (180° tape peel test).

496. "Use of High Temperature and High Humidity to Test the Adhesion of Sputtered Copper to a Polyimide Surface Modified by an AC Nitrogen Glow Discharge," J.B. Ma, J. Dragon, W. Van Derveer, A. Entenberg, V. Lindberg, M. Anschel, D.-Y. Shih, and P. Lauro, *J. Adhesion Sci. Technol.*, 9(4), 487 (1995) (160–170° peel test).

497. "Quantifying the Tape Adhesion Test," G.V. Calder, F.C. Hansen, and A. Parra, in *Adhesion Aspects of Polymeric Coatings*, K.L. Mittal, Ed. (Plenum Press, New York, 1983), p. 569.

498. "Estimation of Paint Adhesion and Comaprison Test of Adhesion Tapes," H. Nuriya, *Jitsumu Hyomen Gijutsu*, 26(2), 86 (1979).

499. "Adhesion of Paints and Similar Coatings. Parameters Determining the Tear-off Method for the Judgement of Adhesion," J. Sickfeld and E. Hosp, *Farbe Lack*, 85(7), 537 (1979).

500. "Semi-quantitative Measurement of Pressure Tape Adhesion," John Ott, *Metal Finishing*, 61 (January 1978).

501. "Strong Adhesion of Vacuum Evaporated Gold to Oxide or Glass Substrates," G.J. Zydzik, L.G. Van Uitert, S. Singh, and T.R. Kyle, *Appl. Phys. Lett*, 31, 697 (1977).

502. "A Technique for Detecting Critical Loads in the Scratch Test for Thin Film Adhesion," J.E. Greene, J. Woodhouse, and M. Pestes, *Rev. Sci. Instrum.*, 45, 747 (1974).

503. "Relationship Between Substrate Surface Chemistry and Adhesion of Thin Films," R.C. Sundahl, *J. Vac. Sci. Technol.*, 9, 181 (1972).

504. "Adhesion Mechanism of Gold Underlayer Film Combinations to Oxide Substrates," K.E. Haq, K.H. Behrndt, and I. Kobin, *J. Vac. Sci. Technol.*, 6, 148 (1969).

505. "Interface Formation During Thin Film Deposition," D.M. Mattox and J.E. McDonald, *J. Appl. Phys.*, 34, 2393 (1963).

506. "On the Cleaning of Surfaces," J. Strong, *Rev. Sci. Instrum.*, 6, 97 (1935).

507. "Evaporated Aluminum Films for Astronomical Mirrors," J. Strong, *Publ. A. S. P.*, 46, 18 (1934).

F.23 THEORETICAL STUDIES

508. "On the Non-linear Elastic Stresses in an Adhesively Bonded T-Joint With Double Support," M. Kemal Apalak, *J. Adhesion Sci. Technol.*, 16(4), 459 (2002).

509. "An Internal Variable Theory of Adhesion," E. Bitterlin and J.F. Ganghoffer, *J. Adhesion Sci. Technol.*, 12(8), 857 (1998).

510. "A Two-Dimensional Stress Analysis of Adhesive Butt Joints of Dissimilar Adherends Subjected to Tensile Loads," T. Sawa, K. Temma, T. Nishigaya, and H. Ishikawa, *J. Adhesion Sci. Technol.*, 9(2), 215 (1995).

511. "Butt Joint Tensile Strength: Interface Corner Stress Intensity Factor Prediction," E.D. Reedy, Jr., and T.R. Guess, *J. Adhesion Sci. Technol.*, 9(2), 237 (1995).

512. "Power-Logarithmic Stress Singularities at Bi-material Corners and Interface Cracks," J.P. Dempsey, *J. Adhesion Sci. Technol.*, 9(2), 253 (1995).

513. "Analysis and Design of Adhesively Bonded Double Containment Corner Joints," M. Kemal Apalak, R. Davies, and Z. Gul Apalak, *J. Adhesion Sci. Technol.*, 9(2), 267 (1995).

514. "Relationship Between Adhesion and Friction Forces," J.N. Israelachvili, Y.-L. Chen, and H. Yoshizawa, *J. Adhesion Sci. Technol.*, 8(11), 1231 (1994) (largely theoretical with some surface force measurements).

515. "Cracks at Adhesive Interfaces," K. Kendall, *J. Adhesion Sci. Technol.*, 8(11), 1271 (1994) (covers variety of tests: peel, wedge, lap shear).

F.24 THERMAL METHODS

516. "Thermal Gradient Adhesion Meter," E.N. Sokolov, *Lakokrus. Mater. Ikh. Primen.*, 45 (1986).

517. "Indicator for Metal Film-substrate Adhesion," D. Guidotti and S.H. Wen, *IBM Technical Disclosure Bull.*, 28(5), 2078 (1985).

518. "Thermographic Detection of Polymer-Metal Adhesion Failures," B.E. Dom, H.E. Evans, and D.M. Torres, in *Adhesion Aspects of Polymeric Coatings*, K.L. Mittal, Ed. (Plenum Press, New York, 1983), p. 597.

F.25 TOPPLE BEAM METHOD

519. "On the Non-linear Elastic Stresses in an Adhesively Bonded T-Joint With Double Support," M. Kemal Apalak, *J. Adhesion Sci. Technol.*, 16(4), 459 (2002) (detailed finite element analysis of T joint).

520. "Measurement of the Adhesion of Silver Films to Glass Substrates," A. Kikuchi, S. Baba, and A. Kinbara, *Thin Solid Films*, 124, 343 (1985).

521. "Some Factors Affecting the Adhesion of Thin Metal Films," D.W. Butler, C.T.H. Stoddard, and P.R. Stuart, in *Aspects of Adhesion*, Vol. 6, D.J. Almer, Ed. (CRC Press, Cleveland, OH, 1971), p. 53.

522. "Thin Film Adhesion: Effect of Glow Discharge on Substrate," C.T.H. Stoddart, D.R. Clarke, and C.J. Robbie, *J. Adhesion*, 2, 270 (1970).

523. "A Simple Film Adhesion Comparator," D.W. Butler, *J. Phys. (E): Sci. Instrum.*, 3, 979 (1970).

F.26 WEDGE TEST

524. "The Role of Interfacial Interactions and Loss Function of Model Adhesives on Their Adhesion to Glass," A. Sharif, N. Mohammadi, M. Nekoomanesh, and Y. Jahani, *J. Adhesion Sci. Technol.*, 16(1), 33 (2002).

525. "Plasma Sprayed Coatings as Surface Treatments of Aluminum and Titanium Adherends," G.D. Davis, P.L. Whisnant, D.K. Shaffer, G.B. Groff, and J.D. Venables, *J. Adhesion Sci. Technol.*, 9(4), 527 (1995).

526. "Effectiveness of Water Borne Primers for Structural Adhesive Bonding of Aluminum and Aluminum-Lithium Surfaces," K.L. Meyler and J.A. Brescia, *J. Adhesion Sci. Technol.*, 9(1), 81 (1995) (wedge test, single lap shear, floating roller peel).

527. "Two Edge-Bonded Elastic Wedges of Different Materials and Wedge Angles Under Surface Tractions," D.B. Bogy, *J. Appl. Mech.*, 93, 377 (1971).

Index

A

acoustic emission methods, references, 401–402
acoustic spectroscopy
  coupling scratch test with, 23
  scratch test negating use of, 27
acoustic waves, for stress measurement, 277; *see also* SAWs (surface acoustic waves) test
acoustoelastic constants, 276
addition, operations for vector, 340
adhesion
  definitions of, 2–5
  stability maps, 333
adhesion, applying fracture mechanics to
  back-of-the-envelope calculations, 180–181
  decohesion number approach, 174–180
  elementary computational methods, 163
  problem of nickel on glass, 183–185
  problem of polyimide on glass-ceramic, 181–183
  references, 185–186
  thin coating on rigid disk, 163–174
adhesion measurement
  contaminants and, 1
  definitions of adhesion, 2–5
  electromagnetic test, 56–60
  indentation debonding test, 17–20
  nomenclature and usage, 5–6
  notes, 68–73
  overview of, 7–8
  peel test, 8–11, 14–17
  references, 400–401
  science of, 2
  SOM theory, 369
  tape peel test, 11–14
  theoretical foundations of, *see* continuum theory
advanced peel test, 193–194
alligator clamp, peel test equipment, 189
aluminum
  fully quantitative peel test, 198
  peel energy for polymides and, 196
  wedge test for aluminum sheets, 41–42
American Society for Testing and Materials (ASTM), 42
ANSYS programming systems, 329

B

arrays
  manipulating matrix, 351–352
  manipulating tensor, 344–347
  transpose of array, 344
arrest energy, 38
automatic mesh generation, 329
axisymmetric delamination, 387

back-of-the-envelope calculations
  fracture mechanics applied to adhesion, 180–181
  nickel on glass, 183–185
  polyimide on glass-ceramic, 181–183
Barkhausen noise, 297–299
beam-bending
  balancing shear force with downward load, 124–125
  deformation, 113–114
  elastic-plastic analysis and, 210–213
  equations for, 372–375
  final solution to, 129–130
  fully quantitative peel test, 200–204
  Hooke's law, 370
  obtaining beam deformations, 126–129
  overcoming limitations of SOM theory, 118–121
  SOM theory and, 112–118, 369
  stress function, 121–123
  tensile force as result of, 114–115
  treating peel strip as, 200–201
beam-bending tests
  advantages/disadvantages of, 43
  applying to coating and film stresses, 251–252
  Brazil nut test, 40–41
  double-cleavage drilled compression test, 39–40
  four-point bend test, 34–36
  overview of, 32
  recommendations, 44
  standard double cantilevered beam test, 35–37
  tapered double cantilevered beam test, 37–40
  three-point bend test, 32–34
  topple beam test, 42–43
  wedge test, 41–42